BLUEPRINTS AND PLANS FO[illegible]

Third Edition

FRANK C. MILLER
WILMA B. MILLER
JOSEPH MORAVEK

PLANS T49067 TO ACCOMPANY BOOK

DELMAR CENGAGE Learning

Australia • Brazil • Japan • K[illegible] United Kingdom • United States

Moulton College

WITHDRAWN from collection

Blueprints and Plans for HVAC, 3rd Edition
Frank C. Miller, Wilma B. Miller, Joseph Moravek

Vice President, Career and Professional Editorial: Dave Garza

Director of Learning Solutions: Sandy Clark

Senior Acquisitions Editor: James Devoe

Managing Editor: Larry Main

Senior Product Manager: John Fisher

Editorial Assistant: Thomas Best

Vice President, Career and Professional Marketing: Jennifer McAvey

Marketing Director: Deborah S. Yarnell

Marketing Manager: Kevin Rivenburg

Marketing Coordinator: Mark Pierro

Production Director: Wendy Troeger

Production Manager: Mark Bernard

Content Project Manager: Michael Tubbert

Art Director: Bethany Casey

Technology Project Manager: Chrstopher Catalina

Production Technology Analyst: Thomas Stover

© 2008 Delmar, Cengage Learning

ALL RIGHTS RESERVED. No part of this work covered by the copyright herein may be reproduced, transmitted, stored, or used in any form or by any means graphic, electronic, or mechanical, including but not limited to photocopying, recording, scanning, digitizing, taping, Web distribution, information networks, or information storage and retrieval systems, except as permitted under Section 107 or 108 of the 1976 United States Copyright Act, without the prior written permission of the publisher.

For product information and technology assistance, contact us at
Professional & Career Group Customer Support, 1-800-648-7450

For permission to use material from this text or product,
submit all requests online at **cengage.com/permissions.**
Further permissions questions can be e-mailed to
permissionrequest@cengage.com.

Library of Congress Control Number: 2008924166

ISBN-13: 978-1-4283-3520-2

ISBN-10: 1-4283-3520-X

Delmar
5 Maxwell Drive
Clifton Park, NY 12065-2919
USA

Cengage Learning products are represented in Canada by Nelson Education, Ltd.

For your lifelong learning solutions, visit **delmar.cengage.com**

Visit our corporate website at **cengage.com.**

Notice to the Reader
Publisher does not warrant or guarantee any of the products described herein or perform any independent analysis in connection with any of the product information contained herein. Publisher does not assume, and expressly disclaims, any obligation to obtain and include information other than that provided to it by the manufacturer. The reader is expressly warned to consider and adopt all safety precautions that might be indicated by the activities described herein and to avoid all potential hazards. By following the instructions contained herein, the reader willingly assumes all risks in connection with such instructions. The publisher makes no representations or warranties of any kind, including but not limited to, the warranties of fitness for particular purpose or merchantability, nor are any such representations implied with respect to the material set forth herein, and the publisher takes no responsibility with respect to such material. The publisher shall not be liable for any special, consequential, or exemplary damages resulting, in whole or part, from the readers' use of, or reliance upon, this material.

Printed in the United States of America
3 4 5 6 7 XX 12 11 10

CONTENTS

WITHDRAWN from collection

Preface xi
New to This Edition xii
About the Authors xiii
Acknowledgments xiii

PART I—BLUEPRINT READING

Chapter 1—Review of Basic Mathematical Procedures 3

Objectives 3
1.1 Introduction 3
1.2 Adding, Subtracting, Multiplying, and Dividing Whole Numbers 4
1.3 Fractions 5
1.4 Determining the Perimeter of a Square or Rectangle 6
1.5 Determining the Area of a Square or Rectangle 6
1.6 Determining the Area of a Triangle 7
1.7 Determining the Area of a Circle 7
1.8 Determining the Circumference of a Circle 8
1.9 Determining Volume 8
1.10 Equations 8
1.11 Additional Problems 9
1.12 Applied Math 9
Answers to Self-Test Questions 11

Chapter 2—Safety 13

Objectives 13
2.1 Introduction 13
2.2 Safety Rules for Technicians on the Job Site 13
2.3 Safety—Federal Law 15
Summary 15
Review Questions 15

Chapter 3—Linear Measurement 17

Objectives 17
3.1 Introduction 17
3.2 The English System and the Metric System 17
3.3 Reduced Scales 18
3.4 The One-Inch Graduation 18
3.5 The One-Foot Graduation 18
3.6 The One-Yard Graduation 19
3.7 Linear Measurement Instruments 20
3.8 Decimals and Fractions 20
Summary 20
Review Questions 21

Chapter 4—Angular Measurement 23

Objectives 23
4.1 Introduction 23

Moulton College R
Class No. 697 MIL C
Acc. No. T43404 T
Learning Resource Centre

4.2 Angular Measurements in a Circle 24
4.3 Four Kinds of Angles 25
4.4 Angles of a Triangle 26
4.5 Using a Protractor 27
4.6 Degrees, Minutes, and Seconds 28
Summary 28
Review Questions 28
Additional Student Exercises: Measuring Angles 29

Chapter 5—Learning to Use the Architect's Scale 31

Objectives 31
5.1 Introduction 31
5.2 The Architect's Scale Rule 31
5.3 Understanding the Architect's Scale Rule 32
5.4 Practicing with the One-Inch Scale 33
5.5 Using the Half-Inch Scale 34
5.6 Using Other Scales 34
5.7 The Most Commonly Used Scales on the Architect's Scale Rule 35
Summary 36
Review Questions 36
Additional Student Exercises: Using the Architect's Scale 37

Chapter 6—Learning to Use the Engineer's Scale 39

Objectives 39
6.1 Introduction 39
6.2 The Engineer's Scale Rule 39
6.3 What Makes the Engineer's Scale Unique 40
6.4 Using the Engineer's Scale Rule 40
6.5 Commonly Used Scales on the Engineer's Scale Rule ... 41
Summary 42
Review Questions 43
Additional Student Exercises: Using the Engineer's Scale 43

Chapter 7—Symbols and Abbreviations 45

Objectives 45
7.1 Introduction 45
7.2 Standardized Abbreviations and Symbols 45
7.3 Variations in Abbreviations 46
7.4 Variations in Symbols 46
7.5 Legend and Symbol Schedules 47
7.6 Abbreviations and Symbols Used Together 47
7.7 Examples of Abbreviations and Symbols 48
Summary 48
Review Questions 48
Additional Student Exercises: Freehand Drawing of Symbols 50

Chapter 8—How Working Drawings Are Created 51

Objectives 51
8.1 Introduction 51
8.2 What Are Working Drawings? 51
8.3 How Are Working Drawings Manually Created? 52
8.4 How Are Working Drawings Created with a Computer? 54
8.5 Drawing Paper or Film Used for Working Drawings 56
8.6 Sepia Drawings 56
8.7 Who Makes Working Drawings? 56
Summary 57
Review Questions 57

Chapter 9—Categories of Working Drawings 59

Objectives 59
9.1 Introduction 59
9.2 Major Categories of Working Drawings 59
9.3 Site Work Section 60
9.4 General Construction Section 60
9.5 Structural Section 61
9.6 Plumbing Section 62
9.7 The Mechanical Plan Section 63
9.8 Electrical Section 63
Summary 64
Review Questions 64

Chapter 10—Sections, Elevations, and Details 67

Objectives 67
10.1 Introduction 67
10.2 Sections 67
10.3 Elevations 69
10.4 Details 72
Summary 72
Review Questions 73
Additional Student Exercises: Extra Chapter 10 Review 73

Chapter 11—Creating Construction Drawings 77

Objectives 77
11.1 Introduction 77
11.2 How Construction Drawings Got the Name "Blueprint" 77
11.3 The Diazo Process for Making Construction Drawings 79
11.4 The Electrostatic Printing Process 81
11.5 Other Methods of Printing Construction Drawings 82
11.6 Microfilm and Computer Disk Storage of Drawings 83
Summary 83
Review Questions 83

Chapter 12—Who Uses Construction Drawings 85

Objectives 85
12.1 Introduction 85
12.2 Plan Review and Approval by the Owner 85
12.3 Review and Approval by Building Construction Agencies 86
12.4 Contractors for Bidding (Pricing) 86
12.5 Equipment Suppliers 86
12.6 Construction Plans for Construction Contractors 86
12.7 As-Built Drawings and Record Drawings Made from Construction Drawings 87
Summary 87
Review Questions 87

Chapter 13—Specifications 89

Objectives 89
13.1 Introduction 89
13.2 Division 1—General Requirements 89
13.3 Division 2—Site Work 90
13.4 Division 3—Concrete 90
13.5 Division 4—Masonry 90
13.6 Division 5—Metals 90
13.7 Division 6—Wood and Plastics 90

13.8 Division 7—Thermal and Moisture Protection 91
13.9 Division 8—Doors and Windows 91
13.10 Division 9—Finishes 91
13.11 Division 10—Specialties 91
13.12 Division 11—Equipment 91
13.13 Division 12—Furnishings ... 91
13.14 Division 13—Special Construction 91
13.15 Division 14—Conveying Systems 91
13.16 Division 15—Mechanical 91
13.17 Division 16—Plumbing 92
13.18 Division 17—Electrical 92
13.19 Responsibility and Liability 92
Summary 92
Review Questions 92
Additional Student Exercises: Matching Specifications 93

Chapter 14—Title Blocks 95

Objectives 95
14.1 Introduction 95
14.2 The Name of the Job 97
14.3 The Architect's and/or Engineer's Name 97
14.4 Sheet Title 97
14.5 Numbering the Sheets in a Set of Drawings 97
14.6 Revisions 98
14.7 Dates on Working Drawings (Construction Drawings) 98
14.8 Names or Initials Found in the Title Block 98
14.9 The Scale Used for the Drawing 98
14.10 The Architect's and Engineer's Seal 98
14.11 The Liability Note 99
Summary 99
Review Questions 100

PART II—FREEHAND SKETCHING AND DRAFTING

Chapter 15—Types and Weights of Lines 103

Objectives 103
15.1 Introduction 103
15.2 Schedule of Lines Used in Drafting and Freehand Sketching 103
15.3 Lines Drawn with Pencils 104
15.4 Object Lines 104
15.5 Hidden Lines 106
15.6 Extension and Dimension Lines 106
15.7 Construction Lines 107
15.8 Projection Lines 107
15.9 Center Lines 108
15.10 Cutting Plane Lines 108
15.11 Break Lines 108
15.12 Border Lines 109
15.13 Schedule of Lines Used in Offices 109
Summary 109
Review Questions 109

Chapter 16—Orthographic Projection 111

Objectives 111
16.1 Introduction 111
16.2 The Viewing Plane 111
16.3 Combining Viewing Planes 113
16.4 Types of Lines Used in Orthographic Projection 114
16.5 Using Orthographic Projection to Construct the Three Basic Views 116

16.6 Selecting the Proper Views 116
16.7 Working Drawings and Orthographic Projection 120
Summary 120
Review Questions 120
Additional Student Exercises: Completing Orthographic Projection Views 121

Chapter 17—Oblique Drawings 125

Objectives 125
17.1 Introduction 125
17.2 How Oblique Drawings Are Created 125
17.3 Using the Architect's Scale for Oblique Drawings 127
17.4 Lines Drawn on Oblique Drawings 127
17.5 Variations in Oblique Drawings 128
Summary 130
Review Questions 130
Additional Student Exercises: Constructing Oblique Drawings 130

Chapter 18—Isometric Drawings 133

Objectives 133
18.1 Introduction 133
18.2 How Isometric Drawings Are Drawn 133
18.3 A Variation of Isometric Drawings 137
18.4 Isometric Piping Diagram 139
18.5 Where Isometric Drawings Are Used 140
Summary 143
Review Questions 143
Additional Student Exercises: Freehand Sketching Isometric Views 143

Chapter 19—Lettering and Notations 149

Objectives 149
19.1 Introduction 149
19.2 Uppercase Lettering 149
19.3 Lowercase Lettering 150
19.4 Lettering Numerals 151
19.5 Wording Notations for Blueprints 153
Summary 154
Review Questions 154
Additional Student Exercises: Practice Forming Letters 154

Chapter 20—Organizing a Drawing Sheet 157

Objectives 157
20.1 Introduction 157
20.2 Creating a Basic Schedule of Plans 157
20.3 A System for Organizing a Job 158
20.4 Organizing Individual Drawing Sheets 158
Summary 158
Review Questions 159
Additional Student Exercises: Constructing Project-Planning Booklets 159

Chapter 21—Shading and Crosshatching 161

Objectives 161
21.1 Introduction 161
21.2 Crosshatching 161
21.3 Poché 162

21.4 Combining Crosshatching and Poché ... 163
21.5 Shading Drawings ... 164
21.6 Procedures for Shading Oblique and Isometric Drawings ... 164
Summary ... 165
Review Questions ... 166
Additional Student Exercises: Shading and Crosshatching ... 166

Chapter 22—Drawing Construction Elevations 169

Objectives ... 169
22.1 Introduction ... 169
22.2 Exterior Elevations ... 169
22.3 Interior Elevations ... 170
22.4 Mechanical Elevations ... 171
22.5 Electrical Elevations ... 171
22.6 Plumbing Elevations ... 171
Summary ... 172
Review Questions ... 172

Chapter 23—Drawing Construction Details 173

Objectives ... 173
23.1 Introduction ... 173
23.2 What Details Should Be Drawn ... 174
23.3 Site Details ... 174
23.4 Structural Details ... 174
23.5 Architectural Details ... 174
23.6 Mechanical Details ... 175
23.7 Plumbing Details ... 175
23.8 Electrical Details ... 176
23.9 Locating Details on the Sheet ... 176
23.10 Drawing Details ... 176
Summary ... 176
Review Questions ... 177

Chapter 24—Freehand Sketching 179

Objectives ... 179
24.1 Introduction ... 179
24.2 Types of Drawings Created by Freehand Sketching ... 179
24.3 Freehand Sketches Are Not Drawn to Scale ... 180
24.4 Lines Used for Freehand Sketching ... 180
24.5 Lettering for Freehand Sketching ... 180
24.6 Locating Notes on Freehand Sketches ... 181
24.7 Locating Drawings on a Page ... 181
24.8 Dimension and Extension Lines ... 181
24.9 Shading and Crosshatching Freehand Drawings ... 181
Summary ... 184
Review Questions ... 184
Additional Student Exercises: Freehand Sketching ... 185

Chapter 25—Drafting with Instruments 187

Objectives ... 187
25.1 Introduction ... 187
25.2 Commonly Used Manual Drafting Instruments ... 187
25.3 Drawing Lines on Working Drawings ... 188
25.4 Time-Saving Drafting Procedures ... 189
25.5 Drawing Heating and Air-Conditioning Components ... 190
25.6 Equipment Room Plan ... 193
Summary ... 195
Review Questions ... 195
Additional Student Exercises: Drafting with Instruments ... 196

PART III: READING AND INTERPRETING ARCHITECTURAL PLANS

Chapter 26—Reading Architectural Plans 203

Objectives ... 203
26.1 Introduction ... 203
26.2 Site Plan and Location Plan ... 203
26.3 Foundation Plan ... 204
26.4 Architectural Floor Plan ... 204
26.5 Wall Sections ... 204
26.6 Roof Framing Plan ... 204
26.7 Transverse and Longitudinal Sections ... 204
26.8 Wall Sections and Details ... 204
26.9 Enlarged Floor Plans and Interior Elevations ... 205
26.10 Schedules, Sections, and Elevations ... 205
Summary ... 205
Review Questions ... 205
Additional Student Exercises: Reading Architectural Plans ... 206

Chapter 27—Ductwork Plans 211

Objectives ... 211
27.1 Introduction ... 211
27.2 What Kind of Ductwork Is Shown on the Plans ... 211
27.3 Shop-Fabricated Sheet Metal Ductwork ... 213
27.4 Using the Duct Factor Chart ... 213
27.5 Measuring the Linear Feet of Ductwork Required ... 214
27.6 Measuring Linear Feet for Duct Fittings ... 214
27.7 Ductwork Takeoff Form ... 215
27.8 Determining the Total Weight of Metal Required for Ductwork ... 217
27.9 Cost for Materials for Shop-Fabricated Ductwork ... 219
27.10 Labor Costs for Shop-Fabricated Ductwork ... 219
27.11 Ductwork Insulation ... 219
27.12 Method for Taking Off Ductwork Insulation ... 219
27.13 Reading the Insulation Chart ... 219
27.14 Insulation Takeoff Form ... 221
27.15 Cost for Insulation Materials ... 221
27.16 Costs of Labor for Installing Insulation ... 221
Summary ... 221
Review Questions ... 223
Additional Student Exercises: Takeoff Practice ... 224

Chapter 28—Reading Mechanical Plans 225

Objectives ... 225
28.1 Introduction ... 225
28.2 Reading the Mechanical Plans ... 226
28.3 Looking for Specific Systems ... 226
28.4 Following the Ductwork ... 226
28.5 The Supply and Return Air Outlets ... 226
28.6 Written Notes and Instructions ... 226
28.7 Other Information Shown on the Mechanical Plan ... 227
28.8 Mechanical Specifications ... 227
Summary ... 227
Review Questions ... 227
Additional Student Exercises: Reading Mechanical Plans ... 228

Chapter 29—Reading Electrical Plans 231

Objectives 231
29.1 Introduction 231
29.2 Reading the Electrical Plans 231
29.3 The Electrical Wiring 232
29.4 Additional Information Shown on an Electrical Plan 232
29.5 Electrical Specifications 232
29.6 Understanding the Plans 232
Summary 232
Review Questions 233
Additional Student Exercises: Reading Electrical Plans 233

Chapter 30—Reading Plumbing Plans 235

Objectives 235
30.1 Introduction 235
30.2 Looking for Specific Systems 235
30.3 Reading the Plumbing Plans 236
Summary 236
Review Questions 237
Additional Student Exercises: Reading Plumbing Plans 237

Chapter 31—Introduction to Load Calculations 239

Objectives 239
31.1 Introduction 239
31.2 Heat Transfer Basics 240
31.3 What You Will Need to Do a Load Calculation 241
31.4 Design Conditions 243
31.5 Building Orientation 243
31.6 Floor Area 244
31.7 Foundation Type 245
31.8 Floor Insulation 245
31.9 Ceiling Area 245
31.10 Ceiling Insulation 247
31.11 Net Wall Area 248
31.12 Wall Insulation 248
31.13 Window Area 249
31.14 Window Construction 249
31.15 Shading Factor 250
31.16 Door Area 252
31.17 Door Construction 252
31.18 Duct Design 254
31.19 Number of Occupants 255
31.20 Infiltration Rate 256
31.21 Ventilation Rate 256
31.22 Miscellaneous Loads 257
31.23 Zoning 259
Summary 260
Review Questions 260
Additional Student Exercises 261

Appendices

Appendix A: Geometric Figures and Formulas 263
Appendix B: Conversion Tables 265
Appendix C: Abbreviations 269
Appendix D: Symbols 273

Glossary 283

Index 291

PREFACE

This text is written for students in a heating, ventilating, and air-conditioning (HVAC) program of a community college or technical school who are planning a career in the building trades. The HVAC technician must be a "jack-of-all-trades" on the job site and in the office. This textbook, therefore, addresses various areas of expertise that may be expected, including:

(1) reading and interpreting architectural, plumbing, HVAC, and electrical plans for the sizing and installation of equipment;

(2) being able to use various kinds of scales and measuring devices and recognizing standard symbols and abbreviations;

(3) following safety procedures in the office and on the job site;

(4) understanding, interpreting, and being able to create isometric, orthographic, and oblique drawings;

(5) working with blueprints for ductwork, piping, electrical wiring, controls, and equipment layout;

(6) being able to do freehand sketching and drafting as needed in the office, in the shop, or at the job site;

(7) recognizing the relationships and responsibilities of owners, architects, engineers, designers, drafters, contractors, and others involved in the planning and construction of any building;

(8) knowing how to operate the various types of printmaking equipment in the office and understanding why some types of prints are better than others for specific uses;

(9) recognizing the advantages of using Computer-Aided Design (CAD) or manual drafting equipment to create plans; and

(10) realizing that the HVAC system is an integral part of the total structure, requiring coordination with the other trades and design personnel.

Numerous illustrations, photographs, and actual building plans have been included for students to refer to as needed. Attention has been given to understandable vocabulary, precise sentences, stated objectives and summaries for each chapter, and review questions and exercises that are realistic and related to current practices in the field. The appendices furnish tables of easy-to-use reference material. The Instructor's Guide contains suggestions for numerous enrichment activities in addition to those offered in the textbook.

The organization of this text is logical and practical. Part I begins with the elements of blueprint reading, progressing from a review of mathematical functions, measurement and scales, and symbols and abbreviations to the process of creating and interpreting working drawings and construction drawings. The components of these drawings, including details, sections, elevations, specifications, and title blocks, are explained and illustrated in separate chapters. Discussion is included on the roles played by the owner, architect, engineer, designer, drafter, contractor, and others involved in the building process.

Part II deals with freehand sketching and drafting. Lines, lettering, shading, and crosshatching are covered in separate chapters. Orthographic projection, isometric drawings, and oblique drawings are discussed and illustrated in detail. Practical suggestions are given about organizing drafting plans, drawing construction elevations, sketching, and manual drafting.

Part III contains numerous exercises for the student, using an adaptation of actual construction plans. The student can read and interpret architectural, mechanical, electrical, plumbing, and ductwork plans just as a technician might be expected to do in the office or on the job site.

A new Chapter 31 has been added to this section. This new chapter applies blueprint reading to heat load calculations. This chapter is a good introduction into using and understanding load calculations.

The text is organized to allow the student (or instructor) to work straight through or to rearrange the order of subject matter as needed. Chapter 1 is a self-test to assess basic mathematical skills needed for the subject matter in the text. The chapters on measurement and on reading scales may seem oversimplified, but the authors and other instructors have noticed that average students often have difficulty in working with scales and measurement.

NEW TO THIS EDITION

This edition has been thoroughly reviewed and updated for changes in the industry since the second edition. Some chapters have minor changes to reflect updated information. Even though the prints and plans industry is upgrading to computer-based information, the paper prints/plans are still widely used in the field, in the office, and when submitted for municipal plan checking for permits. For example, the text makes a reference to a drafting table. Hand-drafting is rarely done today, yet a drafting table is still used to spread out and review paper prints/plans. Sometimes, hand-drafting on a set of prints is done to reflect minor modifications in construction or hand-drafting with the use of red ink to show the actual "as built" drawings.

The third edition still offers the set of popular commercial plans. These plans can be used to teach the topic as well as identify structural, electrical, plumbing, and mechanical components that the HVAC installer is likely to experience in the field. The plans can also be used to develop takeoff materials list or estimates. The paper plans can be used to develop heating and cooling load calculations.

The textbook includes four valuable Appendices:

- Appendix A: Geometric Figures and Formulas
- Appendix B: Conversion Tables
- Appendix C: Common Abbreviations
- Appendix D: Symbols

Finally, the textbook can be used to teach the topic entirely from start to finish or it can be used to teach the HVAC components necessary to install or solve problems in the field.

ABOUT THE AUTHORS

Frank C. Miller graduated from Southern Technical Institute in Marietta, Georgia, and from North Carolina State University in Raleigh, North Carolina. Working with architectural and engineering firms across North Carolina, he designed mechanical systems for buildings ranging from the small bank building included with this textbook to hospitals and multi-story industrial and commercial buildings.

Mr. Miller was an instructor in the HVAC Department of Fayetteville Technical Institute and Central Piedmont Community College. He later served as head of the Technical Careers Department at Central Piedmont Community College. He was a member of a statewide curriculum development committee for HVAC, and he wrote course outlines, planned and conducted training seminars, and helped plan curriculum modifications in HVAC and related programs of study. He was a licensed Plumbing Contractor and a member of ASHRAE and ASPE.

Wilma B. Miller is a graduate of Duke University, majoring in English and Education. She earned a Master's Degree from the University of North Carolina at Charlotte. She also completed an additional year of graduate study from the Metrolina Education Consortium. She has 30 years of teaching experience in the public schools of North Carolina. Early in her career she was a technical editor for Western Electric Publications, editing technical manuals for the U.S. Navy. She is the author of several articles which have been published in educational journals.

Joseph Moravek has a wide range of experiences in the residential and commercial HVAC trade. He has been a Texas licensed mechanical contractor since 1985. He has worked in the field for the City of Houston Parks and Recreation Department and with the City of Houston as a Mechanical Inspector. During his tenure with the City of Houston Inspection Department, he worked as an occupancy inspector, which includes reviewing air-conditioning and building plans to comply with local mechanical codes. He is currently employed as the Corporate Training Coordinator for Hunton Distribution in Houston, Texas.

For 14 years, he was the lead instructor of the HVAC program at Lee College in Baytown, Texas. He has a master's degree in education from the University of Houston. Moravek is a training consultant and has published books related to HVAC topics. If you have any comments regarding this textbook, you can contact him at 713–861–1806 or zmoravek@aol.com.

ACKNOWLEDGMENTS

The authors wish to acknowledge the contributions of many people who have helped in the preparation of this book, including the following: R. L. Rash for technical assistance, for support, and for supplying and preparing architectural plans; Greg C. Miller for help with the sections on symbols and lettering as well as assistance with Computer-Assisted Drafting; Bill Johnson for photography and consultation during the writing of the text; Margaret F. Miller for suggestions for the self-test and technical advice; Laura Miller for scanning and retouching photographs and for help with artwork; W. R. Harris for furnishing information on specifications; Jeff and Jenny Miller, Eleanor Harris, and other family members and friends who have offered encouragement and proofreading assistance during the writing of this book.

We appreciate the help of the following companies that supplied materials and photographs:

Hewlett-Packard Company, San Diego, California

Blu-Ray, Incorporated, Essex, Connecticut

Central Piedmont Community College Bookstore

Duncan-Parnell, Inc., Charlotte, North Carolina

We are also grateful to the following educators who reviewed the manuscript and offered suggestions for improvement:

David C. Alvarez, Western Technical College, El Paso, Texas

Dennis R. Bass, Lincoln Technical Institute, West Palm Beach, Florida

Michael Chandlee, Tennessee Technology Center at Pulaski, Pulaski, Tennessee

Jon J. Cookson, Paul D. Camp Community College, Franklin, Virginia

Art Gaudet, Sarasota County Technical Institute, Sarasota, Florida

Gary Reiman, Dunwoody Institute, Minneapolis, Minnesota

Joe Sutphin, Rustburg, Virginia

Special thanks to Barry Burkan of Apex Technical School in New York City, New York, for performing a technical edit of the manuscript.

PART I

BLUEPRINT READING

CHAPTER 1

Review of Basic Mathematical Procedures

OBJECTIVES

After studying this chapter, you should be able to:

- discuss reasons why heating and air-conditioning technicians must know basic mathematical procedures
- determine what areas of mathematics you need to review before proceeding with this text
- use equations to solve for unknown information

1.1 INTRODUCTION

A successful heating, ventilation, and air-conditioning (HVAC) technician must have a working knowledge of basic mathematical procedures. The technician is called on daily to accurately add, subtract, multiply, and divide whole numbers, fractions, and decimals. Taking measurements on the job requires the ability to calculate the area and volume of rectangles, triangles, and circles. The volume of cylinders and other geometric figures must often be determined. To calculate accurate answers, a technician must formulate and use basic equations.

Self-Test

The following problems are to be used as a self-test of a student's ability to use basic mathematical procedures. Work through all the problems before checking your answers. The correct answers for the odd-numbered problems are at the end of this chapter. The correct answers for the even-numbered problems are in the *Instructor's Guide.*

Formulas for solving problems in Chapter 1 appear in Appendix A. Use Appendix A as a reference, if necessary.

Students who have difficulty correctly answering any question in the self-test should review and learn those procedures where a weakness is evident before continuing with subsequent chapters in this text. This review can be accomplished in any of several ways. An individual may need only to refer to a basic mathematics textbook to refresh one's memory. A computer program for self-study may be available to help you relearn (or learn for the first time) those areas that cause difficulties. A general mathematics course may be well worth the time.

The point is that heating and air-conditioning technicians today need more than good common sense and the desire to do hands-on work in the mechanical field. Technicians are often expected to calculate prices, estimate amounts of materials needed, figure sizes of ductwork and other equipment, and carry on a host of other procedures that require a sound foundation in basic mathematical procedures and calculations.

The self-test begins with the basic functions of addition, subtraction, multiplication, and division, and it progresses into problems of increasing complexity. Remember that careless mistakes can be as costly as those resulting from ignorance of formulas and other procedures.

1.2 ADDING, SUBTRACTING, MULTIPLYING, AND DIVIDING WHOLE NUMBERS

1. $236 + 98$

2. $5490 + 436$

3. $567 - 421$

4. $6783 - 2357$

5. 2896×326

6. 242×32

7. $216 \div 18$

8. $400 \div 25$

9. The air quantities delivered to three areas of a building are as follows: 988 cubic feet per minute (cfm), 2349 cfm, and 6000 cfm. What is the total air quantity (in cfm) delivered to the three areas?

10. An oil company delivers the following quantities of oil to a business that has three tanks: 5689 gallons in Tank 1, 528 gallons in Tank 2, and 100 gallons in Tank 3. What is the total number of gallons of oil delivered to all three tanks?

11. A contractor has two houses under construction. The first house has 4795 square feet, and the second house has 2357 square feet. How much larger is the first house than the second one?

12. A sheet metal shop was billed for 2387 pounds of sheet metal. The shop foreman says that only 2298 pounds of sheet metal were delivered. If the foreman is correct, how many pounds of sheet metal would be needed to complete the order correctly?

13. If 1296 feet of copper tubing will be needed for the plumbing for one building, how many feet of copper tubing will be needed for 15 identical buildings?
14. A rectangular plot of land measures 567 feet frontage and 1234 feet deep. What is the square footage of the plot of land?
15. Two hundred air filters, 18 inches by 24 inches, are needed for a job. The filters are packed in boxes of 10 filters per box. How many boxes of filters should be ordered for the job?
16. The time estimated for completing a drafting job is 385 hours. If a drafter works 35 hours per week on this job, how many weeks will be needed to complete the job?

1.3 FRACTIONS

To convert an improper fraction to a mixed number, divide the numerator by the denominator. For example, to convert $9/2$ to a mixed number: $9/2 = 4\,1/2$.

Convert these improper fractions into mixed numbers:

17. $27/5$
18. $4/3$
19. $18/4$
20. $5/4$

To add or subtract a fraction, you must convert both fractions so they have a common denominator. For example, $1/3 + 1/6$. Convert $1/3$ to $2/6$, add $2/6 + 1/6 = 3/6$, and then reduce to $1/2$.

Use the correct procedures to answer the following:

21. $1/5 + 4/15 + 7/15 =$
22. $2/3 + 1/4 + 3/6 =$
23. $19/24 - 3/8 =$
24. $2\,1/4 - 7/8 =$

To multiply a fraction, multiply the numerators together and then the denominators together. For example, $2/3 \times 4/5 = (2 \times 4)/(3 \times 5) = 8/15$.

Find the answer to these equations by multiplying the fractions:

25. $1/2 \times 1/2 =$
26. $5/6 \times 11/12 =$
27. $1/3 \times 1/2 =$
28. A HVAC technician measures the components of a wall as follows: 4 inches of brick, $3/8$ inches of sheathing, $3\,1/2$ inches of insulation, and $1/2$ inch of Sheetrock. What is the total thickness of the wall?
29. The gasket on a walk-in cooler must be replaced. The door of the cooler measures $34\,1/2$ inches wide and $71\,1/4$ inches high. How many inches of gasket material will be needed to fit around the outside edges of the cooler door?
30. If $1/8$ inch is cut from a section of tubing $1\,1/2$ inches long, how much tubing is left?
31. A cylinder containing 19 pounds of R-22 refrigerant is used to recharge three air-conditioning units. Two pounds are used to fill the first unit, $3\,1/2$ pounds are used for the second unit, and $2\,1/4$ pounds are used for the third unit. How many pounds are left in the cylinder?

32. An oil burner uses 1/2 gallon of fuel per hour. How many gallons of oil will be used over a period of 7 1/2 hours?
33. A HVAC technician spends 3 1/2 hours repairing two units. If he charges $15 per hour for labor, how much will he charge for time spent repairing the 2 units?
34. A service technician spent 7 1/2 hours repairing 3 units. If he spent an equal amount of time on each unit, how many hours did it take to repair each unit?
35. A steel pipe 9 3/4 feet long is to be divided into 3 equal sections. How long will each section be?
36. Into how many 1/4 inch sections can a line 9/16 inches long be divided?

1.4 DETERMINING THE PERIMETER OF A SQUARE OR RECTANGLE

To find the perimeter of a square or rectangle, add the sides together. For example, the perimeter of a 4-foot square is 4' + 4' + 4' + 4' = 16'.

Find the perimeter of these squares:

37. Each side is 6 feet wide.
38. Each side is 4 inches wide.

Find the perimeter of rectangles with these dimensions:

39. Side A = 10 inches

    Side B = 12 inches

    Perimeter =

40. Side A = 6 feet

    Side B = 7 feet

    Perimeter =

1.5 DETERMINING THE AREA OF A SQUARE OR RECTANGLE

To find the area of a square or rectangle, multiply the length by the width. The answer will be in square inches, square feet, etc. This is also stated as FT^2 or sq-FT.

Find the area of the following squares:

41. Each side is 9 meters wide.
42. Each side is 30 inches wide.

Find the area of rectangles with these dimensions:

43. Side A = 42 feet

    Side B = 81 feet

    Area =

44. Side A = 30 inches

    Side B = 24 inches

    Area =

1.6 DETERMINING THE AREA OF A TRIANGLE

To find the area of a triangle, use this formula (Note: Base = L or length):

$$A = \frac{L + H}{2}$$

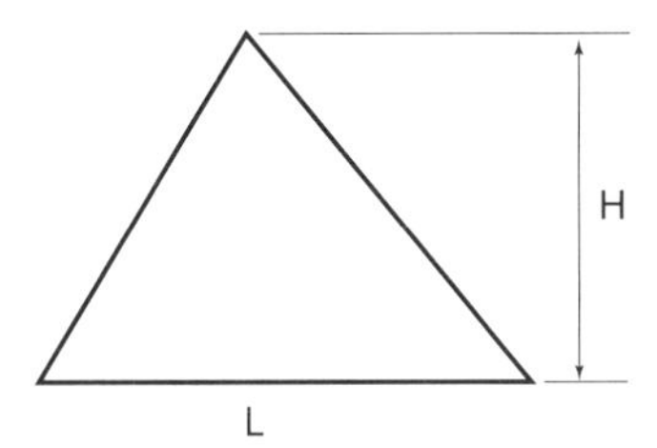

Find the area of triangles with the following dimensions:

45. Base = 5 feet
    Height = 8 feet
    Area =
46. Base = $4^1/_2$ inches
    Height = $9^1/_2$ inches
    Area =
47. Base = 21 feet
    Height = 50 feet
    Area =
48. Base = 17 inches
    Height = 17 inches
    Area =

1.7 DETERMINING THE AREA OF A CIRCLE

To find the area of a circle, use this formula:

$A = \pi \times R^2$

π = pi or 3.14

R = radius; the line from a circle's center to a circle's edge

D = diameter; the line dividing a circle in half

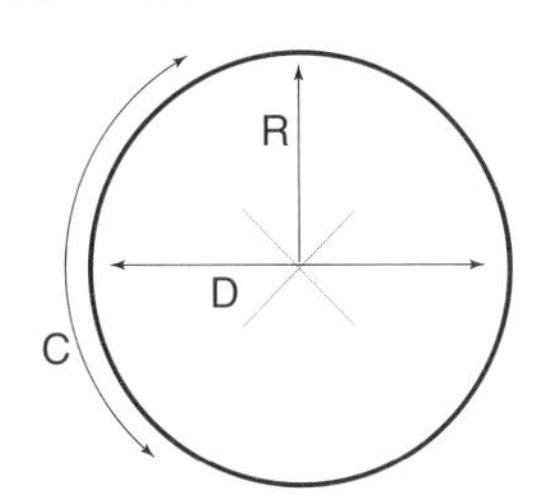

Use the proper equation to determine the area of a circle if the known dimensions are as follows:

49. Radius = 4 feet
50. Radius = $2^1/_2$ inches
51. Diameter = 7 inches
52. Diameter = 40 feet

1.8 DETERMINING THE CIRCUMFERENCE OF A CIRCLE

To find the circumference of a circle, use this formula:

$C = \pi \times D$

π = pi or 3.14

D = diameter; the line dividing a circle in half

Use the proper equation to determine the circumference of these circles:

53. Radius = 3 inches
54. Diameter = 12 inches

1.9 DETERMINING VOLUME

To find volume, use this formula:

$V = \pi \times R^2 \times H$

π = pi or 3.14

R^2 = the radius squared

H = the height of the cylinder

Find the volume of each figure described below (see Appendix A for formula):

55. Cylindrical tank with diameter of 3 feet and length of 5 feet
56. Cylindrical tank with diameter of 1 foot and length of 10 feet
57. Rectangular tank 3 feet wide, 4 feet high, and 8 feet long
58. Rectangular tank 10 feet wide, 7 feet high, and 8 feet long
59. Elliptical tank having a small radius of 2 feet, a large radius of 4 feet, and a length of 5 feet
60. Triangular tank with a 5-foot base, 5 feet long, and 5 feet high (the sloped sides are equal)

1.10 EQUATIONS

Write the equation for the following (Hint: See Appendix A.):

61. Area of a square
62. Area of a circle
63. Perimeter of a rectangle
64. Area of a rectangle
65. Area of a triangle

66. Circumference of a circle
67. Volume of a cylinder
68. Volume of a rectangular container

1.11 ADDITIONAL PROBLEMS

69. A full oil tank holds 300 gallons of Grade 2 fuel oil. During the past 3 months, the amounts of oil burned are 21 gallons, 55 gallons, and 127 gallons. How much oil is left in the tank?
70. One roll of duct insulation insulates 30 linear feet of duct. How many linear feet can be insulated with 65 rolls?
71. There are 144 electrical connectors in a box. How many connectors are there in 9 boxes?
72. A fuel tank holds 200 pounds of fuel. One gallon of fuel weighs 7.1 pounds. How many gallons of fuel are in a full tank?
73. A cylindrical oil tank is 3 feet in diameter and 5 feet long. How many cubic feet of oil will this tank hold?
74. A 100-foot coil of 3/8-inch copper tubing weighs 19.8 pounds. How much does 1 foot of this tubing weigh?
75. A stack of 80 pieces of sheet metal weighs 650 pounds. What is the weight per sheet?

1.12 APPLIED MATH

This chapter has presented many useful formulas for calculating the perimeter, the area, and the volume of various shapes found on blueprints. Review Appendix A for more information on calculating this information.

The purpose of this section is to apply this math to plans that will be useful in the field. Understanding this section will also be useful when completing the exercises in Chapter 31. The answers to this section are not found in the Answers to Self-Test Questions. Use the information provided in this chapter and Appendix A to solve these problems.

76. Find the perimeter of the building in Figure 1.1.
77. Find the gross area of the wall section in Figure 1.2. The wall height is 9".
78. Find the net wall area of the wall section in Figure 1.2.
79. In order to determine the heat gain through a cathedral ceiling shown in Figure 1.3, you will need to calculate its total area (use 1/4" = 1'-0'). What is the total area of this cathedral ceiling? (Hint: Calculate the area of the center rectangle and then add this to the areas of each four adjoining trapezoids.) The view of the ceiling is looking up from inside the room.
80. Calculate the area of the cathedral ceiling shown in Figure 1.3 (use 1/8" = 1'-0').

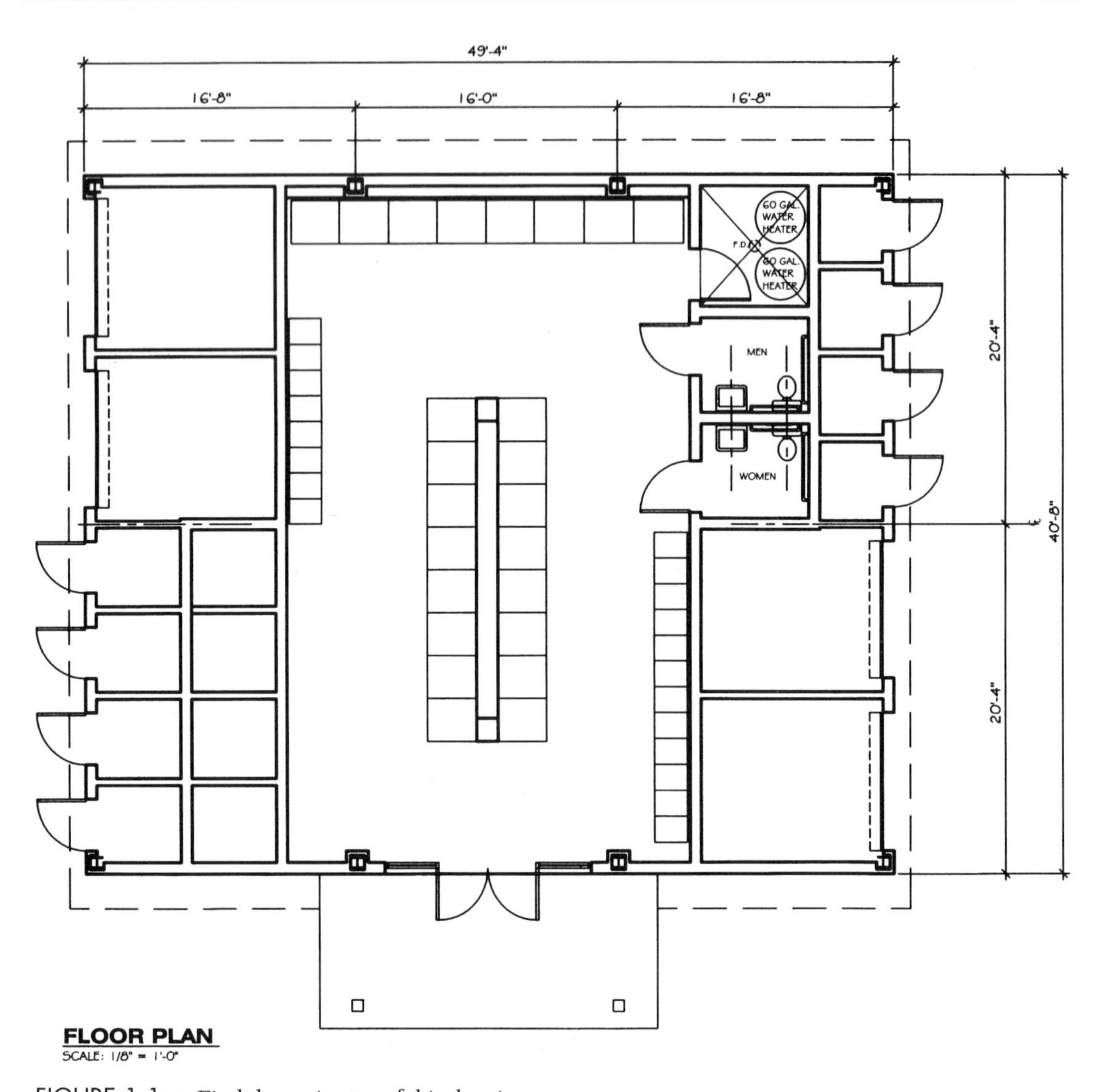

FIGURE 1.1 ■ Find the perimeter of this drawing

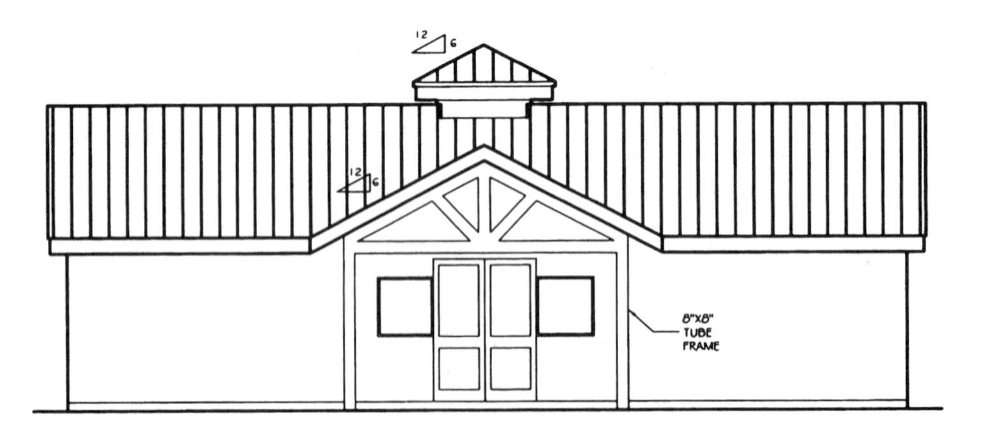

FIGURE 1.2 ■ Find the gross and net area of the front elevation. Hint: The gross area is the total area in square feet. The net area is the gross area minus the area of the windows and doors. Measure carefully

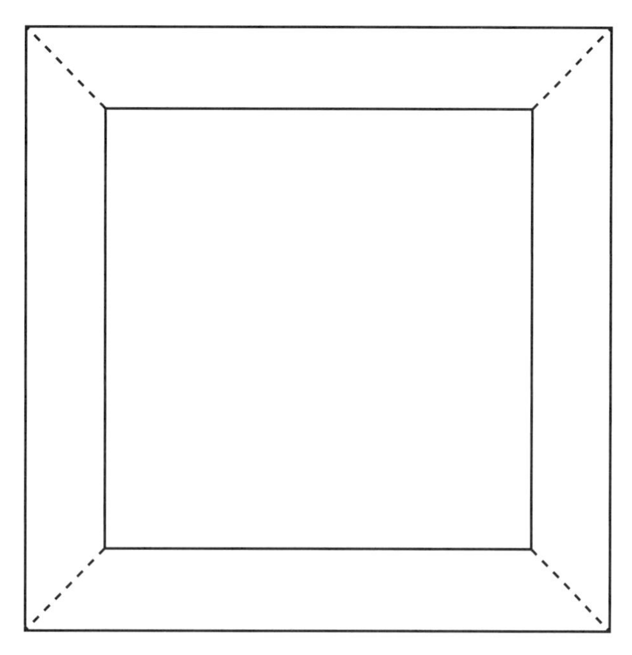

FIGURE 1.3 ■ Measure the gross cathedral ceiling area using a scale of 1/4" = 1'-0" for question 79 and a scale of 1/8" = 1'-0" for question 80

ANSWERS TO SELF-TEST QUESTIONS

(The answers to the odd-numbered questions follow. Even-numbered questions are answered in the *Instructor's Guide*.)

1. 334
3. 146
5. 944,096
7. 12
9. 9337 cfm
11. 2438 square feet
13. 19,440 feet
15. 20
17. 5 2/5
19. 4 1/2
21. 14/15
23. 10/24 or 5/12
25. 1/4
27. 1/6
29. 211 1/2 inches
31. 11 1/4 pounds
33. $52.50
35. 3 1/4 feet

37. 24 feet
39. 44 inches
41. 81 square meters
43. 3402 square feet
45. 20 square feet
47. 525 square feet
49. 50.24 square feet
51. 38.465 square inches
53. 18.84 inches
55. 35.325 cubic feet
57. 96 cubic feet
59. 125.6 cubic feet
61. $A = L \times W$ or $A = 4 \times S$
63. $P = 2L + 2W$
65. $A = L \times H \div 2$
67. $V = 3.14 \times R \times R \times H$
69. 97 gallons
71. 1296 connectors
73. 35.325 cubic feet
75. 8.125 pounds
77. Gross Area = L × H

    = 49'-0" × 9'-0"

    = 441 square feet ± 5 square feet

79. See Appendix A for formulas. The gross area of the cathedral ceiling is found by adding the area of the center square and the four trapezoids that surround the center section. The scale is 1/4" per foot.

    The area of the center square is: 8.5' × 8.5' = 72.25 square feet

    Area of the top and bottom trapezoid is: $1.5' \times \dfrac{8.5' \times 12'}{2} = 76.5$ square feet × 2 = 153 square feet

    Area of the side trapezoids are: $1.5' \times \dfrac{8.5' \times 12'}{2} = 76.5$ square feet × 2 = 153 square feet

    Total Area is 72.25 + 153 + 153 = **378.25 square feet**

CHAPTER 2

Safety

OBJECTIVES

After studying this chapter, you should be able to:

- discuss safety precautions for technicians working in the blueprint room and the drafting room
- discuss safety precautions for technicians on the job site
- tell why safety precautions are necessary for persons working with refrigerants and hazardous materials
- explain why safeguards are needed for persons working with electrical power
- explain what the acronym OSHA means, and discuss the impact of OSHA on safety in the workplace

2.1 INTRODUCTION

In spite of an emphasis on safety in the workplace, U.S. workers suffer approximately two million disabling accidents on the job annually, with some 80 million workdays lost each year. Most employers realize that safety is good business and that profits are greater when employees are safe and healthy. Yet, when owners see the same conditions and practices every day, safety problems become commonplace, and there may be no great urgency to make corrections. Workers must share the responsibility for safe conditions by inspecting the workplace frequently, reporting potential hazards, following manufacturers' instructions when using tools or machinery, and acting responsibly on the job.

Safety should be on a technician's mind at all times. Common sense and safety precautions must be observed in the blueprint room, in the drafting room, and on the job. The listing of safety rules in this chapter is not in any order of importance and is certainly not complete. The intent of this chapter is to remind the technician that accidents happen even in an environment not considered to be hazardous—it only takes the technician becoming careless or thoughtless on the job.

2.2 SAFETY RULES FOR TECHNICIANS ON THE JOB SITE

1. Keep the work area clean and orderly.
2. Wear proper protective gear, such as gloves and safety glasses, when working in extreme temperatures.
3. Be aware that pressures above and below atmospheric pressure can cause injury. (High pressures can pierce and cut the skin or inflict damage from objects that become airborne. A vacuum can cause a blood blister on the skin.)
4. Wear gloves and safety glasses when working with refrigerants.
5. Ensure that proper ventilation is present when working with refrigerants. (Refrigerants, being heavier than air, displace air. If adequate ventilation is not present, the technician may lose consciousness or become ill.)

6. Do not store refrigerants near an open flame, such as the pilot on a water heater or furnace. Refrigerant does not burn; however, when it is exposed to a flame, a toxic and corrosive gas is given off that deteriorates metal in the vicinity. This gas burns the eyes and hampers breathing.
7. When working with hand tools, use the proper tool for the job. Using the improper tool can cause injury to the technician, to the tool, and to the object being worked on.
8. Never release refrigerants into the atmosphere. Refrigerant reclaim systems are available, and federal law requires that all refrigerants be recovered rather than be released into the atmosphere to destroy the ozone layer.
9. Never pour refrigerant oil on the ground; it pollutes the earth. Send used refrigerant oils to a recycling company for treatment.
10. Transferring refrigerants from one cylinder to another should be done by an authorized person at the refrigerant manufacturing plant. Throwaway cylinders are meant to be used one time and then disposed of. It is illegal to refill these throwaway cylinders.
11. Shut off electrical power at the distribution entrance panel before installing equipment. A locking device should be put on the disconnect device by the installing technician so that someone else cannot turn the power back on inadvertently.
12. Ensure that all electrical circuits to air-conditioning equipment are grounded.
13. All electric hand tools used by a technician should be grounded or double insulated. Ground-fault circuit interrupters (GFCI) are required on temporary construction site power supplies.
14. Take special precautions when using a torch around combustible materials. Use a protective shield to prevent the heat from damaging surrounding materials and combustibles.
15. Never try to solder or weld a pipe or tubing that is sealed. Heat causes the contents of the pipe or tubing to expand, which may cause bodily injury.
16. When working around rotating equipment, do not wear neckties, gloves, or loose-fitting clothing that could become caught in the equipment, resulting in injury or death. Tie back long hair and secure it close to the head.
17. Make sure all nuts are tight on motor couplings before starting the motor.
18. Never attempt to stop a motor or other belt-driven equipment by gripping the belt.
19. Use your legs, not your back, to lift heavy objects. Always keep your back straight when lifting objects.
20. Personnel subject to lifting heavy weights should wear steel-toed shoes.
21. When using recommended chemicals in the air-conditioning field, know the chemicals and follow the manufacturers' recommendations for use. **Do not use chemicals except as recommended by the manufacturer.** Maintain a file of Material Safety Data Sheets (MSDS).
22. When working with electricity, consider every wire energized until proven de-energized. Verify that all circuits serving the working area are turned off and locked out. Tag the disconnect with your name, contact information, and date. This is true with *any* voltage. This is known as the Lockout/Tagout procedure.
23. Wear protective glasses or shields when working with liquid sprays or objects that might become airborne.
24. When replacing a three-phase motor, verify the rotation of the motor before connecting the drive.
25. Avoid "horseplay" in the work area.
26. Be sure that flame-source heating systems have ample combustion air and are properly vented.
27. Never use an open flame where the air has heavy concentrations of refrigerants. Doing so would cause phosgene gas, a colorless respiratory irritant with an unpleasant odor, to be created.

2.3 SAFETY—FEDERAL LAW

The federal government has recognized the fact that unsafe practices in the workplace are hazardous to workers and waste valuable human and material resources. In an effort to create a safer working environment, the federal government set up the Occupational Safety and Health Administration (OSHA) in 1970 to establish a set of regulations that companies must follow.

Regulations enforced by OSHA include those dealing with protective garments, fire prevention, and exposure to asbestos, lead, and other potentially dangerous substances. OSHA tries to educate workers and employers about hazards in the workplace. OSHA representatives visit companies and manufacturing plants unannounced to check on safety conditions for the workers. OSHA can issue citations for safety violations and can levy fines on the company. These fines can be heavy enough to put a small company with significant safety violations out of business.

Although executives make frequent complaints about paperwork, expensive changes required that may or may not improve safety for workers, and too much governmental intrusion into private industry, OSHA continues to make a significant impact on industries. In an effort to decentralize authority and to have an even more far-reaching effect, the federal government has encouraged states to develop their own health and safety programs to be approved by OSHA. The intent of government is to make workers and management realize that safe conditions and practices must have a high priority in the workplace.

SUMMARY

- The rules of safety and common sense should be followed by the technician in the office and on the job.
- Food and drink should not be present in blueprinting areas.
- Fumes and moving parts related to the printing process can be dangerous. Proper ventilation is required.
- Safety equipment is necessary when working in extreme temperatures or with hazardous equipment on the job.
- Materials should be safely stored.
- Take precautions when working with electricity to prevent hazardous conditions.
- Numerous legal and practical considerations for persons working with refrigerants have been established.
- A federal agency commonly known as OSHA enforces regulations dealing with safety issues.
- Safe conditions and practices should be a cooperative goal of labor, management, and government agencies.

REVIEW QUESTIONS

1. Explain why proper ventilation is necessary when a technician is working with refrigerants.
2. Why is it especially dangerous to store refrigerants near an open flame?
3. How should refrigerants be disposed of? Why?
4. What is the purpose of temporarily installing a locking device at the electrical distribution entrance panel?
5. List two instances in which proper clothing or protective gear is essential for safety.
6. Define OSHA and briefly discuss three of the duties of OSHA.
7. Why are business executives sometimes critical of OSHA?
8. In your opinion, who is responsible for seeing that proper safety practices are followed in the workplace? How can a company set up an effective safety program?

CHAPTER 3

Linear Measurement

OBJECTIVES

After studying this chapter, you should be able to:

- list three reasons why technicians need to understand scales and measurements
- define linear measurement
- differentiate between units of measurement in the English system and the metric system
- understand and read a 12-inch scale
- describe the relationship between inches, feet, and yards
- list and define the major instruments used in linear measurement

3.1 INTRODUCTION

To be successful on the job, the HVAC technician must know how to read standard scales and take accurate measurements. Dimensions and measurements must be correct for the calculation of heating and cooling loads. Correct dimensions are also necessary for manufacturing and installing ductwork, piping, and equipment.

Scales (also called scale rules) are measuring devices or instruments that enable persons to take measurements and then repeat these exact measurements at a later time and in another location. Scales have been designed so that they can easily be carried on the job.

Examples of the most common instruments for measuring linear distances are the ruler, the folding rule, and the steel tape rule. These devices are divided into standard graduations that can be read at any time. In other words, dimensions taken on the job are recorded and then reproduced in the shop and in the drafting room.

To establish and maintain accurate measurement standards for industry, science, and commerce in the United States, the National Bureau of Standards (NBS) was established in 1901. NBS is now known as the National Institute of Standards and Technology (NIST). This agency compares and coordinates its standards with those of other countries in the world. All scales and measuring devices have been calibrated in accordance with primary standards of the NBS, verifying and ensuring standard units of measure in the United States and throughout the world.

3.2 THE ENGLISH SYSTEM AND THE METRIC SYSTEM

The word linear is defined as "of or pertaining to a line or lines." Linear measurement is the measurement between two points along a straight line. This measurement can be the distance from one end of an object to the other or any distance along the way. It can be the distance from one wall to the other, the width of a door, the height of a ceiling, or the width of a piece of ductwork. Any time a person uses a folding rule, a 12-inch ruler, a steel tape, or a yardstick to measure a distance, the person is taking a linear measurement.

Two linear measuring systems—the English system and the metric system—are commonly used in the world today. The English system has as basic units of measurement the inch (in.), foot (ft), yard (yd), and mile (mi). Symbols for feet and inches are often used in the building industry. Using these symbols, 6 feet would be written as 6'-0".

The basic unit of the metric system is the meter. Prefixes express multiples of the meter. Common metric prefixes are *deci* (10), *centi* (100), and *milli* (1000). One meter equals 10 decimeters, 100 centimeters, and 1000 millimeters. Commonly used abbreviations in the metric system include the following: millimeter (mm), centimeter (cm), decimeter (dm), and meter (m).

Although most countries use the metric system, workers at all levels of the building trades in the United States have almost always preferred the English system. In 1975, Congress passed the Metric Conversion Act, calling for a voluntary change from the English system to the metric system; however, very little change in that direction occurred. Only a few companies that manufacture heating and air-conditioning equipment in the United States use the metric system, and architects and engineers continue to use the English system for drawing and dimensioning plans.

Since metric dimensions are occasionally used in heating and air conditioning, metric conversion tables are included in Appendix B to help students calculate conversions if necessary. We will use the English system throughout the body of this text.

3.3 REDUCED SCALES

Because it is not practical to draw plans full scale, the technician must be able to read and use reduced scales on construction plans. The reduced scale is used to measure distances; therefore, this kind of measurement is linear measuring. The architect's scale (Chapter 5) and the engineer's scale (Chapter 6) are used to reduce the scale of drawings.

3.4 THE ONE-INCH GRADUATION

The basic graduation for a measuring instrument in the English system is the inch. In Figure 3.1A, a rectangle has been drawn to represent one inch. In Figure 3.1B, the inch is divided into two equal parts. Each part of the inch is one-half (1/2) inch.

In Figure 3.1C, the two 1/2-inch segments have been divided into equal parts. With four equal parts of the original inch, each segment represents one-fourth (1/4) inch.

In Figure 3.1D, each 1/4-inch segment has been divided in half, creating eight equal parts called one-eighth (1/8) inch segments.

On some scales, the 1/8-inch segments have been divided in half to form 1/16-inch segments. The inch can be divided into 32 or 64 segments (also called a scale of 1/32 or 1/64). These divisions, however, are smaller than needed for most heating and air-conditioning work.

On the 1-inch scale, the longest division lines represent the 1-inch graduation. Slightly shorter lines represent the 1/2-inch graduation, even shorter lines represent the 1/4-inch graduation, and so on. By noticing the differentiation between the various lengths of division lines, one can easily determine the inch, the 1/2-inch, the 1/4-inch, the 1/8-inch, and the 1/16-inch graduations.

3.5 THE ONE-FOOT GRADUATION

In the English system, 12 inches are equal to 1 foot. When a measuring instrument is longer than 1 foot (12 inches), the markings for each foot are usually longer and more noticeable than the graduation marks for each inch (Figure 3.2).

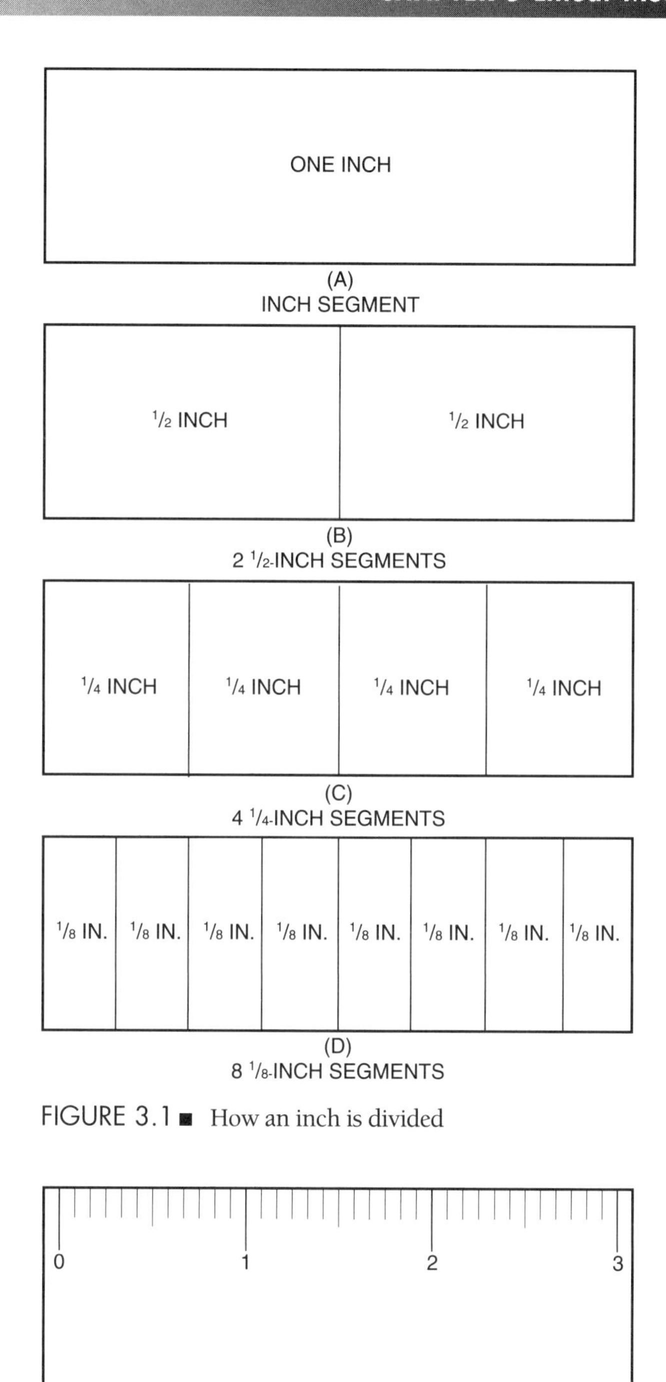

FIGURE 3.1 ■ How an inch is divided

FIGURE 3.2 ■ A 3-foot rule

3.6 THE ONE-YARD GRADUATION

The measuring instrument containing three 1-foot graduations is called the yardstick. The 1-foot graduations are subdivided into the 1-inch, the 1/2-inch, the 1/4-inch, and the 1/8-inch graduations. In most cases, yardsticks are made of wood and are used in the home as a general-use instrument. Yardsticks are hard to carry on the job, and they are not commonly used in the heating and air-conditioning industry.

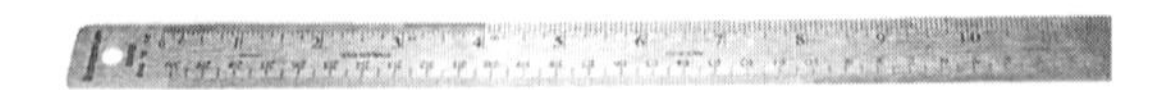

FIGURE 3.3 ■ 1-foot steel rule *(Photo by Bill Johnson)*

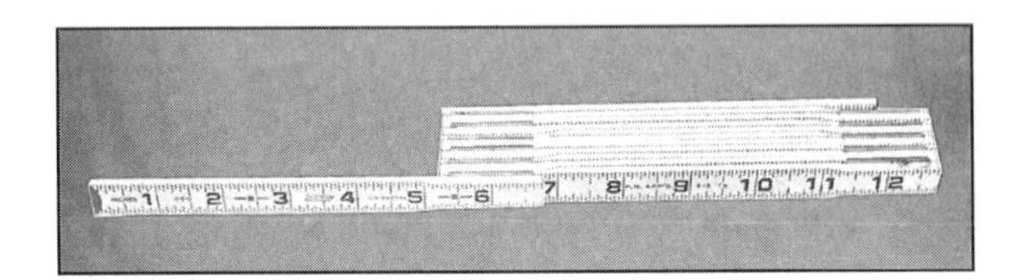

FIGURE 3.4 ■ 6-foot folding rule *(Photo by Bill Johnson)*

FIGURE 3.5 ■ Steel tapes in different lengths *(Photo by Bill Johnson)*

3.7 LINEAR MEASUREMENT INSTRUMENTS

The most common linear measurement instruments used in the heating and air-conditioning industry are:

- 1-foot scale—usually steel construction (Figure 3.3)
- 6-foot folding rule—usually wood construction with metal pivoting joints every 6 inches (Figure 3.4)
- Steel tape—flexible steel tape varying in length from a few feet to 100 feet (Figure 3.5)

3.8 DECIMALS AND FRACTIONS

Occasionally, a technician needs to be able to convert a number from a decimal to a fraction (or from a fraction to a decimal). For this reason, tables of commonly used fractions and their decimal equivalents have been included in Appendix B.

To review briefly, remember the following: To change a fraction to a decimal form, divide the numerator (top number) by the denominator (bottom number). The fraction $^{2}/_{5}$, therefore, becomes .40 (forty hundredths) as a result of this mathematical conversion.

Changing a decimal to a fraction can also be accomplished in a few simple steps. If the decimal is stated in hundredths (.40), place the numeral 40 without the decimal over the numeral 100 ($^{40}/_{100}$); if the decimal is stated in thousandths (.004), place the numeral 4 without the decimal over 1000 ($^{4}/_{1000}$); and so on. Consult the tables in Appendix B for further information on decimal-fraction equivalents.

SUMMARY

- Measurements are very important in the heating and air-conditioning field. Bids are determined from this information.
- The two major systems for linear measurements are the English system and the metric system.
- Most measurements in heating and air conditioning use the English system.
- Linear measurement refers to measurements made along a straight line.
- The graduations on a measuring instrument are standard within the United States, and the divisions and lengths are named and specified by definition.
- Accurate dimensions are necessary for the accurate calculation of heating and cooling loads, the provision of proper space for air-conditioning units, and the proper installation of ductwork and piping.

REVIEW QUESTIONS

1. Briefly discuss why a heating and air-conditioning technician needs to understand and be able to take accurate measurements.
2. What U.S. government agency is responsible for maintaining uniform standards of measurement?
3. In your opinion, why is it important that uniform standards of measurement be maintained?
4. Name the two major measuring systems used in the United States.
5. Define linear measurement.
6. Why does a heating and air-conditioning technician need to understand reduced scales?
7. Explain in a paragraph or diagram how a 1-inch graduation is subdivided into smaller segments.
8. List and describe three commonly used measuring devices.
9. Figure 3.6 contains a series of lines. Using a 12-inch scale rule, determine the length of each line.

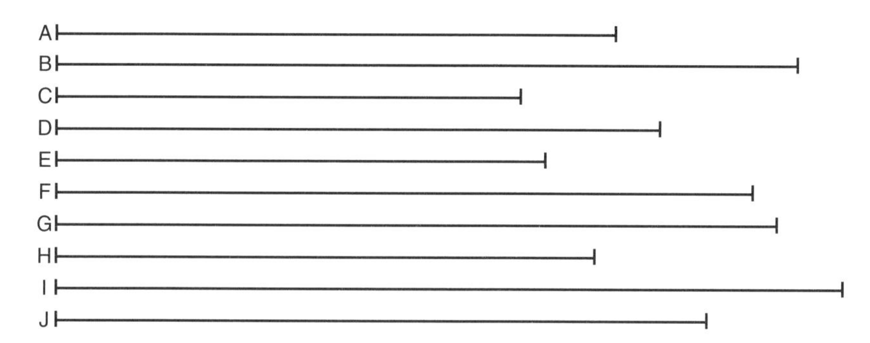

FIGURE 3.6 ■ Exercise for measuring lines

CHAPTER 4

Angular Measurement

OBJECTIVES

After studying this chapter, you should be able to:

- demonstrate an understanding of angular measurement
- describe the four angles formed at the center of a circle that has been divided into four equal parts
- define and illustrate three other types of angles
- describe the relationship between degrees, minutes, and seconds
- identify angles in a triangle
- determine the third angle in a triangle when the other two angles are known
- explain how there are 180 degrees in a straight line
- read and use a protractor to measure angles

4.1 INTRODUCTION

In addition to understanding linear measurement, the HVAC technician must know how to measure circles and other geometric figures. Buildings have circular spaces. Ductwork is often circular in shape. Unusual angles are created by the architectural design of many modern buildings. Triangular shapes are created when a room has a cathedral ceiling (Figure 4.1).

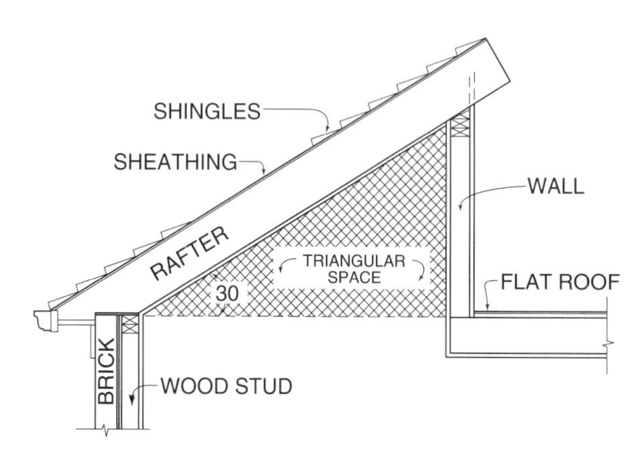

FIGURE 4.1 ■ Triangular shape in a building (crosshatched)

Regardless of the shape of the space involved, the HVAC technician must know how to determine areas and volumes in order to determine heating and cooling loads for sizing equipment and for air distribution. (Refer to Appendix A for information on calculating area and volume.)

4.2 ANGULAR MEASUREMENTS IN A CIRCLE

A circle is a closed curve on a plane with all points on the curve the same distance from the center of the circle. A circle contains 360 degrees (Figure 4.2). In other words, a circle can be divided into 360 sections that could be cut from the center of the circle. Each of these 360 sections equals one angular degree. When a circle is divided into four quadrants, each quadrant has 90 degrees.

A useful term in angular measurement is vertex. The point where two straight lines intersect is called the vertex (Figure 4.3). Lines extending from the vertex are called *sides*. Spaces between the lines, which taper to a point at the vertex, are called *angles*. These spaces (angles) are measured in degrees. If a vertical line passes through the center of a circle, 180 degrees can be measured between the line to the right of center and the line to the left of center (see again Figure 4.2).

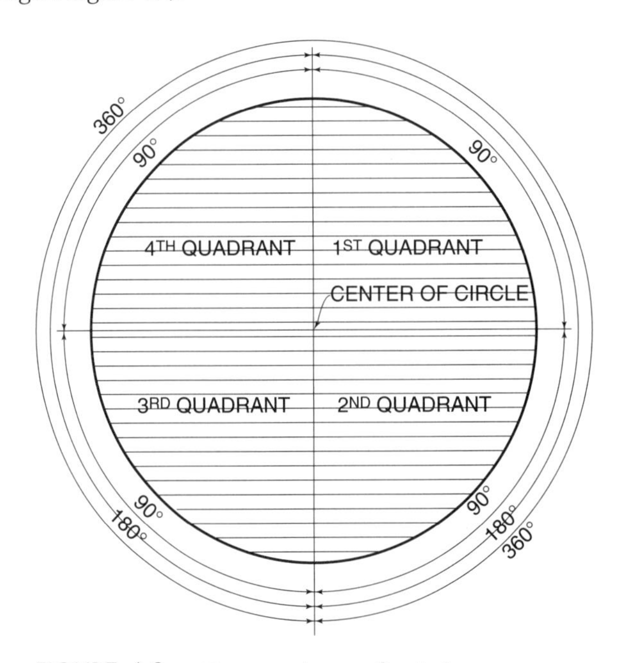

FIGURE 4.2 ■ Four quadrants of a circle

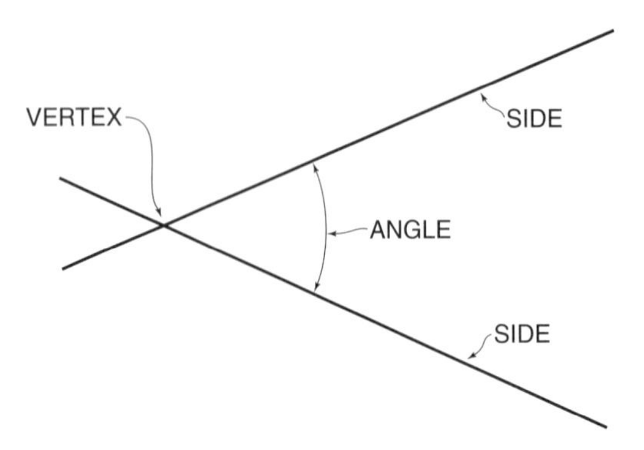

FIGURE 4.3 ■ Vertex and sides of an angle

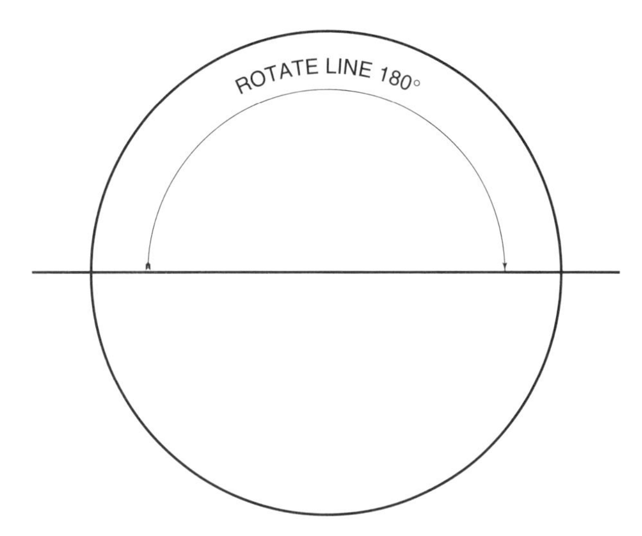

FIGURE 4.4 ■ A straight angle

If a *horizontal* line is drawn through the center of a circle, the portion of the line to the left of center is 180 degrees from the portion of the line to the right of the center. *A straight line, therefore, contains 180 degrees* (Figure 4.4).

4.3 FOUR KINDS OF ANGLES

Four types of angles are the *right angle*, the *acute angle*, the *obtuse angle*, and the *straight angle*.

If a vertical line and a horizontal line cross at the center of a circle as illustrated in Figure 4.5A, the circle is divided into four equal parts. Each of the four equal angles in the circle has 90 degrees (one-fourth of the 360 degrees in the whole circle). Each 90-degree angle is called a right angle.

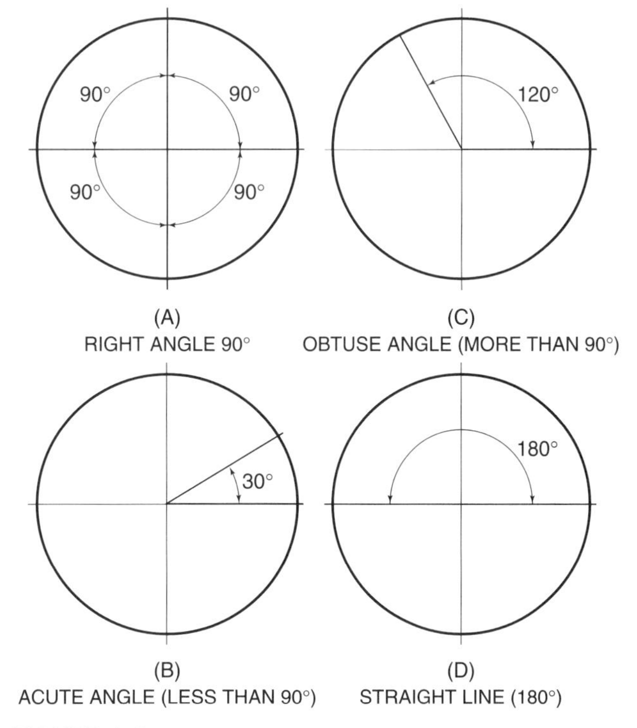

FIGURE 4.5 ■ Four kinds of angles

An angle of less than 90 degrees is called an acute angle (Figure 4.5B). An angle greater than 90 degrees is called an obtuse angle (Figure 4.5C). An angle between two horizontal lines drawn end to end through the center of the circle is equal to 180 degrees and is called a straight angle or a straight line (Figure 4.5D).

4.4 ANGLES OF A TRIANGLE

A triangle is a geometric figure with three sides and three inside angles. A triangle can be created by breaking a straight line in two places and then joining the ends together. The figure that has been created has three sides and three inside angles. Since the figure started as a straight line, the sum of all three angles is equal to 180 degrees (Figure 4.6).

If the sum of two known angles is subtracted from 180 degrees, the difference is the number of degrees in the unknown angle. This rule holds true regardless of the shape of the triangle. In Figure 4.7, angles A + B + C = 180 degrees. *The sum of two enclosed angles subtracted from 180 degrees equals the third angle.* Hence, the equations:

$A + B + C = 180$ degrees

180 degrees $- (A + B) = C$

180 degrees $- (B + C) = A$

180 degrees $- (C + A) = B$

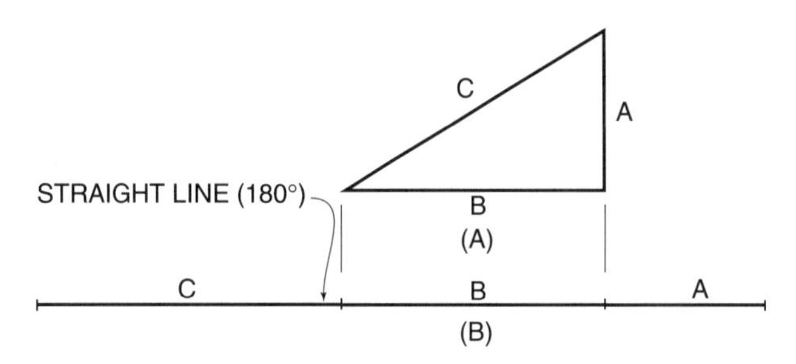

FIGURE 4.6 ■ How a triangle is formed

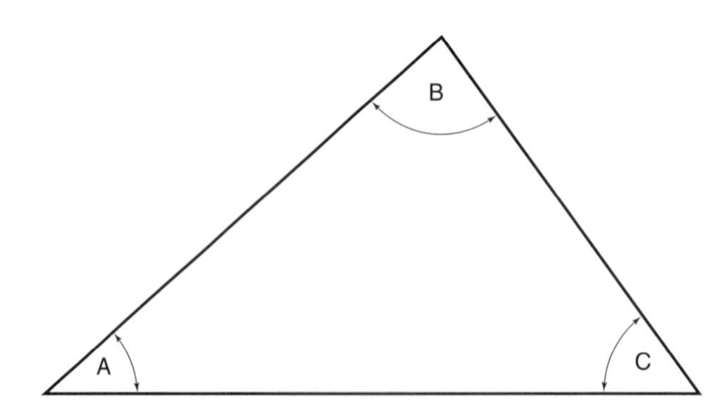

FIGURE 4.7 ■ The three angles of a triangle

4.5 USING A PROTRACTOR

A protractor is an instrument consisting of a half-circle with the midpoint (center) marked on the horizontal portion (base) of the protractor. This midpoint, also called the reference point, is clearly marked on the protractor. The half-circle shape of the protractor is divided into degrees, which are usually numbered from right to left and from left to right, so that angles can be measured from the right side or the left side (Figure 4.8).

To use the protractor, place it over the angle to be measured, with the base (horizontal) portion of the protractor directly on the side of the triangle and the point of the angle at the midpoint on the base of the protractor. The angle in Figure 4.9 measures 30 degrees.

With the protractor base directly on the side of the angle and the point of the angle directly under the midpoint, the adjacent side (line) of the triangle intersects the degree scale on the half-circle. The number of degrees of the angle can be read on the degree scale.

If the triangle is too small for the adjacent side to cross the degree scale of the protractor, extend the adjacent side so that the extended line intersects the degree scale.

Special care must be taken in the use of a protractor. The base must be placed directly over the side of the angle being measured, and the point of the angle must be directly under the midpoint of the base of the protractor. The drafter must be accurate in reading the degrees on the half-circle at the point where the adjacent side intersects. For accuracy in measuring angles, it is very important to take all these precautions and to hold the protractor firmly in place without letting it slip.

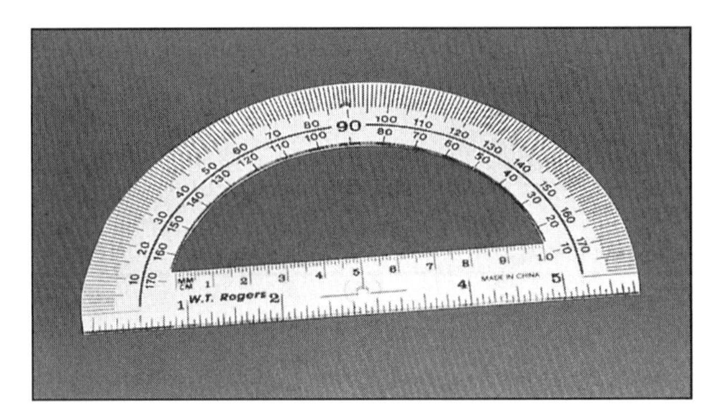

FIGURE 4.8 ■ A protractor *(Photo by Bill Johnson)*

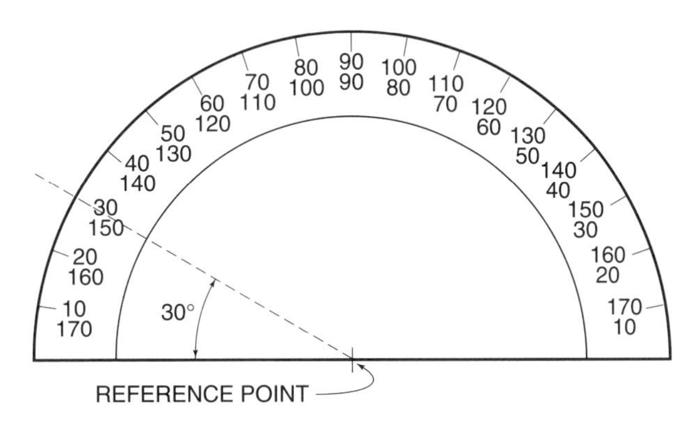

FIGURE 4.9 ■ Using a protractor

4.6 DEGREES, MINUTES, AND SECONDS

Sometimes, a measurement requires more accuracy than can be obtained by measuring whole degrees. To allow this accuracy, each degree can be subdivided into sixty sections called *minutes*. Each minute can be further subdivided into sixty sections called *seconds*. (Remember that minutes and seconds as they refer to angular measurements are different from minutes and seconds as measurements of time.) Minutes and seconds divide parts of a circle into very small segments. For example, in a right angle (90 degrees), there are 5400 minutes (90 degrees × 60 minutes per degree) and 324,000 seconds (90 degrees × 60 minutes × 60 seconds per degree).

A small variation in an angular measurement makes a great difference. This difference is especially obvious in navigation for air, sea, and space travel. For example, the distance across the United States is approximately 3000 miles. If a navigator makes a 1-degree mistake in charting the destination across the country, the vehicle will miss the mark by more than 52 miles!

It should be noted that computers and advanced technology can be used to figure angular measurements to extremely minute (very small) numbers. Space technology and high-tech equipment require measurements smaller than the human eye can discern. However, extremely small measurements are rarely needed by the HVAC technician who relies on instruments like the protractor to take measurements in whole degrees.

SUMMARY

- A circle is a closed curve on a plane with all points on the curve the same distance from the center of the circle.
- The vertex is the point where two intersecting lines meet.
- The lines extending out from the vertex are called sides.
- The space between the lines extending out from the vertex is the angle between those lines.
- A straight line (or straight angle) has 180 degrees.
- A right angle has 90 degrees.
- An acute angle has fewer than 90 degrees.
- An obtuse angle has more than 90 degrees.
- A triangle has three sides and three inside angles.
- A protractor is an instrument used for measuring angles.
- Angles can be subdivided into tiny segments called minutes and seconds.

REVIEW QUESTIONS

1. Why must a heating and air-conditioning technician know how to measure circles and other geometric figures in his or her job?
2. List and define the following terms as they refer to circles:

   A. Vertex

   B. Sides

   C. Angle

   D. Straight angle

   E. Right angle

   F. Acute angle

   G. Obtuse angle
3. Define a circle. How many degrees are contained in a circle?
4. Define a triangle. How can a straight line be changed into a triangle?
5. If two of the angles of a triangle are known, how can the number of degrees in the third angle be determined?

6. Define a protractor. What is meant by the reference point of a protractor?
7. Why is it essential that the drafter be very careful in using the protractor?
8. How many minutes are in one degree?
9. How many seconds are in one minute?
10. In your opinion, why is it usually not necessary for a heating and air-conditioning technician to measure angles in minutes and seconds?

ADDITIONAL STUDENT EXERCISES: MEASURING ANGLES

1. Using a protractor, measure the two angles shown in Figure 4.10 and then record your answers below.

   Answers:

   Angle A = _____ degrees

   Angle B = _____ degrees

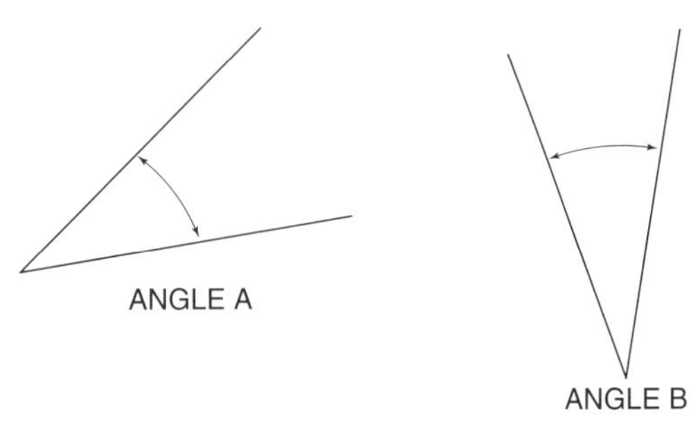

FIGURE 4.10 ■ Exercise 1

2. Using a protractor, determine all the angles in each triangle in Figure 4.11 and then record your answers below.

   Answers:

   A = _____ degrees

   B = _____ degrees

   C = _____ degrees

   D = _____ degrees

   E = _____ degrees

   F = _____ degrees

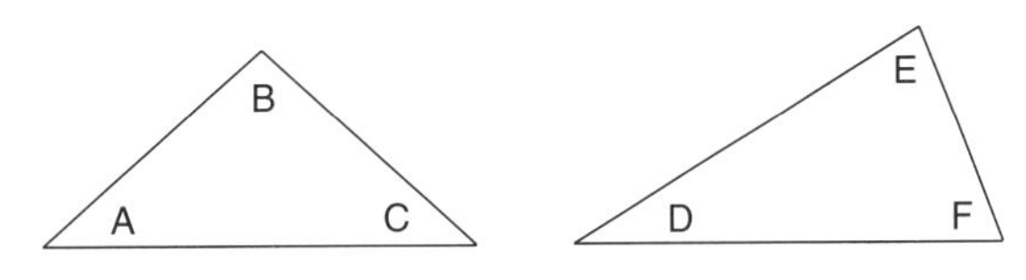

FIGURE 4.11 ■ Exercise 2

CHAPTER 5

Learning to Use the Architect's Scale

OBJECTIVES

After studying this chapter, you should be able to:

- describe several sizes and kinds of architect's scale rules
- demonstrate the ability to use a 1-inch scale
- explain why a drafter needs to be able to read and use an architect's scale
- state several criteria used by a drafter in order to select the proper scale for a drawing
- demonstrate the ability to use the $^1/_8$-, $^1/_4$-, $^1/_2$-, and $^3/_4$-inch scales

NOTE: Refer to an architect's scale while studying this chapter.

5.1 INTRODUCTION

In drafting and freehand sketching, it is important for the entire drawing to be in proportion. With freehand sketching, drawings are made to *near scale*. In the drafting room, drawings are drawn *to scale*. Drawing a building to *full scale* would be unthinkable because the finished drawing would be the same size as the building. Even if plans could be drawn and printed to full scale, they would be too bulky for use on the construction site. To keep drawings and plans in proportion and at a practical size, architects, engineers, and drafters depend on scale rules (often simply called scales) for measurement. With these scales, drawings can be reduced to a size suitable for the construction site or the shop. Also, since the drawings are *to scale* (even though they are not *to full scale*), all lines on a plan are in proportion. Choosing the scale is important because the architect, engineer, or drafter tries to draw plans as large as possible within the confines of the page (the larger the plan, the easier it is to read). Many reduced scales are on the architect's scale rule. This chapter will discuss the ones that are most commonly used.

5.2 THE ARCHITECT'S SCALE RULE

Of the various types of scale rules in the drafting room, the architect's scale rule is probably used most often. It is also frequently carried by those in the air-conditioning field. The best known is a triangular scale rule 12 inches long that has a total of eleven scales (Figure 5.1).

Another version of the triangular scale rule is only 6 inches long. This 6-inch scale rule also contains eleven scales.

A flat scale rule 6 inches long is very popular because it contains eight of the most commonly used scales, and it easily fits into a shirt pocket (Figure 5.2). Manufacturers sometimes give these flexible, plastic, 6-inch

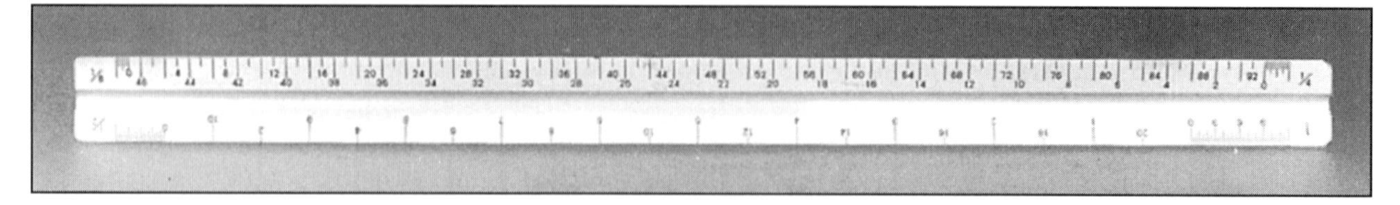

FIGURE 5.1 ■ 12-inch architect's scale *(Photo by Bill Johnson)*

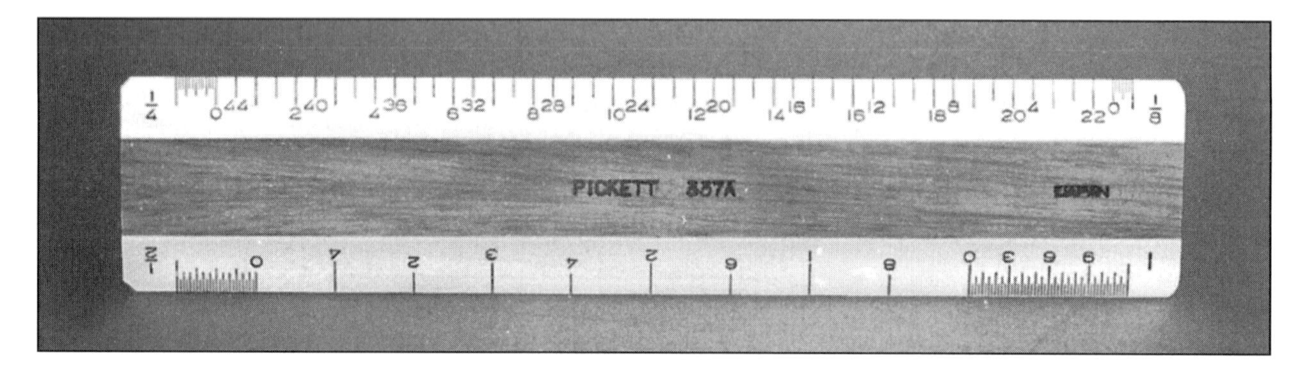

FIGURE 5.2 ■ 6-inch architect's scale *(Photo by Bill Johnson)*

scale rules with advertisements printed on them to architects, engineers, and drafters. Some architects, engineers, and drafters prefer 12-inch, flat scale rules that have the same scales as the 6-inch versions.

5.3 UNDERSTANDING THE ARCHITECT'S SCALE RULE

At first glance, the architect's scale rule appears complicated and confusing because so many different scales are shown on the same measuring device. On the 12-inch, triangular rule, two sides of the rule hold four different scales each, and the other side holds three additional scales. Some of these scales are read from left to right, and others are read from right to left. The size of each scale is indicated at the end of the rule.

The scale size of a drawing is actually a ratio of the size of the drawing compared to the size of the building (or another object being drawn to scale). A scale of 1/4" = 1'-0" means that each one-fourth inch on the drawing represents one foot on the building itself. Other frequently used scales are as follows:

1/8" = 1'-0"

1/2" = 1'-0"

3/4" = 1'-0"

1" = 1'-0"

The architect makes a note on the blueprint to denote which scale was selected for that particular drawing.

5.4 PRACTICING WITH THE ONE-INCH SCALE

To learn how to read and use the architect's scale rule, you will need the following items: a triangular, 12-inch architectural scale rule, a piece of paper, and a pencil. (A flat architectural scale rule could be used if the triangular rule is not available.) Then, follow these step-by-step directions:

1. Place the architectural scale rule in front of you so that the "1" on the rule is on your left and the "1/2" is on your right. The numeral "1" stands for the 1-inch scale, meaning that each inch represents 1 foot on the original building (or other object). You read the 1-inch scale from *left to right*; the markings from "1" to "10" represent measurements for the 1" = 1'-0" scale.
2. Next, look at the "1/2" at the right-hand end of the scale. Reading (backward) from *right to left*, notice that the 1/2" = 1'-0" scale has small markings from "2" to "20." On this scale, each one-half inch represents 1 foot of the original building or object. You will use the "1/2" scale later.
3. Now concentrate on the 1-inch section located at the left of the scale between the "1" and the "0" (Figure 5.3). Next to the "1" signifying the 1-inch scale are tiny subdivisions and the numerals "9," "6," "3," and "0." This section of the rule is used for lines that represent less than 1 foot in measurement. Each numeral within the small section represents that many inches using the 1-inch scale; that is, the "9" stands for a 9-inch measurement, the "6" stands for 6 inches, and so on. By counting the tiny marks between the numbers, you can also figure unlabeled inches, half-inches, and quarter-inches on this scale.
4. Now you are ready to draw lines using the 1" = 1'-0" scale. To draw a line representing a measurement of 6 feet, for example, count past the 1, 2, 3, and so on to find the numeral 6 on the "1" scale. Then, draw a line between the "6" and the "0," being careful to drag the pencil and not tear the paper. (Usually, a right-handed drafter will move the pencil from left to right, and a left-handed drafter will move from right to left.) The line you have drawn should be 6 inches long, representing 6 feet on a scale of 1" = 1'-0".

To draw a line that represents 6 feet and 3 inches ($6\frac{1}{4}$ feet), use the same scale to draw a line from the "6" to the "0" and continue past the "0" to the small "3" (which represents 3 inches more). The line represents 6 feet and 3 inches ($6\frac{1}{4}$ feet) on a scale of 1" = 1'-0".

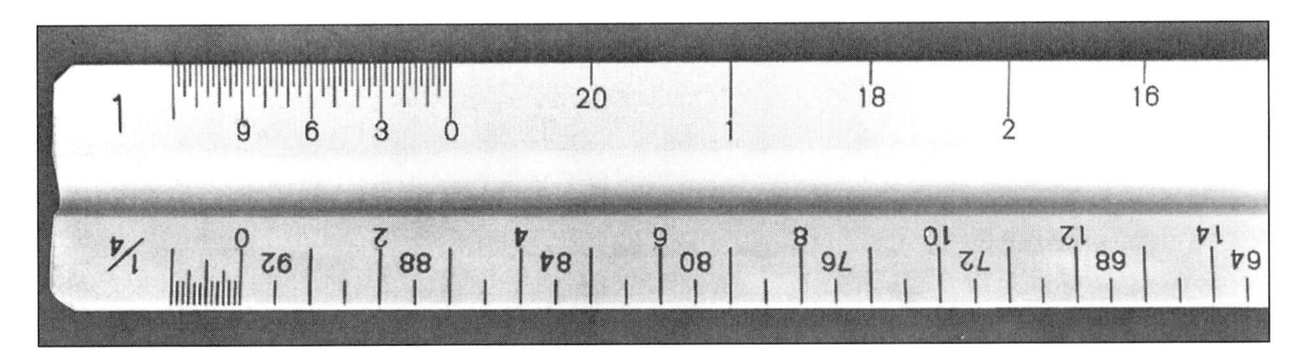

FIGURE 5.3 ■ 1-inch scale on the architect's scale *(Photo by Bill Johnson)*

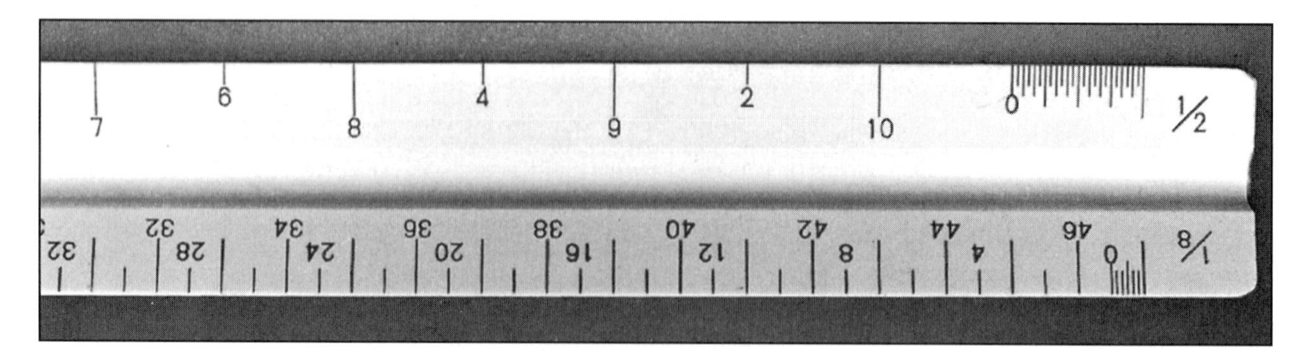

FIGURE 5.4 ■ 1/2-inch scale on the architect's scale *(Photo by Bill Johnson)*

5.5 USING THE HALF-INCH SCALE

If you understand the "1" scale as described above, it should not be difficult to read the "1/2" scale too. To read the 1/2-inch scale, use the same face of the triangular scale, but this time, you need to read from *right to left* (Figure 5.4). Again, follow the step-by-step procedure:

1. Look at the right-hand end of the architectural scale rule at the fraction "1/2," denoting the 1/2" = 1'-0" scale. Each one-half inch on this scale represents 1 foot on the actual building or object.
2. Notice the marks just to the left of the "1/2" designation. This small section shows 1 foot (12 inches) broken down into inches and fractions of an inch. In Figure 5.4, the section up to the "0" has 24 tiny lines so that you can accurately draw measurements containing inches or fractions of an inch. Each of the smallest division lines represents one-half inch. The longer divisions represent whole inches, and the longest divisions represent 3 inches each.
3. On your 12-inch architectural rule, read from *right to left* every other number from "2" (for 2 feet) to "20" (for 20 feet). The line you draw will extend between the "0" at the right-hand end of the scale and the designated number. (For this step, Figure 5.4 is not adequate, since it shows only a small segment of the rule.)
4. Now you are going to use the "1/2" scale to draw a line representing 12 feet. Start at the "0" on the right-hand end of the scale and read backward by twos to the numeral "12." Draw the line between the "12" and the "0." Your line is 6 inches long, and it represents 12 feet when the 1/2" = 1'-0" scale is used.
5. To draw a line representing 12 feet and 9 inches (12 3/4 feet), draw the line from the numeral "12" to the "0" just as you did before and then continue on for three major subdivisions to the right of the "0," representing 9 inches more than the 12 feet.

5.6 USING OTHER SCALES

The other scales function in the same way as the ones you have just used: The small area subdivided at each end of the rule ending with a "0" represents fractional parts of 1 foot, with the number of subdivision lines indicating what fraction is represented. The small area always represents 12 inches, and each mark inside may stand for 1 inch, one-half inch, or another fractional part, depending on what scale you are working with.

You can use other scales accurately by referring to the correct position on the triangular architectural scale. You must be careful to read the numbers from the correct end of the scale, since two different scales occupy the same space on the face of the triangular scale and the numbers for the different scales overlap.

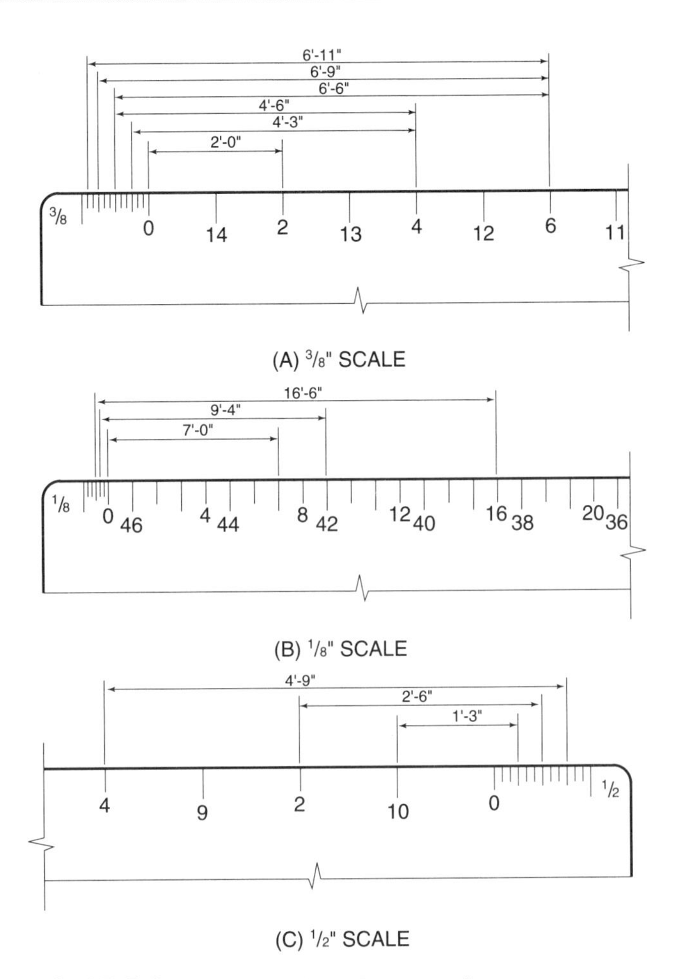

FIGURE 5.5 ■ Reading the architect's scale

The scale on the left reads *left to right*, and the scale on the right reads *right to left*. Figure 5.5 shows measurements that have been made with the 3/8" scale, the 1/8" scale, and the 1/2" scale.

5.7 THE MOST COMMONLY USED SCALES ON THE ARCHITECT'S SCALE RULE

The drawing paper or drawing film used by architects and engineers is usually 24" × 36" or 36" × 42". The 24" × 36" drawing paper or film is almost always used for drawing plans for residential and small commercial buildings. If the building is 100 feet by 200 feet, the architect and engineer can put the entire floor plan on a 24" × 36" plan with a scale of 1/8" = 1'-0". (Figure 5.4 contains the 1/8" = 1'-0" scale along with the 1/2" = 1'-0" scale, but it can be best read by inverting the page.) A building 20 feet × 50 feet can be easily shown on a 24" × 36" sheet when the 1/4" = 1'-0" scale is selected.

For larger buildings, the architect usually chooses the 1/8" = 1'-0" scale. In some cases, smaller scales are used, but the plans become very difficult to read with any accuracy.

Depending on the size of the building and the size of the drawing paper (or drawing film), the architect and engineer usually draw floor plans of buildings using the 1/8" = 1'-0" or the 1/4" = 1'-0" scales. Details, diagrams, and sections are sometimes drawn to larger scales, such as the 3/4" scale, the 1" scale, and even the 1 1/2" scale.

SUMMARY

- It would not be practical to draw building plans to full scale.
- The 12-inch, three-sided architect's scale rule has eleven scales.
- The "0" on each scale separates the inches from the feet on the architect's scale.
- Several kinds and types of architect's scale rules exist; however, the standard scales should be the same on all rules.
- The smaller the scale, the smaller the drawing.
- The most commonly used scales on the architect's scale are the 1/8" scale and the 1/4" scale.
- Details, diagrams, and sections are usually drawn to a larger scale, such as 3/4", 1", and 1 1/2" scales.

REVIEW QUESTIONS

1. Name all the scales found on the 12-inch, three-sided architect's scale.
2. What are the two most commonly used scales on the architect's scale rule?
3. Discuss the factors considered by an architect or engineer when choosing a scale for drawing a floor plan on a sheet.
4. With an architect's scale rule, measure each line in Figure 5.6 using the 1/8", the 1/4", the 1/2", the 3/4", and the 1" scales for measuring each line. (You will have five different answers for each line measured, with a different answer for each scale used.)
5. With an architect's scale rule, measure each line in Figure 3.6 using the 1/8", the 1/4", the 1/2", the 3/4", and the 1" scales for measuring each line.

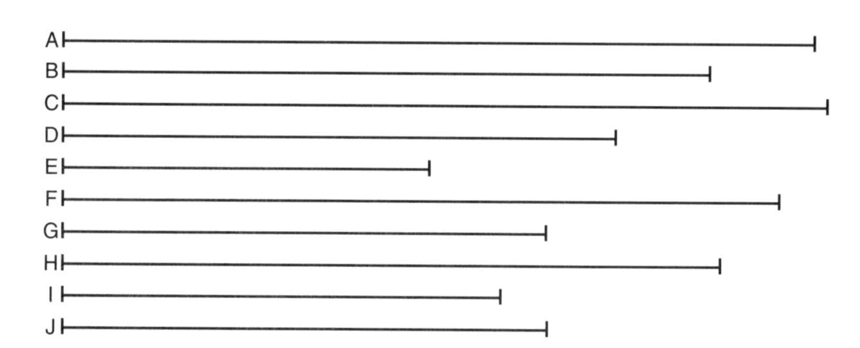

FIGURE 5.6 ■ Reading the architect's scale

ADDITIONAL STUDENT EXERCISES: USING THE ARCHITECT'S SCALE

Using the 1/8" scale and the 3/8" scale, give the indicated dimensions for Figure 5.7.

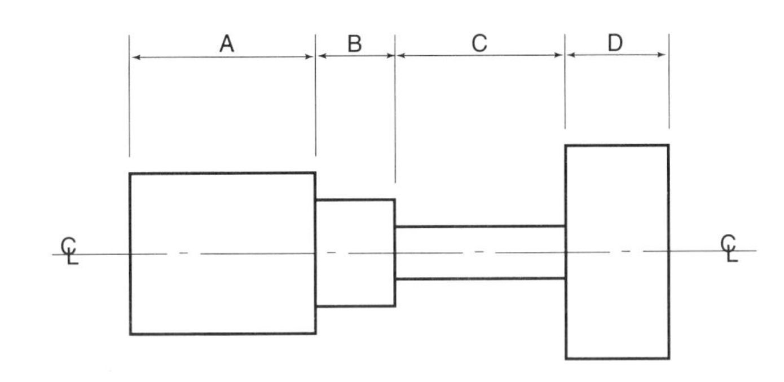

USING THE ARCHITECTURAL SCALE, GIVE THE INDICATED DIMENSIONS USING THE 1/8 SCALE AND THE 3/8 SCALE.

SCALE	DIMENSIONS			
ANSWER	A	B	C	D
1/8 SCALE				
3/8 SCALE				

FIGURE 5.7 ■ Exercise 1

CHAPTER 6

Learning to Use the Engineer's Scale

OBJECTIVES

After studying this chapter, you should be able to:

- discuss the purpose of using the engineer's scale rule
- give examples of types of plans for which the engineer's scale rule is commonly used
- describe three different kinds of engineer's scale rules
- explain and demonstrate how the 10 scale is used
- list other commonly used scales and demonstrate how to use them

NOTE: Refer to an engineer's scale while studying this chapter.

6.1 INTRODUCTION

The engineer's scale rule and the architect's scale rule are very similar. The purpose of the engineer's scale rule is to reduce the size of a plan to scale so that the plan can be put on workable-size drawing paper or drawing film. The civil engineer often uses the engineer's scale rule because it easily adapts to making drawings of large tracts of land that are to be developed.

With the engineering scale rule, almost any reduced scale can be used as long as it relates to tens, hundreds, thousands, and so forth. Site plans and plot plans are usually drawn using the engineer's scale. HVAC technicians use site plans and plot plans to locate air-conditioning piping and equipment on the site. For this reason, technicians must know how to use the engineer's scale rule.

On smaller jobs, architects often use the engineer's scale to develop the site plan and show the grading plan for the site. If the site is small, a smaller scale on the *architect's rule* (like 1/16"= 1'-0"or even 1/32" = 1'-0") may be selected. In most cases, however, the architect prefers the engineer's scale for developing small scale drawings.

6.2 THE ENGINEER'S SCALE RULE

The engineer's scale rule is designed so that large dimensions can be reduced to a small scale. Several types of engineer's scale rules exist. The most common type is the 12-inch, triangular scale rule (Figure 6.1). Each of the three sides has two scales. The triangular engineer's scale is also available in a 6-inch model. Some drafters find this 6-inch rule easier to use than the 12-inch version.

FIGURE 6.1 ■ 12-inch engineer's scale *(Photo by Bill Johnson)*

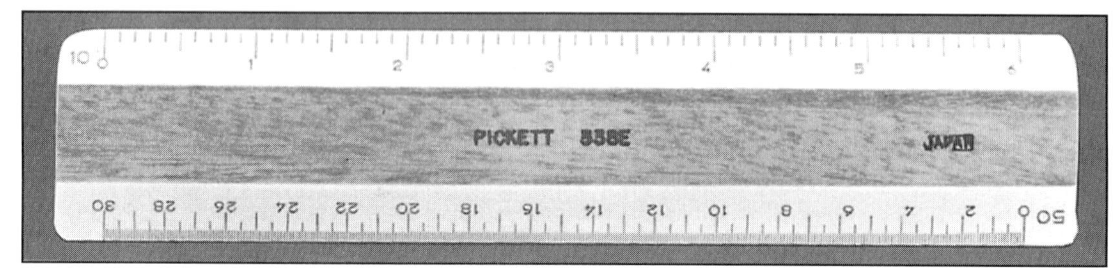

FIGURE 6.2 ■ 6-inch engineer's scale *(Photo by Bill Johnson)*

Some engineers and drafters prefer to use a 12-inch, flat (or two-sided) engineer's scale rule. These flat rules have only two scales on each side; however, these four scales are the most commonly used scales in the engineering field. The flat rule is also available in a 6-inch model, which is popular because it is easily carried in the shirt pocket (Figure 6.2). Like the flat 12-inch rule, it is limited to two scales on each side rather than two scales on each side of the triangular rule.

Equipment manufacturers and other retailers and businesses distribute flexible plastic engineer's scale rules with company names on them for advertisement. These rules are similar to the flat rule described above. They are popular because they are easy to carry and usually obtained free.

6.3 WHAT MAKES THE ENGINEER'S SCALE UNIQUE

The architect's scale is divided into units of 2, 4, 6, 9, 12, and so forth, which relate to feet and inches very readily. The engineer's scale is divided into units of 10, 20, 30, 40, 50, and 60. One must be careful not to mix the engineer's scale with the architect's scale. Even on the 10 scale, where each 1-inch segment is numbered, the inch segment on the engineer's rule is divided into tenths (with ten division marks per inch). Figure 6.3 shows the divisions on a partial 10 scale. On a standard foot rule, each inch is separated into eight divisions, signifying halves, quarters, and eighths of an inch.

6.4 USING THE ENGINEER'S SCALE RULE

The triangular, 12-inch engineer's scale has six scales (two on each side). The scales are numbered 10, 20, 30, 40, 50, and 60. Each scale is then subdivided so that each inch has ten major subdivisions on the 10 scale, twenty major subdivisions on the 20 scale, thirty major subdivisions on the 30 scale, and so on for the other three scales. Some scales have additional subdivisions to make the rule easier to use. The person who understands how to use one scale can use the other scales because the procedure is the same.

The 10 scale is often used by architects and engineers. The scale is identified by the number 10 at the left end of the rule, and inches are indicated by numbers reading left to right (Figure 6.3).

Each inch is divided into ten equal parts. On this scale, 1 inch could equal 10 feet (usually stated as 1" = 10'-0"), and each subdivision between the inch marks would equal 1 foot. The scale could also be

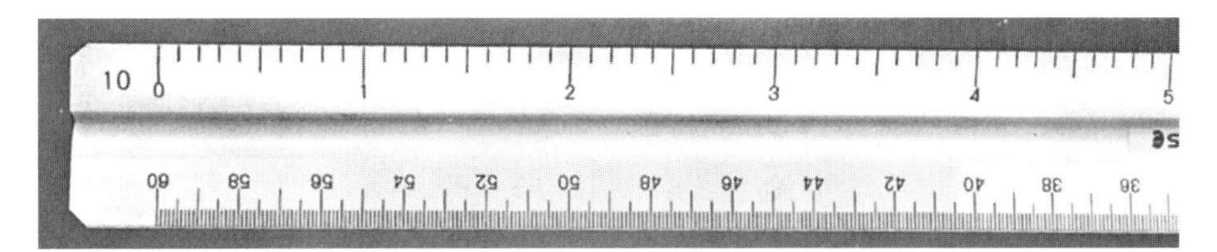

FIGURE 6.3 ■ 10 scale on the engineer's rule *(Photo by Bill Johnson)*

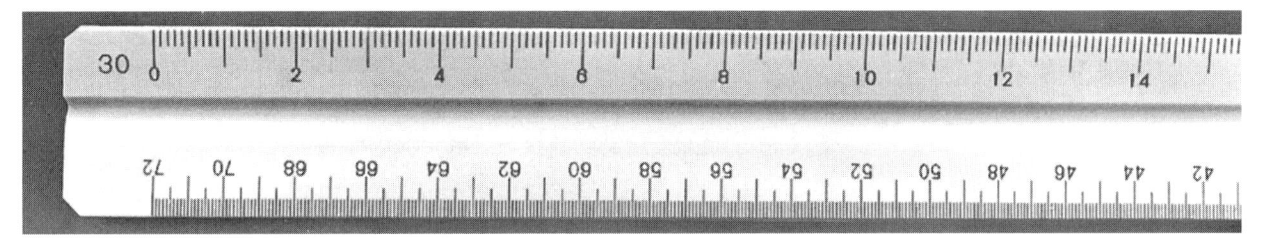

FIGURE 6.4 ■ 30 scale on the engineer's rule *(Photo by Bill Johnson)*

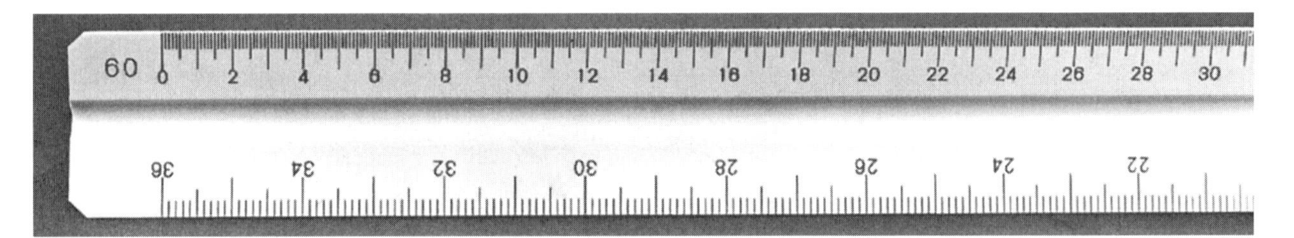

FIGURE 6.5 ■ 60 scale on the engineer's rule *(Photo by Bill Johnson)*

used so that 1 inch equals 100 feet, with each subdivision between the inch marks representing 10 feet. A scale of 1 inch equals 1000 feet means that each subdivision between the inch marks equals 100 feet. The scale could even be used for 1 inch equals 10,000 feet or 100,000 feet. The principle for reading and using the scale is the same regardless of the length represented.

The 30 through 60 scales are set up the same way (Figures 6.4 and 6.5). Although numbering is not indicated for each inch on these scales, reading the scales is not difficult. For example, the 30 scale has major divisions numbered 6, 8, 10, 12, and so forth, up to 36. The first inch could represent 30 feet, even though the numbering on the scale is 3. The first inch is halfway between the 2 and 4 (or 3, even if it is not numbered). This scale can be used to represent 1 inch equals 30 feet, 300 feet, 3000 feet, and so on.

6.5 COMMONLY USED SCALES ON THE ENGINEER'S SCALE RULE

The most commonly used scales on the engineer's scale are the 10, 20, and 30 scales. The architect and the engineer can usually put all the information needed by the HVAC technician on a site plan drawn to a 1" = 10'-0", a 1" = 20'-0", or a 1" = 30'-0" scale. Site plans that are used by HVAC technicians are usually localized to the building rather than to large tracts of land. Figure 6.6 illustrates the way the 10 scale can be used for many different units of measurement.

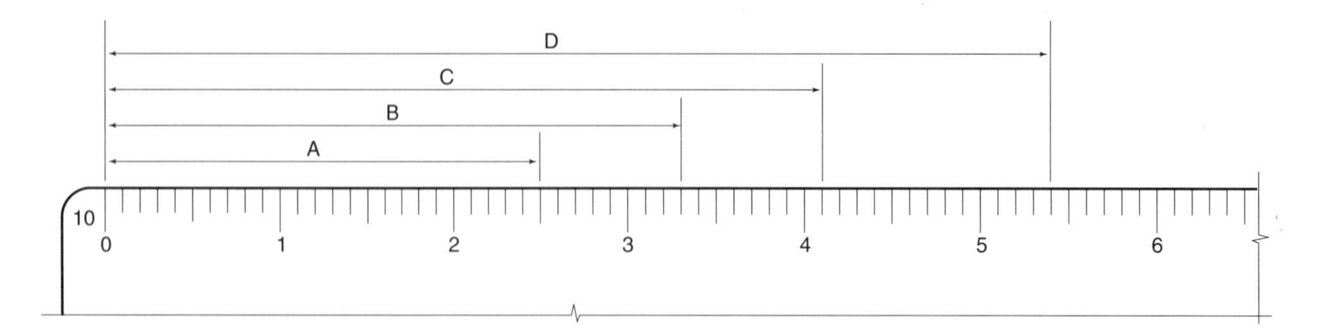

Line	Scale	Distance
A	1" = 10'	25'
B	1" = 10'	33'
C	1" = 10'	41'
D	1" = 10'	54'
A	1" = 100'	250'
B	1" = 100'	330'
C	1" = 100'	410'
D	1" = 100'	540'

Line	Scale	Distance
A	1" = 1000'	2500'
B	1" = 1000'	3300'
C	1" = 1000'	4100'
D	1" = 1000'	5400'
A	1" = 10,000'	25,000'
B	1" = 10,000'	33,000'
C	1" = 10,000'	41,000'
D	1" = 10,000'	54,000'

FIGURE 6.6 ■ Reading the engineer's scale

SUMMARY

- It is not practical to draw large tracts of land to a large scale. It is hard to fit such drawings on standard sheets of drawing paper or drawing film.
- Architects often use the engineer's scale to draw site plans.
- The 12-inch, triangular engineer's scale rule has a total of six scales.
- Each scale on the engineer's scale rule can be used for at least four scales: 10s, 100s, 1000s, and 10,000s.
- Several types of engineer's scale rules exist, but the scales are the same. Some of the scales described in this chapter do not appear on all engineer's scale rules.
- The most commonly used scales on the engineer's scale are the 10, 20, and 30 scales.
- The larger the scale number, the smaller the drawing will be.

REVIEW QUESTIONS

1. Explain when and why an architect uses an engineer's scale rule.
2. Explain when and why HVAC technicians use the engineer's scale rule.
3. What scales on the engineer's scale rule are used most often? Why?
4. Can the architect's scale and the engineer's scale be interchanged on the same drawing? Explain.
5. Using the engineer's scale, if 1 inch equals 100 feet, each subdivision of the inch represents _______________ feet.
6. Using the engineer's scale, if 1 inch equals 1000 feet, each subdivision of the inch represents _______________ feet.

ADDITIONAL STUDENT EXERCISES: USING THE ENGINEER'S SCALE

1. Using the indicated scales in Figure 6.7, determine the distance of each line. Record your answers in the space provided.

A
B
C
D
E

Scale	Distance		
A, B, & C at 1" = 10'			
E, D, & C at 1" = 20'			
A, B, & D at 1" = 30'			
C, D, & E at 1" = 40'			
E, D, & C at 1" = 200'			
A, B, & C at 1" = 60'			
B & C at 1" = 400'			

FIGURE 6.7 ■ Exercise 1

2. Using the engineer's scale, measure the property lines and total the overall perimeter of Figure 6.8. Record your answers in the space provided.

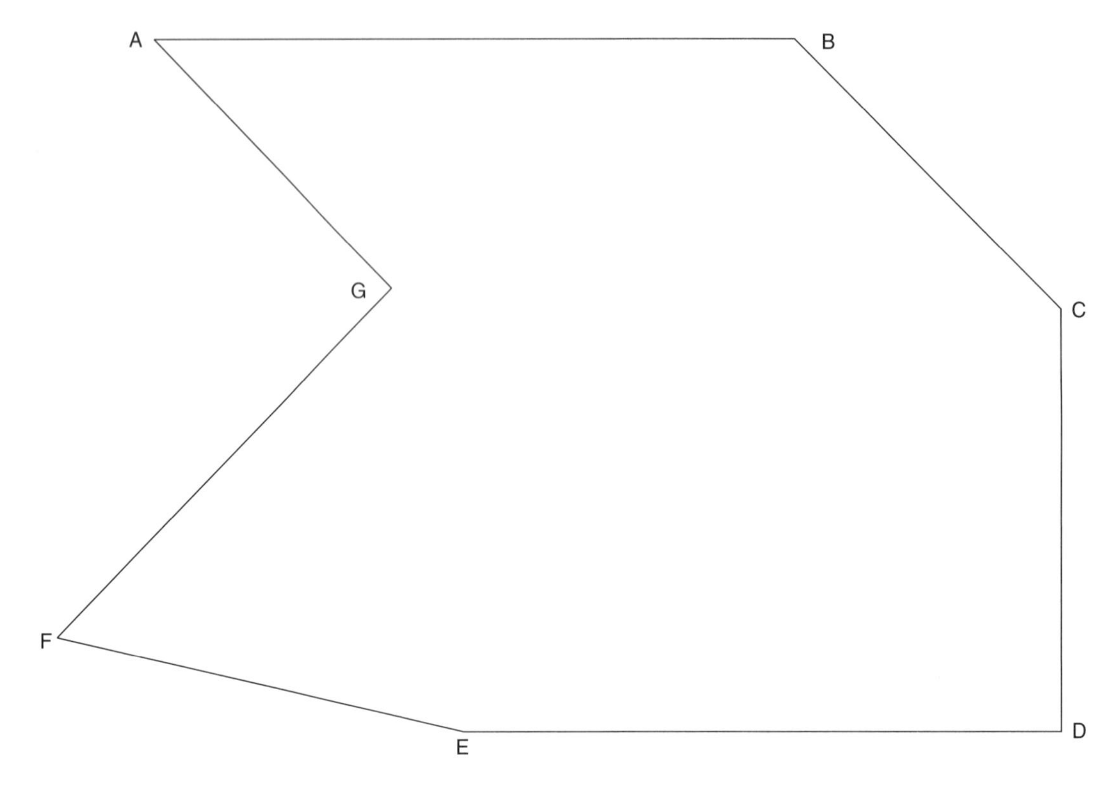

	Line A–B	Line B–C	Line C–D	Line D–E	Line E–F	Line F–G	Line G–A
1" = 10'-0" Scale							
1" = 20'-0" Scale							
1" = 30'-0" Scale							

FIGURE 6.8 ■ Exercise 2: Measuring a plot of land

3. Using the bank building in sheet number 1, redraw the perimeter with a scale of 2"= 20'-0'.
4. How many different scales are found on the M-1 drawings?
5. Refer to M-1. What are the facing dimensions of the fan section located in the 2 mechanical space?

CHAPTER 7

Symbols and Abbreviations

OBJECTIVES

After studying this chapter, you should be able to:

- define and give examples of abbreviations
- define and give examples of symbols
- define and give examples of acronyms
- discuss the use of nonstandard symbols and abbreviations on drawings
- discuss the advantages of using a Computer-Aided Design (CAD) library of symbols
- explain the purpose of the "Legend and Symbol Schedule"
- find examples of abbreviations and symbols used together

7.1 INTRODUCTION

In blueprint reading, abbreviations and symbols are used for similar reasons: to save time for the drafter and to save space on the plan. Abbreviations are key letters of words that are used to denote the complete word. When abbreviations are used to replace whole words on a drawing, the drawing is less cluttered and easier to read. Symbols are graphic representations of the building materials and components required to complete a building. Using symbols on a drawing allows the drafter to concretely and succinctly communicate the information needed by the person reading the plans. The drafter and the person reading the drawings are responsible for becoming familiar with standard abbreviations and symbols in order to communicate clearly with each other.

7.2 STANDARDIZED ABBREVIATIONS AND SYMBOLS

Standardized abbreviations and symbols for the various trades have been developed by organizations like the American National Standards Institute (ANSI) and the American Society of Heating, Refrigeration, and Air-Conditioning Engineers (ASHRAE). These standard abbreviations and symbols are generally used by architects and engineers; however, architects and engineers sometimes create their own symbols and abbreviations to represent materials and equipment on drawings. If symbols and abbreviations are not standard, they are usually noted on the drawing.

Some abbreviations for materials, equipment, and titles are *acronyms*. An acronym is an abbreviation formed by using the first letter of each word. Some examples of acronyms are cfm (cubic feet per minute), fpm (feet per minute), BTU (British thermal unit), and gpm (gallons per minute).

7.3 VARIATIONS IN ABBREVIATIONS

Some single-letter abbreviations represent several different words. The letter "R" stands for *radius*, *Rankine*, *road*, *room*, and *thermal resistance*. Generally speaking, the location of the single-letter abbreviation indicates what that particular abbreviation stands for.

Some words have more than one abbreviation. The word *DOWN* can be written as *DWN* or *DN*. Either abbreviation is correct, but the drafter should choose one and then use it throughout a drawing.

Some abbreviations (single letter or several letters) have a period after the last letter to eliminate confusion for the reader. It is not necessary to place a period after an acronym or between the letters in an acronym.

7.4 VARIATIONS IN SYMBOLS

When symbols are drawn by hand, templates are usually used (Figure 7.1). Templates are commercially prepared patterns made of sheets of plastic with the shapes of various symbols cut from the plastic. To insert a symbol at the proper place on the drawing, the drafter draws around the cut-out portion of the template. If the symbol is a complicated one, two or more templates may be required.

Computer-Aided Design (CAD) systems have a library of symbols from which the exact symbol can be selected and plotted on the drawing. This library contains symbols for various materials and pieces of equipment in each section of drawings (architectural, plumbing, mechanical, and electrical sections).

In some cases, there is no standard symbol for a material or piece of equipment. For example, there is no standard symbol for *EARTH* when shown on the plan view of a drawing. When there is no standard symbol, the drafter notes on the drawing what the material or equipment is.

Some symbols for the same material are shown differently on different drawings. For example, when brick is shown on the elevation views of a building, the drafter usually shows the brickwork as it will look when the wall is complete. The mortar joints and the brick are shown in true perspective. When drawn on the floor plan, this brick wall is shown with crosshatching (parallel lines drawn close together at a 45-degree angle). It is important for the reader of the drawings to keep these variations in mind (Figure 7.2).

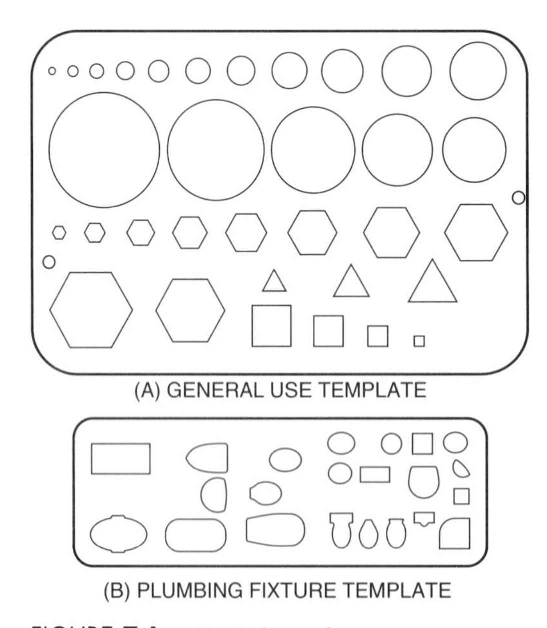

FIGURE 7.1 ■ Typical templates

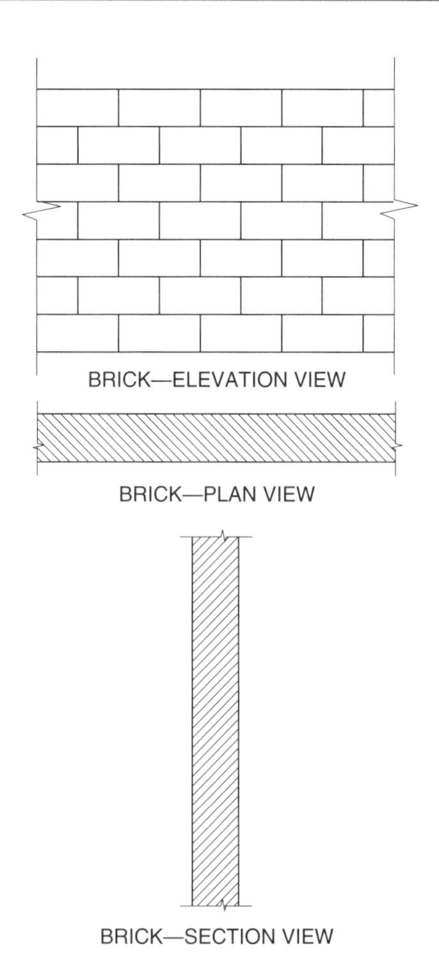

FIGURE 7.2 ■ Architectural symbols for brick

7.5 LEGEND AND SYMBOL SCHEDULES

In most cases, the architect and engineer include a listing of symbols used on the drawings called the "Legend and Symbol Schedule." It is usually placed on the first drawing of each section of the set of drawings (architectural, mechanical, plumbing, and electrical). When this schedule is present, each symbol used in that section is shown on the schedule, whether it is a standard symbol or one created by the drafter.

7.6 ABBREVIATIONS AND SYMBOLS USED TOGETHER

Abbreviations and symbols are combined on working drawings to relate information to the builder. Abbreviations and symbols are used jointly to describe the sidewall supply register in Figure 7.3. The arrow indicates the discharge of supply air from the register. The air quantity is shown on the arrow with the abbreviation cfm (cubic feet per minute). A circle, square, or other symbol is placed at the arrowhead to indicate which register to look for in the "Register, Grille, and Diffuser Schedule." This kind of combination of symbols and abbreviations is often found on mechanical plans.

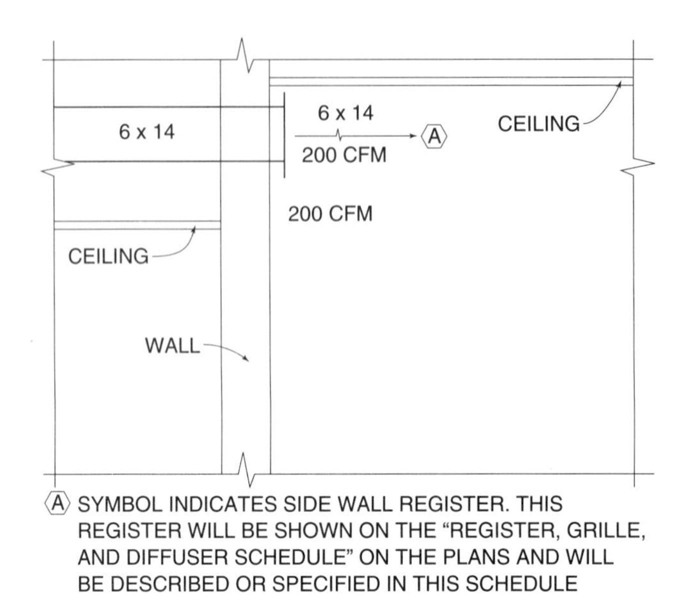

FIGURE 7.3 ■ Use of symbols and abbreviations

7.7 EXAMPLES OF ABBREVIATIONS AND SYMBOLS

Some of the standard abbreviations and symbols used in blueprint reading are listed in the appendices at the end of this book. Appendix C contains standard abbreviations, and Appendix D contains standard symbols. Students should study both appendices now and refer to them later as needed. The construction drawings folded in the back cover demonstrate the use of abbreviations and symbols on actual construction drawings.

SUMMARY

- Abbreviations and acronyms are used on drawings to relate information in a shortened, easy-to-read form.
- Standard abbreviations have been developed for words that are used often in the construction field.
- Some single-letter abbreviations are used to represent several different words.
- If a word has more than one abbreviation, the same abbreviation should be used throughout a set of drawings.
- Symbols and abbreviations are often used together to relay information to the reader.
- Symbols are graphic representations of the building materials and components that are shown on construction drawings.
- Templates are used as patterns for manually drawn symbols.
- Computer-Assisted Design (CAD) systems have a library of standard symbols to choose from.
- A symbol schedule on the drawings explains what each symbol represents. This schedule is especially helpful if nonstandard symbols have been used.

REVIEW QUESTIONS

1. Why are abbreviations and acronyms used on drawings?
2. Define *symbols*. Why are symbols used on drawings?

3. What is a template?
4. Why do architects and engineers sometimes use nonstandard symbols?
5. What information is placed on the "Legend and Symbol Schedule"?
6. Using Appendix C at the end of the textbook, give the abbreviations for the following terms:

   A. Basement
   B. Ceiling
   C. Cross-section
   D. Fire extinguisher
   E. Glass
   F. Men's restroom
   G. Pounds per cubic foot
   H. Thermometer
   I. Air conditioning
   J. Bathtub
   K. Cement floor
   L. Cold water
   M. Drinking fountain
   N. Fireproof
   O. Hot water heater
   P. Shutoff valve
   Q. Velocity
   R. Access panel
   S. British thermal unit
   T. Conduit
   U. Fabricate
   V. Gallons per minute
   W. Elevator
   X. Pressure-reducing valve
   Y. Vent
   Z. Alternating current
   AA. Breaker
   BB. Current
   CC. Direct current
   DD. Solder
   EE. Volt

7. The following abbreviations are familiar to architects and engineers. Explain what each one stands for:

   A. ARCH
   B. AGA
   C. ASME
   D. BS

E. NFPA
F. UL
G. ASHRAE
H. ASRE
I. AIEE

ADDITIONAL STUDENT EXERCISES: FREEHAND DRAWING OF SYMBOLS

DIRECTIONS: Draw freehand the following symbols that can be found in Appendix D.

1. Gate valve
2. Pressure-reducing valve
3. Pipe union
4. Backdraft damper
5. Pipe turning down
6. Butterfly valve
7. Angle gate valve
8. Compressed air pipe
9. Fuel gas pipe
10. Floor drain
11. Hot water heater
12. Pipe sleeve on a pipe
13. Three-way wall switch
14. Electrical lighting panel (distribution panel)
15. Power panel (electrical service panel)
16. Electrical motor
17. Electrical isolating switch (circuit breaker)
18. Turning vanes in a round-throated elbow
19. Electric heater in ductwork
20. Plan view, rectangular duct turned up
21. Flexible duct connection
22. Plan view, stud partition wall
23. Plan view, wall with concrete block and brick
24. Finished grade line, as shown on site plan
25. Deflector, behind a register or grille

CHAPTER 8

How Working Drawings Are Created

OBJECTIVES

After studying this chapter, you should be able to:

- define working drawings
- list tools and equipment used to create working drawings
- define CAD
- describe the process by which working drawings are manually created
- describe the process by which working drawings are created with a computer
- list advantages and disadvantages of using a computer to create working drawings

8.1 INTRODUCTION

When an owner decides to construct a building, the following procedures are followed:

1. Financing is arranged as necessary for the construction.
2. An architect and engineer are hired to design the building.
3. Discussions take place with the architect and engineer so that the building as designed will meet the owner's needs.

The plans resulting from this collaboration are called working drawings.

8.2 WHAT ARE WORKING DRAWINGS?

Working drawings are a set of plans containing graphic information and written instructions for a construction project. These plans give dimensions and details as to how different materials and construction members are to be joined in the construction. Working drawings are the *original drawings*.

Working drawings may be drawn by hand or with the aid of a computer. Until recent years, all working drawings had to be drawn manually. Today, many companies use computers to create working drawings.

Working drawings are drawn on translucent drawing paper that is too fragile for use on the construction site because the paper will not hold up under heavy use by construction workers. For this reason, copies of working drawings are made to use on the construction site as the building is being constructed. These copies are called "blueprints" or "construction drawings." How blueprints (construction drawings) are made from working drawings will be discussed in Chapter 11.

8.3 HOW ARE WORKING DRAWINGS MANUALLY CREATED?

Hand-drawn working drawings are drawn to scale using T-squares, parallel bars, or a drafting machine (Figure 8.1). T-squares and parallel bars are used to draw horizontal lines only. Vertical and angular lines are drawn using triangles in conjunction with T-squares and parallel bars.

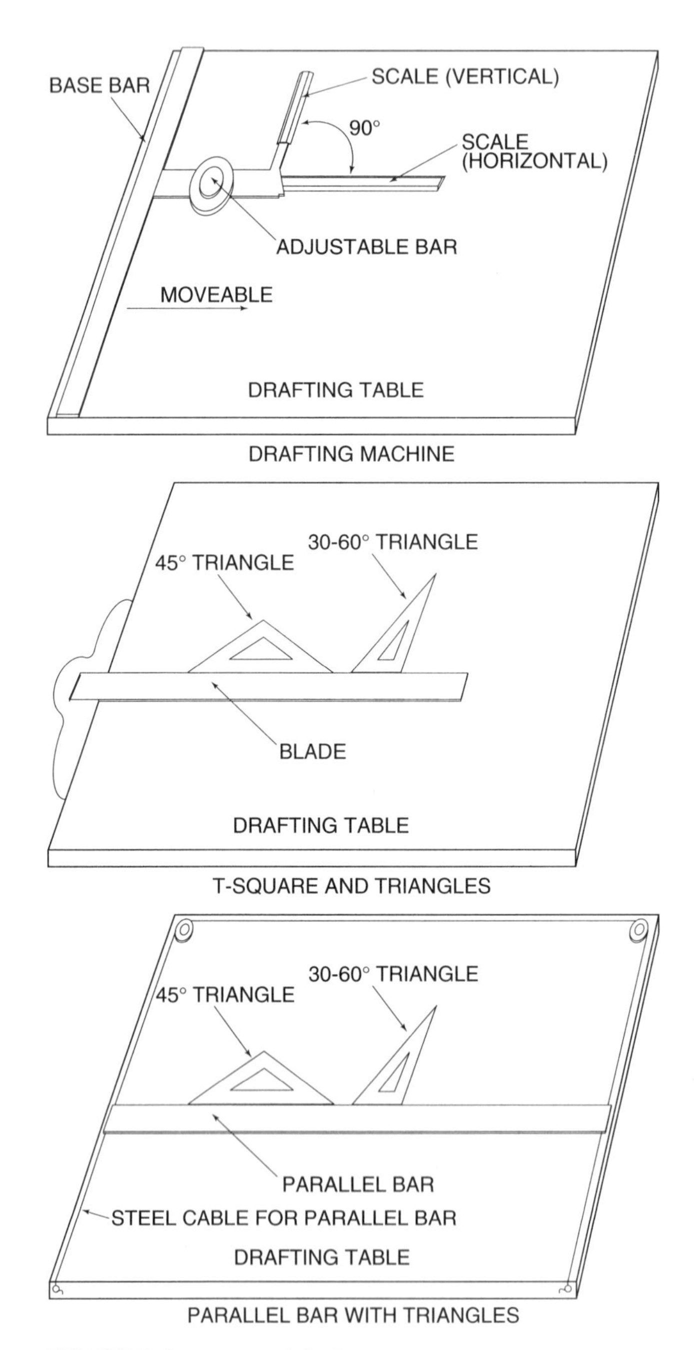

FIGURE 8.1 ■ Manual drafting equipment

Instead of using T-squares and parallel bars, the drafter may choose to use a drafting machine. When a drafting machine is used, the rotating head can be adjusted to draw horizontal, vertical, and angular lines without the use of triangles.

Compasses and other drafting instruments are used to draw circles and arcs. Dividers and scale rules are used to measure distances and lengths (Figure 8.2). Review Chapter 5 for information on reading the

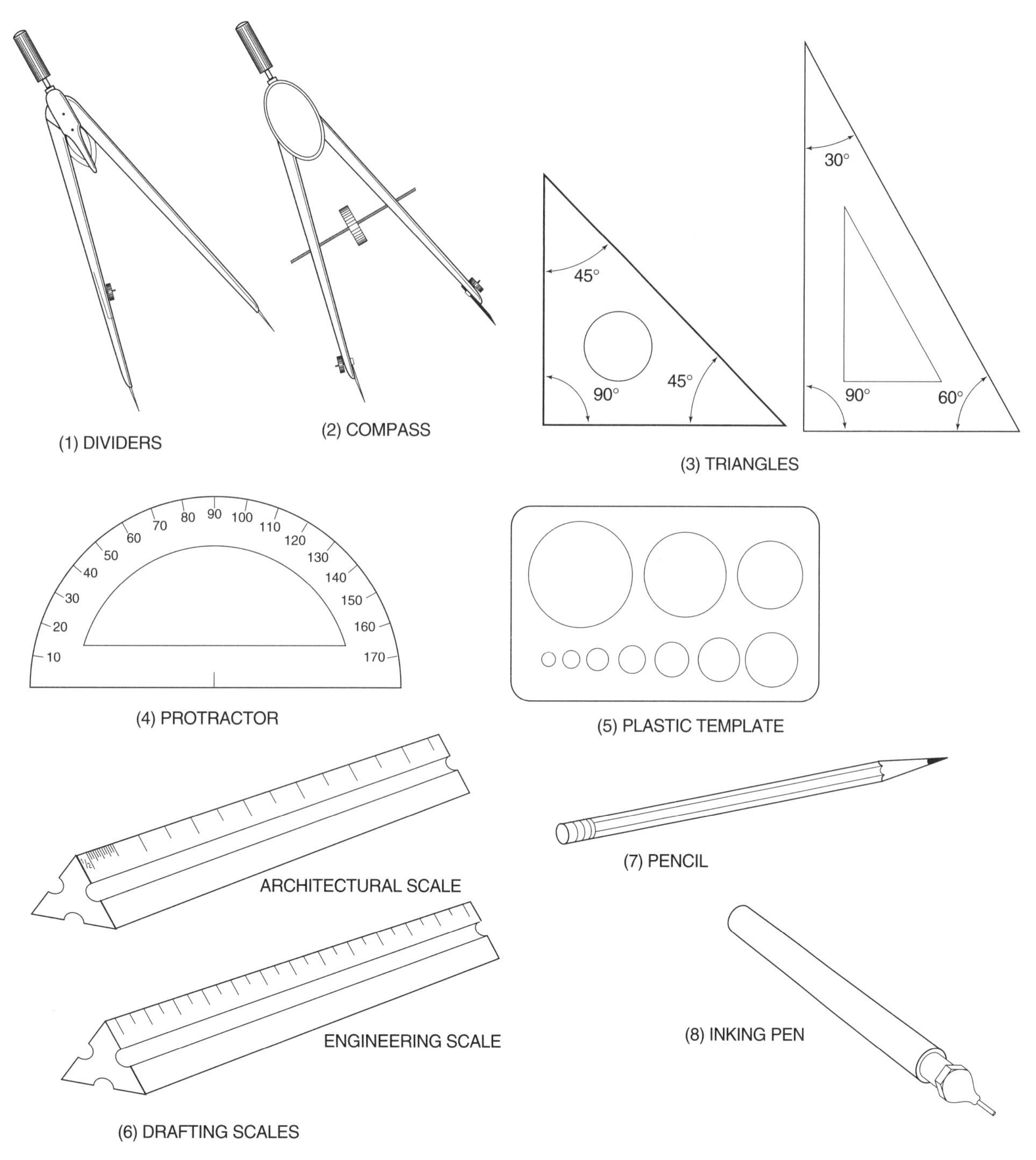

FIGURE 8.2 ■ Manual drafting instruments

architect's scale and Chapter 6 for information on reading the engineer's scale. Chapter 4 discusses how to use a protractor.

Working drawings are made on translucent drawing film using a pencil or pen and ink. The plan size can range from about 8½" × 11" to 36" × 42". Some plans are even larger; however, a plan much larger than 36" × 42" is hard to handle on the job. The most common size for working drawings in residential and small commercial work is 24" × 36".

8.4 HOW ARE WORKING DRAWINGS CREATED WITH A COMPUTER?

Drafters and designers who have had special training often use computers to create working drawings. This method is called "CAD" (Computer-Aided Design or Computer-Aided Drafting). The CAD operator draws segments of drawings on the computer monitor (screen) and puts these segments together to produce a final drawing.

The advantages of CAD include accuracy, consistency, and speed. Another major advantage is the ease of making changes. Line weights, lettering, and symbols are consistent and neat. Using CAD allows the designer to create several design alternatives at the same time, since the repetitive aspects of designing can be handled easily and quickly. In most cases, a well-trained CAD operator has more time for creative efforts and less time is expended on the actual preparation of drawings.

The computer-drafting machine is composed of a software program for computer-assisted drafting and hardware consisting of a keyboard and related input equipment (mouse), a processing unit, and a monitor (Figure 8.3). The processing unit receives information from the input devices (the keyboard with function keys and the mouse). Memory is built into the processing unit to store information. Information is also stored on computer disks (Figure 8.4).

These are two major sizes of disks. The first is the 3½" disk (Figure 8.4A), and the other is a compact disc, known as a CD (Figure 8.4B). The 3½" disk is encased in plastic and has a metal window-like device to protect the disk when it is not in the computer. When the disk is inserted into the computer, the window is opened and the computer records information onto the disk. The CD is more popular because of its reliability and capacity to store more information in comparison to the 3½" disk.

When a plan has been completed on the computer, it is printed on translucent paper or drawing film with a printer or plotter. The printer draws on the translucent paper or film with ink drafting pens. The printer is capable of drawing with several line weights and many colors if desired. The printer prints the information received from the computer. Paper is held in place in the printer, and pens are guided across the paper by the mechanism to complete the drawing. The pens are moved horizontally across the paper, and the paper is moved vertically. With the two movements geared together, circles and other curved lines can be drawn (Figure 8.5).

When CAD machines first came on the market, they were so expensive that only large companies could afford them. They are still expensive; however, the price has dropped enough that more companies are able to buy and use them.

Major advantages of having working drawings drawn by a CAD system are that line weights are more uniform, symbols are always the same, and lettering is precise and easy to read. Another advantage to CAD is that alternative designs can be created quickly and easily. Some companies use a system whereby color keys designate which designers did which drawings (architectural, mechanical, electrical, plumbing, etc.). Overlaying the various colors of sheets of drawings assures better coordination between departments. If two or more items have been designed by different departments to occupy the same space in the building, potential problems become apparent and can be corrected. The quality of drawings done manually depends on the technique and neatness of the drafter. If the drafter is inexperienced or careless, lettering may not be uniform and the quality of drafting may be poor.

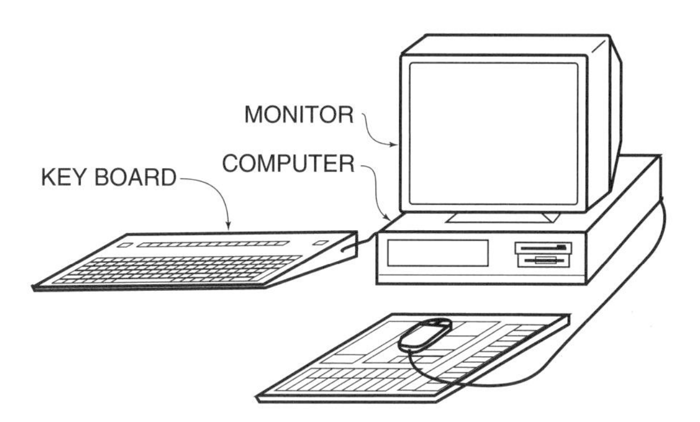

FIGURE 8.3 ■ The computer drafting machine

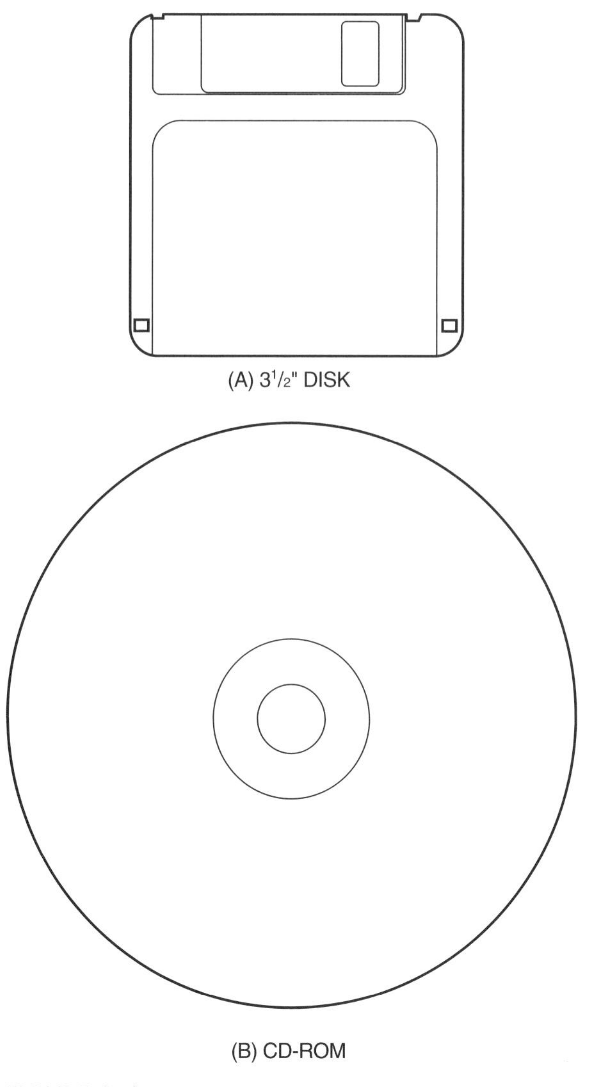

FIGURE 8.4 ■ 3 1/2" and 5 1/4" computer disks

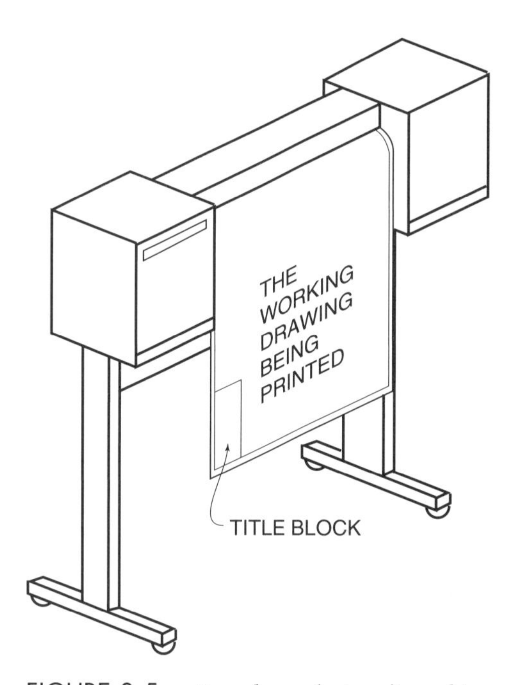

FIGURE 8.5 ■ Pen plotter (printer) machine

8.5 DRAWING PAPER OR FILM USED FOR WORKING DRAWINGS

Several drawing materials are used for working drawings. One expensive drawing sheet is a high grade of linen. This linen is tightly woven and treated with a film that makes the linen smooth and translucent like drawing paper. The finish of the linen has a texture that allows a graphite pencil or drawing ink to adhere to the surface. This drawing material is not widely used because of the expense of the material. When erased several times, the surface becomes slick, and graphite and ink will not adhere to it. One advantage of using linen sheets is that they are fairly resistant to deterioration when stored in a dry place.

A second drawing material is rag content drawing paper. This paper is translucent and has a finish that allows a graphite pencil or ink to adhere to the surface. The paper is rather fragile because it is thin and brittle from the finishing process. Many grades of rag content drawing paper are available, but even the better quality of material is moderately priced. Rag content drawing paper starts deteriorating rather quickly when stored in a dry place. The sheets become brittle and fall apart after storage for several years.

In recent years, a special plastic drawing film has been introduced. This film is translucent and is a tough plastic-like material that will not tear. It is water-resistant and does not deteriorate for many years. The surface of the film is treated to receive graphite pencil or ink marks. A special wax-like pencil lead has been marketed that works well on Mylar® film and will not smudge like graphite lead.

8.6 SEPIA DRAWINGS

Chapter 11 (Section 11.3) discusses the Diazo printing process and describes sensitized print paper. Sensitized print paper results from a chemical treatment that is applied to translucent drafting paper. When working drawings are placed over the sensitized paper, a print of the working drawings is made on the drafting paper. This drafting paper is a sepia, which can be drawn on with pencil or ink. Sepias can be printed as working drawings by using the Diazo printing process.

Sepias save time for the drafter, who is not required to trace the floor plan before adding the various systems to be installed in the building. The sepia serves as the floor plan, and the mechanical, plumbing, or electrical system can be drawn directly on the sepia.

8.7 WHO MAKES WORKING DRAWINGS?

The architect and the engineer are responsible for designing a building and preparing working drawings for it. The architect and engineer are registered professionals, as required by the state government.

Registered architects and engineers have a seal that is required on all drawings that they are responsible for. This seal must be on the plans before the building can be constructed. (Figure 14.3 is an example of an engineer's seal. The architect's seal is similar.)

After completing their college requirements, architects and engineers must pass a comprehensive series of written examinations and verbal interviews conducted by the state professional examining board. After passing these examinations, they are licensed to practice architecture or engineering in that state. They receive a copy of the bylaws required by that state and have the right to administer the professional seal.

Each state has an examining board; however, the written examinations are prepared and graded by the national examination board. Since the examination is conducted on a national level, a registered architect or engineer can usually obtain a license and seal in another state by reciprocity. The architect and engineer can do design work in other states if they have applied for and received a professional license in the state where the project is to be constructed.

The architect is responsible for building design and for overseeing construction. The engineer is responsible for systems installed in the building, such as the plumbing, air-conditioning, heating, and electrical systems.

Designers and drafters who work under the supervision of the architect usually draw the working drawings. The designers, working closely with the architect and the owner, make sketches to show how workers should construct the building. These sketches are given to the drafter, who puts the information on the final drawing.

Like the architect, the engineer usually supervises one or more designers and drafters. The engineer and the designers work with the architect to design systems (air conditioning, heating, plumbing, and electrical) that will serve the building properly and fit in the space set aside for that purpose in the building. The drafter takes the sketches made by the engineer and the designer to create the finished working drawing.

For related information, refer to Chapters 9 and 11. Information on freehand sketching and drafting techniques is located in Chapters 24 and 25.

SUMMARY

- Working drawings are the original drawings of plans for a building.
- Copies of the working drawings are called "blueprints" or "construction drawings."
- Working drawings can be created manually or with a computer.
- T-squares, parallel bars, drafting machines, and compasses are used to create working drawings by hand.
- Computer-Aided Design (CAD) drawings have the advantage of accuracy, consistency, and speed.
- The architect is responsible for the building design and for overseeing construction of the building.
- The engineer is responsible for systems installed in the building.
- Engineers and architects are registered professionals who are assisted by designers and drafters.
- Working drawings are often drawn on linen, rag content paper, or Mylar®.
- Working drawings can be put over sensitized paper and printed as sepia prints.

REVIEW QUESTIONS

1. Define working drawings.
2. What is the difference between working drawings and blueprints?
3. Tell specifically how the following instruments are used to create working drawings:
   A. T-squares
   B. Parallel bars
   C. Drafting machine
   D. Compass
   E. Dividers
4. Why are building plans rarely put on sheets larger than 36" × 42"?
5. What are the components of a CAD system?
6. What are the advantages of the preparation of plans using a CAD system?
7. List and briefly discuss each step by which a CAD plan is produced.
8. Briefly discuss the duties of the following in the preparation of working drawings:
   A. Architect
   B. Engineer
   C. Designer
   D. Drafter
9. Define a "sepia." Why are sepias sometimes used for working drawings?
10. Describe some of the kinds of drawing paper used for working drawings, and list advantages and disadvantages for each kind.

CHAPTER 9

Categories of Working Drawings

OBJECTIVES

After studying this chapter, you should be able to:

- identify and describe the major categories of working drawings
- discuss the types of plans included under the site work section
- discuss the types of plans included under the general construction section
- explain the types of information found in the mechanical section
- explain the types of information found in the electrical section
- locate specific items of information in the various subsections of each category of working drawings

9.1 INTRODUCTION

Working drawings are the original drawings prepared by the architect and engineer with the help of their drafters. Chapter 8 discussed the process by which working drawings are created. This chapter is concerned with the categories of working drawings and why working drawings are placed in categories.

9.2 MAJOR CATEGORIES OF WORKING DRAWINGS

Major categories of working drawings are:

1. Site work section
2. General construction section
3. Structural section (can be combined with the general construction section)
4. Mechanical section
5. Plumbing section (can be combined with the mechanical section)
6. Electrical section

If the construction project is small, several categories can be combined. If the construction job is very large, additional subsections are added to help separate responsibility among the construction contractors.

9.3 SITE WORK SECTION

This section relays information to the contractor who is to do the site work on the construction site. The site plan may consist of a single sheet, several sheets, or many sheets grouped together. The site work section usually consists of the following:

1. **Demolition Site Plan**—shows what exists on the site and what is to be removed (buildings, trees, and unwanted roads, for example).
2. **Soil Boring Site Plan**—shows the locations of test borings to determine types of soils on the site, the locations of undesirable rock formations, and the locations of undesirable soil.
3. **Grading Plan**—shows where soil is to be removed or added to make the site suitable for the construction being planned. This plan also takes into account the drainage of storm water from the site.
4. **Site Location Plan**—shows the locations and dimensions of the buildings on the site.
5. **Landscaping Site Plan**—shows the locations of plantings and lists the names of plants, trees, shrubs, and grasses to be planted.
6. **Irrigation Site Plan**—shows the irrigation system, including piping, the locations of sprinkler heads, and related equipment.
7. **Utilities Site Plan**—shows all service utilities, including domestic water, sanitary sewer, electrical, fuel gas, telephone, storm drainage, cable TV, and any other service that is required for the building.

If needed, additional site plans may be added by the architect or engineer to the seven site plans listed above. In many cases, a separate site plan is prepared to show the outside work for a trade area. For example, the site plan for plumbing site work may be included in the plumbing category of plans. Mechanical and electrical categories may also include separate site plans.

9.4 GENERAL CONSTRUCTION SECTION

The general construction category of working drawings includes building construction plans. The general contractor usually does the construction work on small jobs; however, on large jobs, responsibilities may be delegated to other contractors.

For example, on a small job, workers hired by the general contractor are usually responsible for the structural or framing portion of the building. On a large job, structural or framing work is done under a separate contract by workers hired by the structural contractor.

A set of general construction drawings is broken down into the following major types of plans:

1. **Elevation Plans**—show what the building looks like from various viewpoints, including the front elevation, right side elevation, left side elevation, and rear elevation. (See Sheet A-2 of the construction drawings at the end of this textbook for examples.)

   In some cases, the building has wings projecting at an angle from the main building. This view of the building will be distorted if it is not parallel to the viewing plane. In such cases, an elevation view is usually drawn to show the wing to scale as seen with a viewing plane parallel to the face of the building. (See Chapter 16 for a detailed explanation of viewing planes.) Elevation views show the relationship of different components, such as windows, doors, different textures of building materials, and so forth.

2. **Foundation or Basement Plans**—show the locations of the foundations for outside walls and columns of the building. (See Sheet 2 of the construction drawings at the end of this textbook for an example of a foundation plan.) If there is a basement, the basement plan shows the size of the basement along with the locations of column footings and foundations for the outside walls.

3. **Floor Plans**—are usually grouped in a sequence beginning with the plan for the first floor and going up a floor at a time to the top of the building. Each floor plan shows the construction features necessary for the completion of work on that floor. Dimensions are given and symbols for different construction members are incorporated into each floor plan. Elevations, sections, and details are noted on the floor plans. Also noted are instructions telling where these elevations, sections, and details are shown in the set of plans. (See Sheet A-1 of the construction drawings at the end of this textbook for examples of floor plans.)
4. **Roof Plans**—complete the series of plans for the top portion of the building. The roof plans show all equipment to be located on the roof, roof openings required, roof drains (if roof drains are required), walkways for servicing roof-mounted equipment, and the pitch (or slope) required for the roof. Also indicated are sections and details showing how the roof equipment is to be mounted and how special architectural features are to be installed. The roof plan indicates where these sections and details are located on the plans.
5. **Elevations, Sections, and Details**—shown on the sheets that follow the roof plan. These drawings, usually done to a large scale, enable a construction worker to understand just how the materials and equipment fit together to meet the requirements of the design. Larger scale drawings allow detailed information and instructions to be included. (See Sheet A-8 of the construction drawings at the end of the textbook for examples of sections.)
6. **Equipment Schedules**—several equipment schedules are always provided on construction jobs, including the following:
   a. Door schedule—shows all doors used on the job. The door schedule shows the door size (dimensions), the door type, materials used to construct the door, hardware (hinges and door locks), the door finish, the door frames, and other information needed for ordering doors.
   b. Window schedule—like the door schedule, information required for ordering the windows includes the size, type of glass, type of window, information as to the window frame, and so forth.
   c. Room finish schedule—each room in the building is included on this schedule. Some items listed are the type of ceiling, baseboard, wall treatment (paint or wall covering), and floor finish (hardwood, treated, painted, carpet, etc.). Special notations give instructions on the finishes in each room.
   d. Symbol schedule—shows symbols and describes them.
   e. Abbreviations schedule—lists the abbreviations used in the general construction area along with an explanation of each abbreviation.
   f. Other schedules—include kitchen equipment, hospital equipment, laundry equipment, and special equipment installed in the building under the general construction contract.

Refer to Sheet A-8 of the construction drawings at the end of the textbook for examples of equipment schedules.

9.5 STRUCTURAL SECTION

The structural plans consist of the following:

1. **Foundation/Basement Plans**—give the sizes and locations for all footings and foundations for the load-bearing wall, the outside walls, and the columns.
2. **Floor Framing Plans**—show the structural members that support each floor of the building along with the size and spacing of these structural members.
3. **Roof Framing Plans**—show the structural framing to support the roof, including size and spacing of structural members. (See Sheet A-8 of the construction drawings at the end of this textbook for examples.)

4. **Elevations, Sections, and Details**—includes how structural members are to be joined, types of bolts, types of rivets, welds, clips required, and other important information. These drawings are generally drawn to a larger scale to provide information in more detail. (Structural details are shown on Sheet A-8 of the construction drawings at the end of this textbook.)
5. **Schedule Plan**—schedules on this plan include the following:
   A. Column schedule—includes information about the columns, such as their locations, the number of them, the types of materials used, and their sizes
   B. Schedule of miscellaneous steel or structural members
   C. Welding symbols schedule
   D. Bolt/nut and rivet schedule
   E. Symbol schedule
   F. Abbreviation schedule

On small jobs, the architect can show structural work on the architectural plans. Some architects include structural plans at the end of the General Construction section. Most architects group structural plans together in a separate category, particularly on a large job.

9.6 PLUMBING SECTION

The plumbing section is sometimes included as a separate part of the mechanical section. In this chapter, plumbing will be treated as an entirely separate section. The plumbing section usually includes the foundation or basement, the floor plan, the roof plan, elevations, sections and details, and equipment schedules.

Plumbing work outside the building is shown on the site plan. This site plan is usually bound with the other site plans as described in Section 9.3 above. Sometimes, the plumbing site plan is bound as the first sheet of the plumbing section.

In addition to the plumbing work site plan, the plumbing section usually includes the following:

1. **The Foundation Plan**—shows piping that is installed under the first floor. The piping is located in a crawl space (if there is one) or under the floor slab of a concrete floor on grade. Piping consists of hot and cold domestic water, sanitary sewer or drainage piping, and special services piping, such as fuel gases, medical gases, or process piping.
2. **Basement Plan**—piping serving the basement (underfloor piping) is shown on a basement floor plan if there is a basement; however, notes on the plan specify whether the piping is under the basement floor or at the basement ceiling level.
3. **Floor Plans**—each floor plan shows plumbing fixtures and equipment served by the plumbing system. Each piece of equipment and/or plumbing fixture has a symbol referring the print reader to the schedule section to see exactly what equipment/fixture is specified.

   Piping serving the plumbing equipment/fixtures is usually shown on the floor plan where the piping is to be installed. In other words, if the piping is located above the second floor ceiling and serves equipment/fixtures on the third floor, the piping is drawn on the second floor, showing all runouts and riser piping that turn up to the third floor. The plan is noted to say that piping shown on this plan is to be located above the second floor ceiling. In some cases, the engineer/drafter shows the piping for the equipment/fixtures on the same floor plan on which the equipment/fixtures are shown. To prevent confusion, the engineer/drafter adds notes that explain where the piping is to be installed.
4. **Roof Plan**—all plumbing work required on the roof or through the roof is shown on the roofing plan. This includes plumbing vents through the roof, roof drains (if required), gutters (if required), and special plumbing equipment to be mounted on the roof.

5. **Elevations, Sections, and Details Plans**—these drawings, usually drawn to a larger scale, show details and give information necessary for proper installation of the plumbing system, equipment, and fixtures. An isometrically drawn piping diagram is not uncommon for the detail sheet. (See Chapter 18 for a detailed explanation of isometric diagrams.)
6. **Equipment and Fixture Schedules**—schedule sheets usually contain the schedules for plumbing fixtures, plumbing equipment, piping, piping connections, plumbing symbols, and plumbing abbreviations.

Sheet P-1 of the construction drawings at the end of this textbook shows the plumbing floor plan, details, and the equipment and fixture schedule for a small bank and trust company.

9.7 THE MECHANICAL PLAN SECTION

The mechanical section, like the plumbing section, may include a site plan at the beginning that shows mechanical work to be done outside the building.

In addition to this site plan, the mechanical section includes the following:

1. **Foundation and/or Basement Plans**—show mechanical work to be done in the crawl space, under the slab on grade, or in the basement.
2. **Floor Plans**—one plan per floor shows mechanical work required for each floor. Ductwork is shown on the floor where it is to be installed. For example, ductwork located above the second floor ceiling and serving the second floor is shown on the second floor plan. If ductwork is located above the ceiling of the second floor to serve a space on the third floor, the ductwork is drawn on the second floor and is noted as being installed on the second floor. Piping for the heating and cooling system is drawn and noted in the same manner.
3. **Elevations, Sections, and Details**—located on sheets that follow the floor plan sheets. These large scale drawings show how units, ductwork, and piping are installed. The drawings also show how the units are to be connected with piping, ductwork, water, steam, and so forth.
4. **Schedules**—mechanical schedule sheets such as those listed below relay information to the mechanical worker:

   A. Unit or equipment schedules

   B. Register, grilles, and diffuser schedules

   C. Duct insulation schedule

   D. Piping schedule

Sheet M-1 of the construction drawings at the end of the textbook shows mechanical plans for a small bank and trust company, including ductwork, piping installation, and sections and details.

9.8 ELECTRICAL SECTION

Electrical work shown on the site plan can be the first sheet of the electrical section. The electrical section also includes the following:

1. **Foundation/Basement Plan**—shows electrical equipment (panel boards, switches, electrical outlets, etc.) required for the basement, located in the crawl space, or located under a slab on grade.
2. **Floor Plans**—show the electrical equipment required for each floor, including receptacles, lighting fixtures, and necessary electrical connections to equipment furnished by other contractors.
3. **Roof Plan**—electrical wiring, equipment, and electrical connections to equipment on the roof are shown on the electrical roof plan.

4. **Elevations, Sections, and Details**—these are detailed drawings showing how the electrical equipment is installed. Special instructions and information are relayed to the electrical worker through these drawings.
5. **Schedule Plan**—contains schedules for the electrical devices, including the following:

   A. Lighting fixture schedule

   B. Panel board schedule

   C. Conduit and raceway schedule

Sheet E-1 of the construction drawings at the end of this textbook shows the electrical floor plan, the symbol schedule, the lighting fixture schedule, and a panel board schedule for a small bank and trust company.

The HVAC technician should be familiar with the electrical plans. The locations of equipment electrical disconnects will be important when installing HVAC equipment. Codes also require lighting in certain attic spaces and a GFCI-protected circuit near ground level and rooftop air-conditioning equipment.

SUMMARY

- Working drawings are original drawings prepared by the architect and engineer to give information to the contractor on a construction job.
- Working drawings are usually classified in the following categories: site work, general construction, structural, mechanical, plumbing, and electrical.
- Site work plans give details on the preparation of the site for construction.
- General construction plans include the foundation plans; floor plans; roof plans; elevations, sections, and details; and equipment schedules.
- Structural plans give information to contractors on the foundation, framing, and other structural information needed for erecting the building.
- Plumbing plans are sometimes shown as part of the mechanical section.
- The mechanical section contains information about piping, ductwork, and systems to be installed in the building.
- The electrical section contains details on panel boards, electrical outlets, and electrical fixtures to be installed in the building.

REVIEW QUESTIONS

1. Define working drawings.
2. Briefly discuss each of the following types of site work plans:

   A. Demolition site plans

   B. Soil boring site plans

   C. Grading plans

   D. Site location plans

   E. Landscaping site plans

   F. Irrigation site plans

   G. Utilities site plans

3. What types of elevation plans are usually included in the general construction section?
4. What is the sequence for floor plans, and what information is found on them?
5. What kinds of information are shown on the roof plans?

6. Why are elevations, sections, and details usually drawn to a larger scale than other parts of a set of plans?
7. What is the purpose of the door schedule?
8. What information is found in the structural section?
9. What types of piping may be shown on the foundation plan of the plumbing section?
10. How does the architect/engineer indicate on the floor plans that piping above the second floor ceiling serves fixtures on the third floor?
11. Where should a person look on the mechanical plans for specific information on how to install units, ductwork, and piping?
12. What is the purpose of the schedule plan in the electrical section?
13. Describe where each of the following would be found:
    A. Which trees are to be removed
    B. The location of columns of a building
    C. The slope of the roof
    D. The kind of wall covering
    E. The size and style of doors
    F. How the roof is to be framed
    G. If piping is to be located under a floor slab
    H. Whether gutters are required
    I. Where ductwork serving a particular floor is to be installed
    J. What shrubs and trees are to be planted

CHAPTER 10

Sections, Elevations, and Details

OBJECTIVES

After studying this chapter, you should be able to:

- recognize and define sections, elevations, and details
- discuss reasons why sections, elevations, and details are shown on plans
- use sections, elevations, and details to find specific information about building construction

10.1 INTRODUCTION

Sections, elevations, and details are an integral part of any set of construction drawings. They are used by architects and engineers to relay information that cannot be shown on the floor plans. This information helps building construction workers understand how the building is put together and how to construct it as designed. HVAC technicians use the information for calculating heating and cooling loads and for designing and installing systems that will fit into the allotted spaces.

Sometimes, the architect and engineer put sections, elevations, and details on separate sheets; however, on small jobs, these drawings are put on unused spaces of the construction drawings for the various trades. Architectural plans contain details pertaining to construction; mechanical plans have details relating to the HVAC system; and so on for each trade involved in the project.

10.2 SECTIONS

Section drawings are drawn to scale, showing internal portions of objects. Section drawings are developed through the use of a cutting plane.

The cutting plane is an imaginary single plane similar to a pane of glass. This cutting plane is used to cut (visually) through a wall, a mechanical unit, or another object. In Figure 10.1, the cutting plane is inserted into an apple. In Figure 10.2, one-half of the apple has been removed and the inside of the apple can be seen through the cutting plane. This view of the apple is the section view.

Sections show how the components inside a drawing relate to each other. Construction sections enable the builder to see how different members of a building are joined and their relationship to each other. Figure 10.3 shows a section of a cornice. All the various construction members are shown and named. Sometimes, dimensions are included on section drawings.

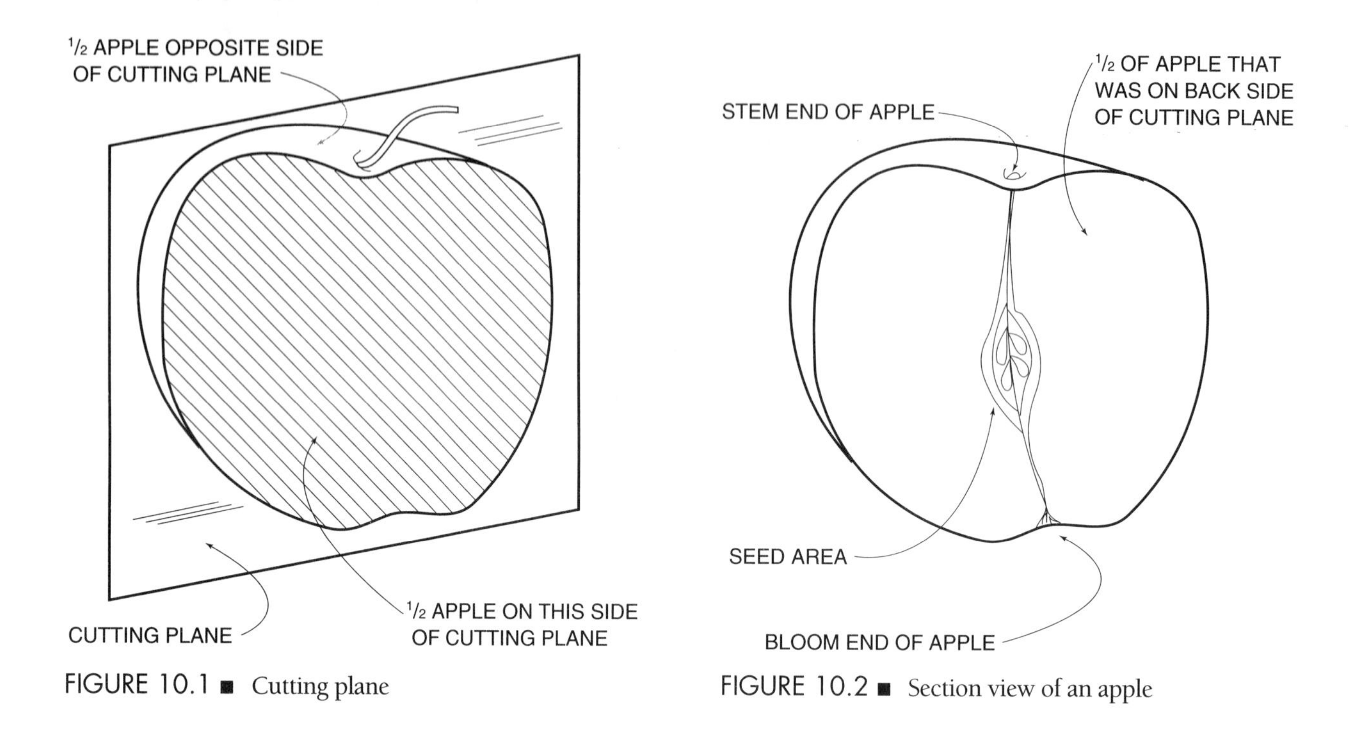

FIGURE 10.1 ■ Cutting plane

FIGURE 10.2 ■ Section view of an apple

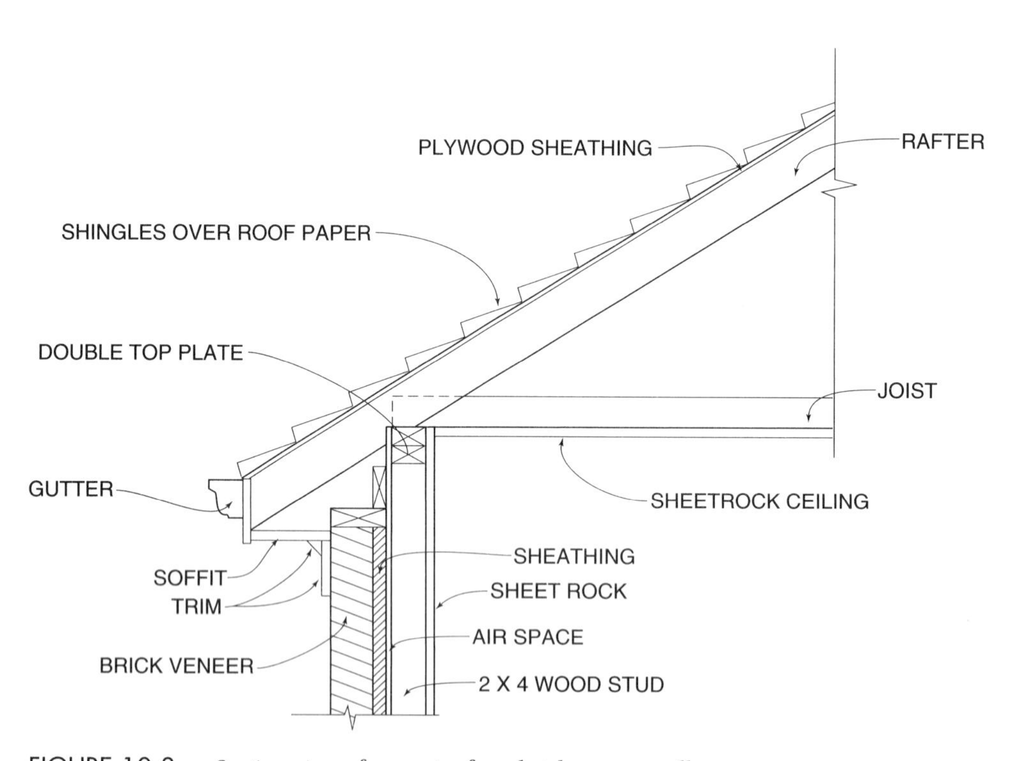

FIGURE 10.3 ■ Section view of a cornice for a brick veneer wall

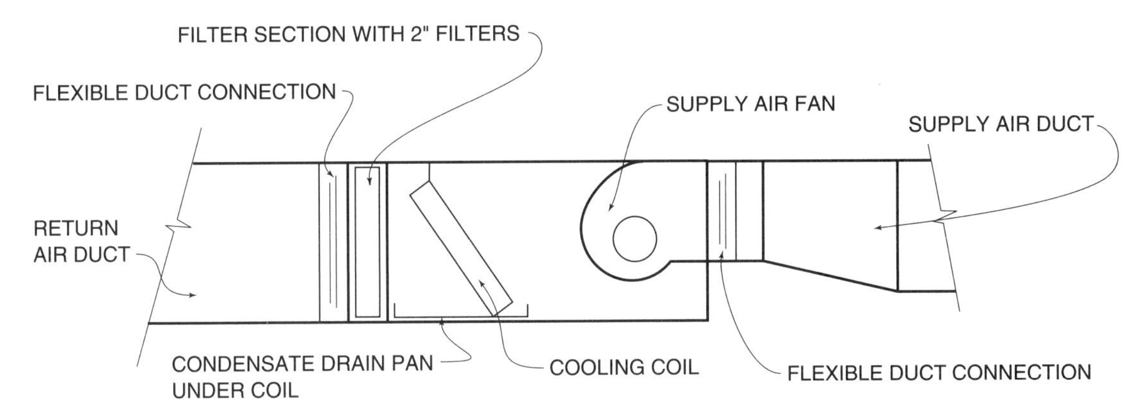

FIGURE 10.4 ■ Section of an air-conditioning unit

Figure 10.4 is a section view of an air-conditioning unit. The duct connection, supply fan, cooling coil, filter section, and return air duct connection can be seen. The internal condensate pan is shown along with flexible duct connections on the return and supply sides of the unit.

Sections are drawn for any portion of work in each trade to clarify how a unit functions or how components are joined. A wall section with footings, insulation, flooring, and other components is shown in Figure 10.5. Since wall sections are usually drawn to scale, dimensions have been omitted for clarity. Figure 10.6 is a drawing of a partial wall section with stucco on the outside of the wall and paneling on the inside.

Arrows or special markings representing the cutting plane identify the point at which sections are taken. Figure 10.7 is a plan view of typical locations where sections are cut. Two types of symbols indicate the cutting plane. The first is a heavy line drawn into the object with a 90-degree line, an arrowhead, and the letter "A" shown at the point of the arrowhead. This symbol indicates the location of the cutting plane and the direction one should look to see the section. Using this symbol, Figure 10.7 identifies the section as Section A-A. If there are other sections, they will be identified as Section B-B, Section C-C, and so on.

A second type of symbol indicating the cutting plane is also shown on Figure 10.7. A heavy line is drawn outside the object to a circle that has a number above the center line and a combination letter and number under the line. An arrowhead formed on the circle indicates which way one should look to see the section. The number above the center line indicates the number of the section. The combination letter and number under the line gives the sheet number where the section is drawn.

Using the combination symbol, Figure 10.7 identifies the section as Section 3 shown on Sheet M-1. Sheet A-2 of plans at the end of this textbook demonstrates sections that are cut through building elevations. Sheet A-5 of the same set of drawings contains longitudinal sections and transverse sections through the building. Drawings of this type allow a construction worker to visualize what the building will look like and provide information that helps the HVAC designer locate equipment.

10.3 ELEVATIONS

Elevations are scaled drawings showing what the building should look like when construction is completed. Elevations are drawn on a single plane, showing no depth. They supply information that cannot be given on the plan view of an object. Elevation drawings denote the locations of windows and doors, the types of wall finishes to be used, the exterior trim, the location of exterior cooling and heating equipment, and other vital information. Typically, building elevations are shown for each face of the building (front side, left side, right side, and rear). Figure 10.8 shows a simple building elevation. Several examples of exterior building elevations are shown on Sheet A-2 at the end of this book.

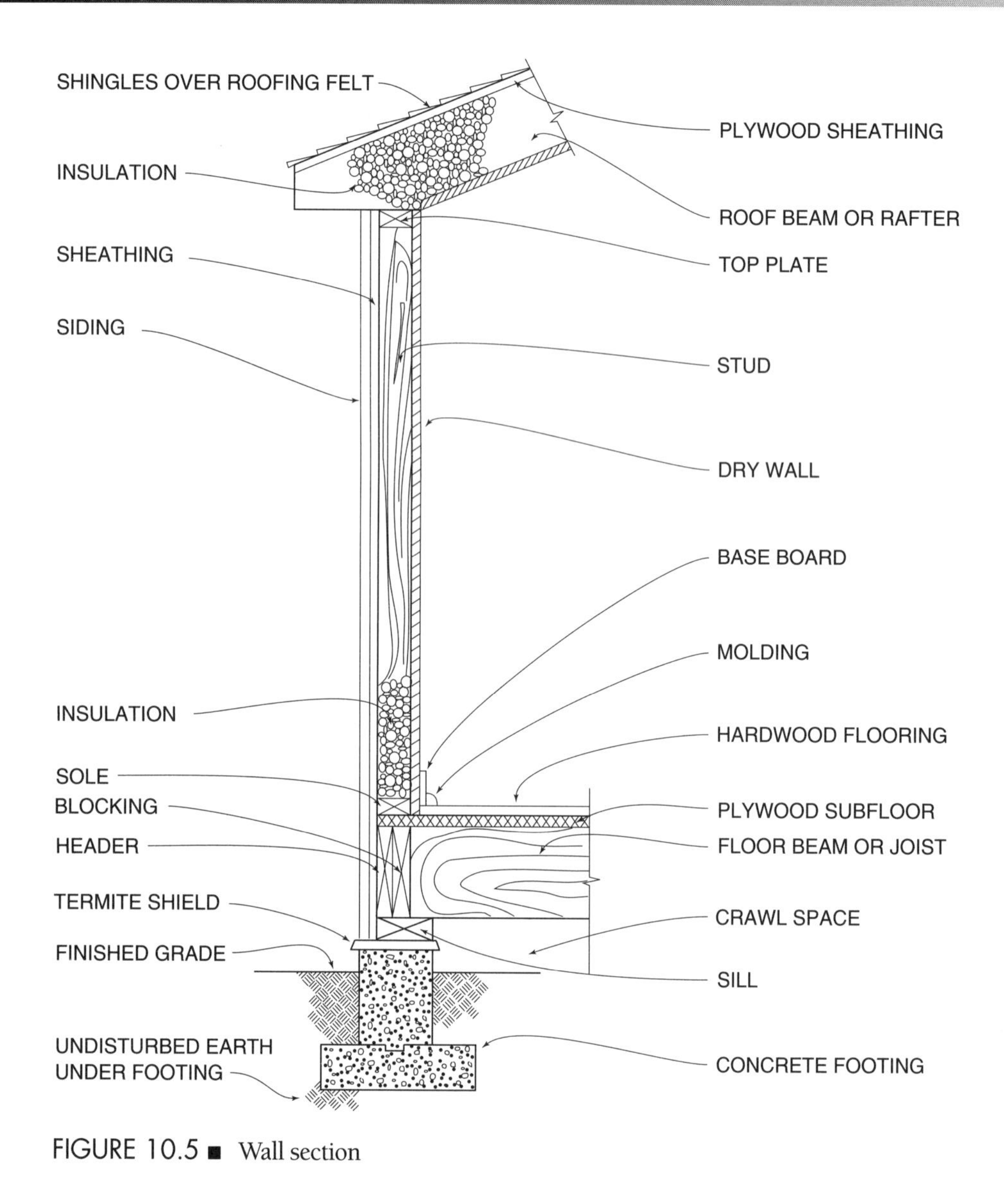

FIGURE 10.5 ■ Wall section

Larger elevations can be drawn starting with the plan view of an object. The finished plan view is taped on the sheet where the elevation is to be drawn. Light extension lines are drawn between the plan view and the elevation location. These extension lines transfer the necessary horizontal distances from the plan view to the elevation view. A drafter uses the architect's scale to draw the height dimensions on the elevation. This method saves time and allows the drafter to draw the elevation drawings accurately. (See Figure 10.12 for a plan view of a condensing unit with extension lines drawn between the plan view and the elevation view.)

Interior wall elevations are sometimes drawn to show the wall from floor to ceiling and from adjoining wall to adjoining wall. The drafter labels these drawings as wall elevations. (See Sheet A-7 of the plans at the end of this textbook for examples of wall elevations. These drawings are different from the section drawings shown on Sheet A-5 because the section covers the entire building and a wall elevation covers only one wall inside the building.)

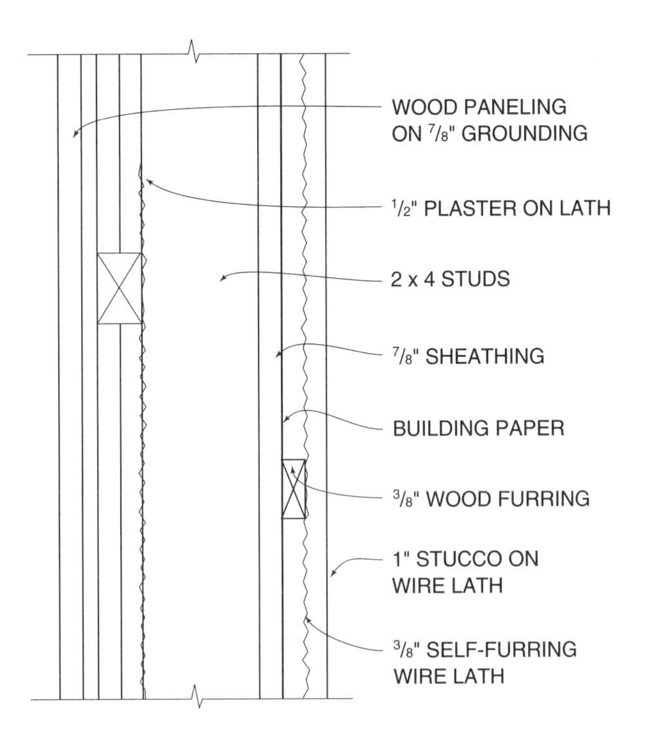

FIGURE 10.6 ■ Partial wall section

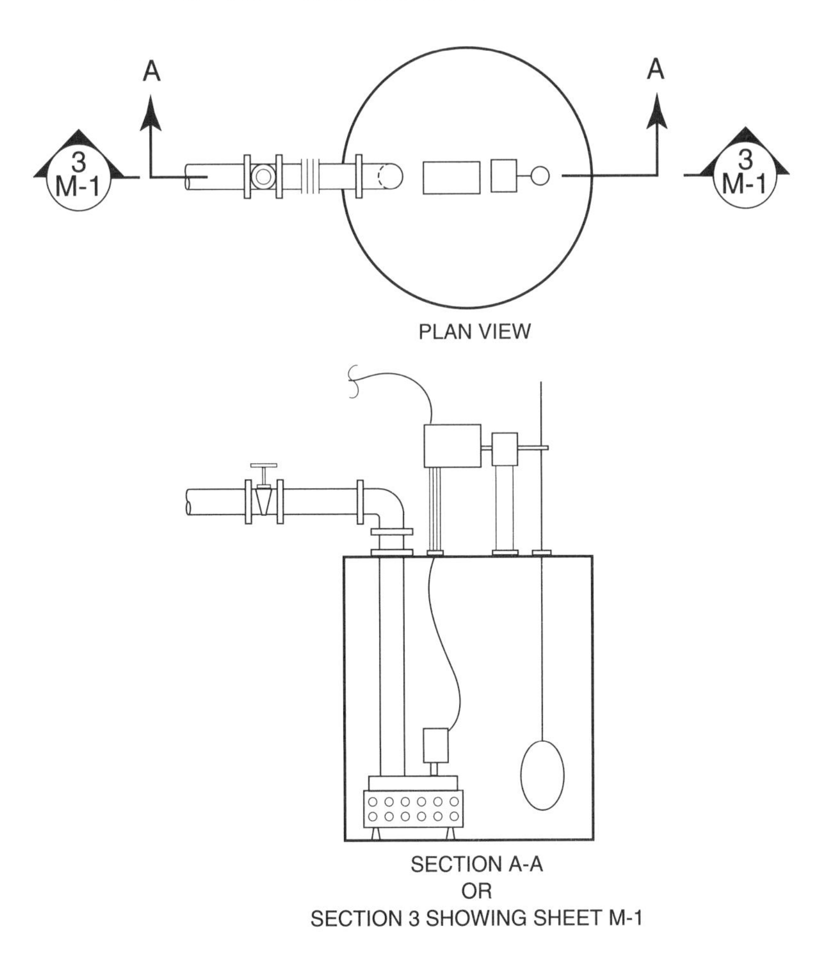

FIGURE 10.7 ■ Two ways to show the location of a section

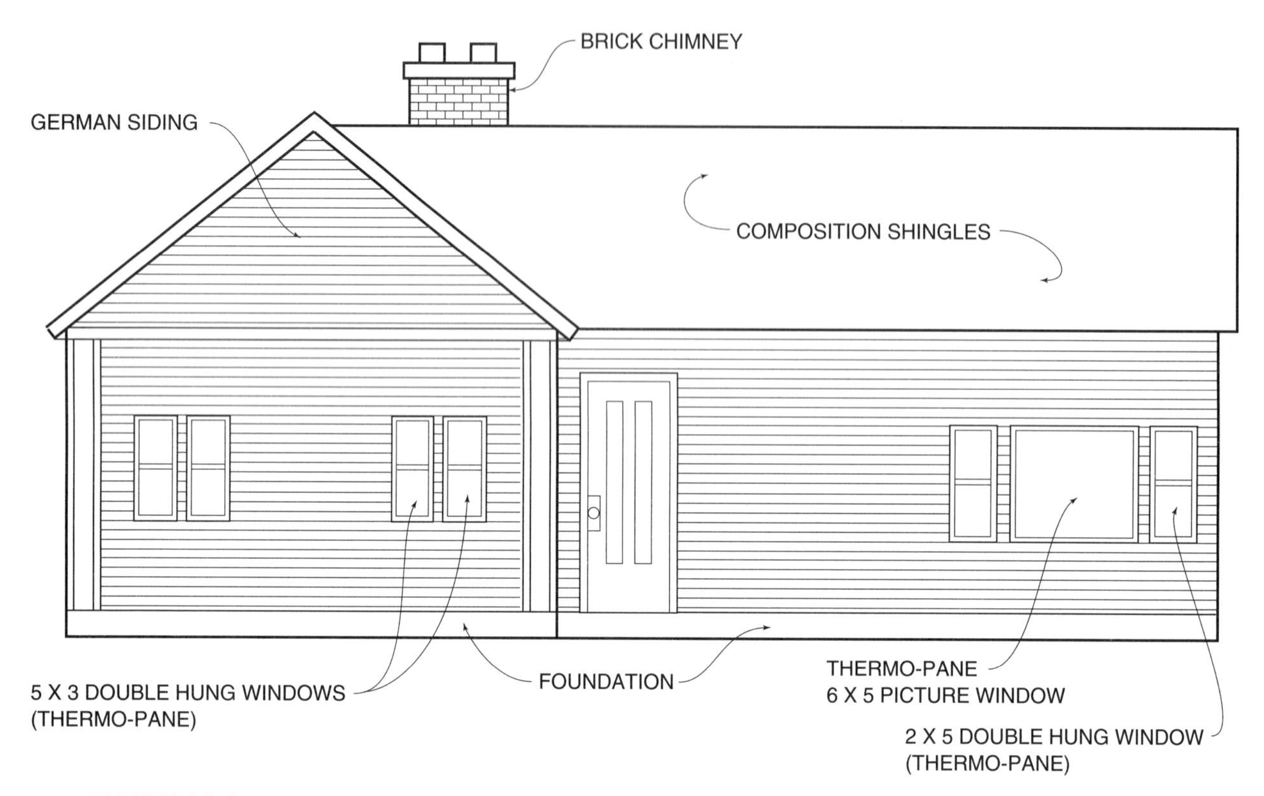

FIGURE 10.8 ■ Front elevation of building

10.4 DETAILS

At times, the architect and engineer need a drawing that gives a three-dimensional appearance. These drawings are labeled "Detail" on the plans under the title of the detail. Details are not usually drawn to scale. For clarity, the drafter notes "Not to Scale" or gives the scale used. Some examples of such drawings are shown in Figures 10.9 through 10.13. Each of these details gives a realistic view of the object along with specific necessary information.

NOTE: See also Chapter 17, "Oblique Drawings," and Chapter 18, "Isometric Drawings."

SUMMARY

- Details, sections, and elevations are included on plans to relay information from the architect to persons who will be studying the construction drawings.
- Details, sections, and elevations may be on a separate page or on unused spaces on the construction drawings.
- A detail shows how materials and components are joined, and it is not usually drawn *to scale*. Details are often drawn using the isometric or oblique drafting method.
- A section drawing shows the internal portion of an object and is usually drawn *to scale* using a cutting plane. Sections illustrate how different components of a building are joined and how they relate to each other.

- An elevation shows in detail what an object actually looks like and is usually drawn *to scale*. Elevations usually show the different views of a building and give detailed information about its parts, including locations and (sometimes) dimensions.

REVIEW QUESTIONS

1. Define a detail, and describe its purpose.
2. Define a section, and describe its purpose.
3. Define a cutting plane, and show an example of a drawing using a cutting plane.
4. Define an elevation, and describe its purpose.
5. How does a drafter indicate whether a detail has been drawn to scale or not to scale?
6. List examples of information usually found on elevation drawings.

ADDITIONAL STUDENT EXERCISES: EXTRA CHAPTER 10 REVIEW

NOTE: Refer to Figures 10.9 through 10.13 to answer the following review questions.

1. Figure 10.9 is a detail of a hot water unit heater. How many globe valves are shown?
2. Describe the water flow through the unit heater in Figure 10.9.

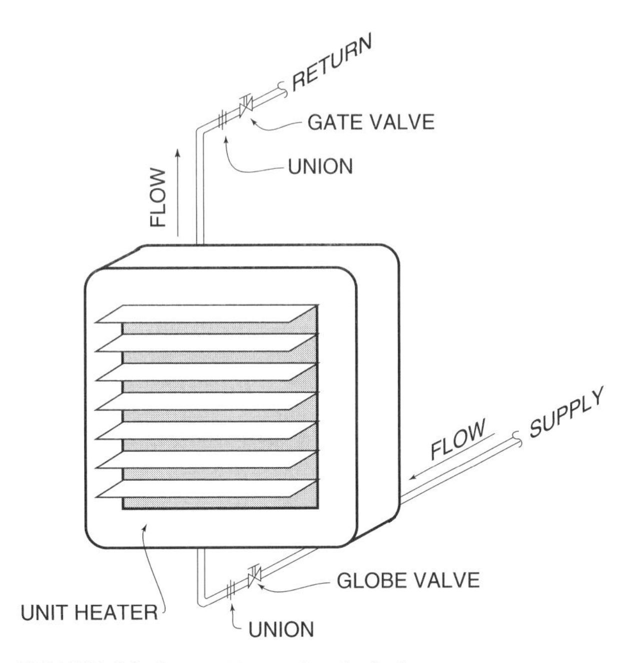

FIGURE 10.9 ■ Oblique detail of a hot water unit heater

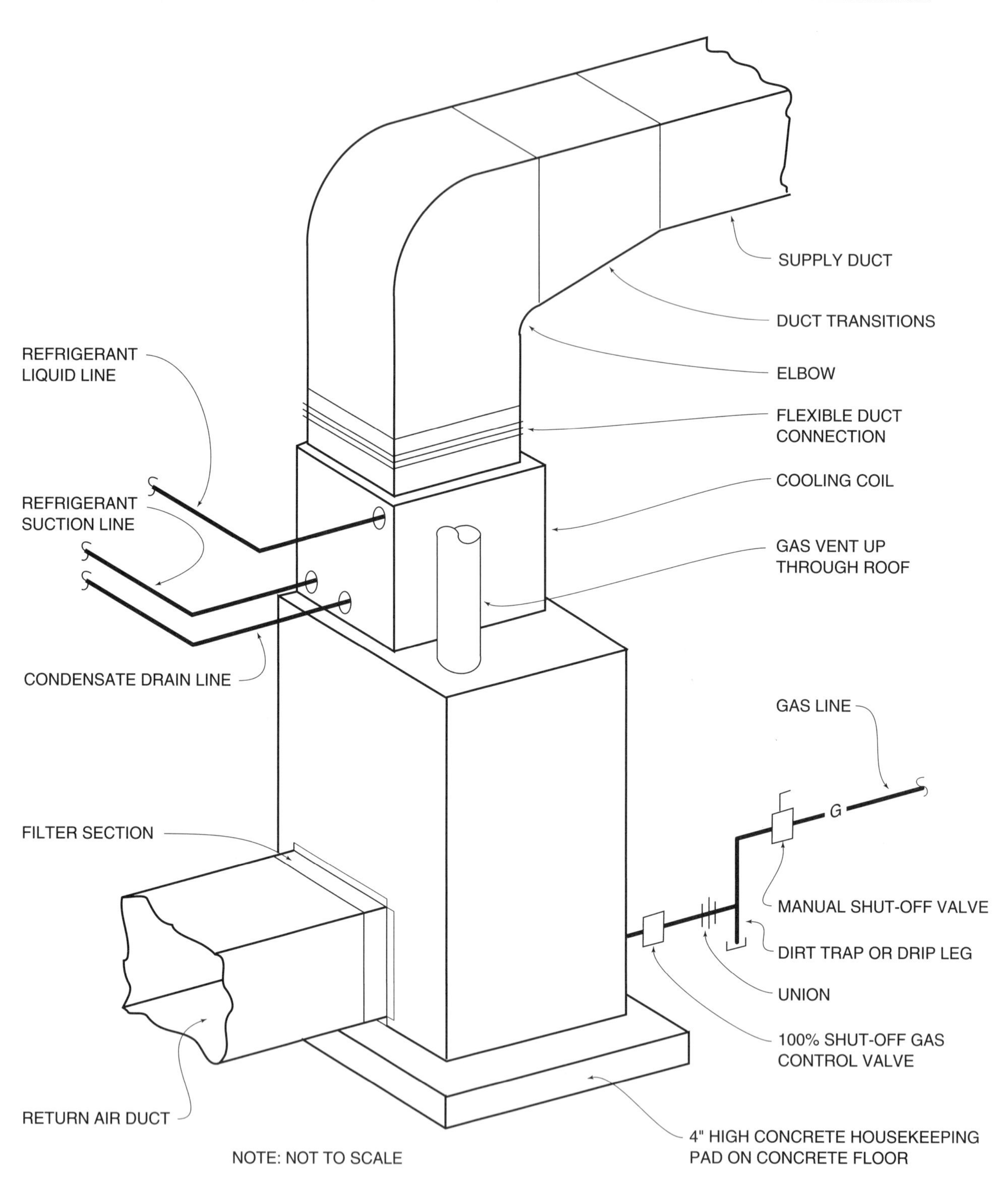

FIGURE 10.10 ■ Detail of connections to a gas furnace in an AC unit

3. In Figure 10.10, is the filter located inside the furnace or outside the furnace?
4. How high is the concrete housekeeping pad in Figure 10.10?
5. Is any provision made for venting air from the coil in Figure 10.11?
6. Name the types of valves shown in Figure 10.11.
7. In Figure 10.12, what information on the elevation view is not shown in the plan view?
8. How high is the mounting pad in Figure 10.13?
9. List the information shown in Figure 10.13 that is not shown in Figure 10.12.
10. Looking at Figures 10.12 and 10.13, which detail do you prefer? Give reasons for your choice.

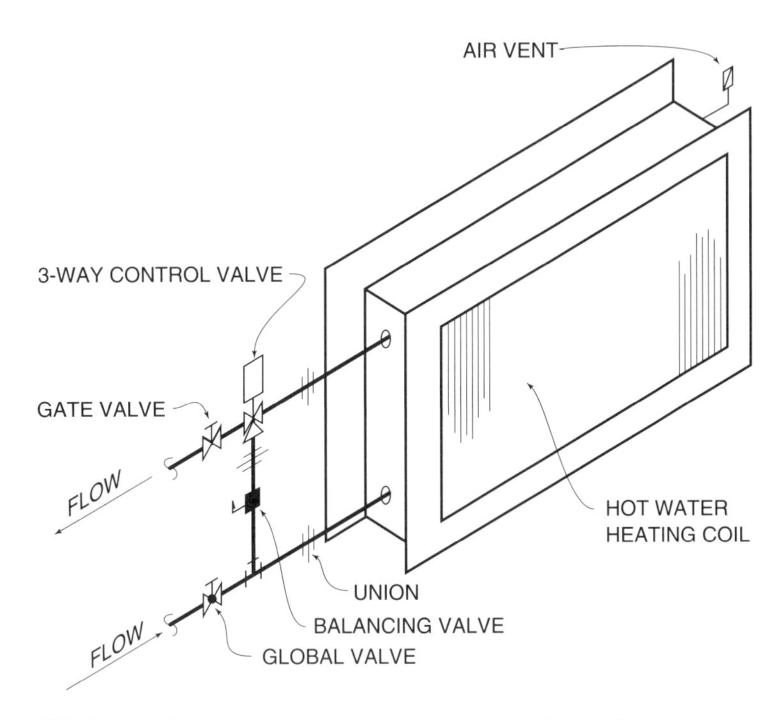

FIGURE 10.11 ■ Detail of hot water heating coil with three-way control valve

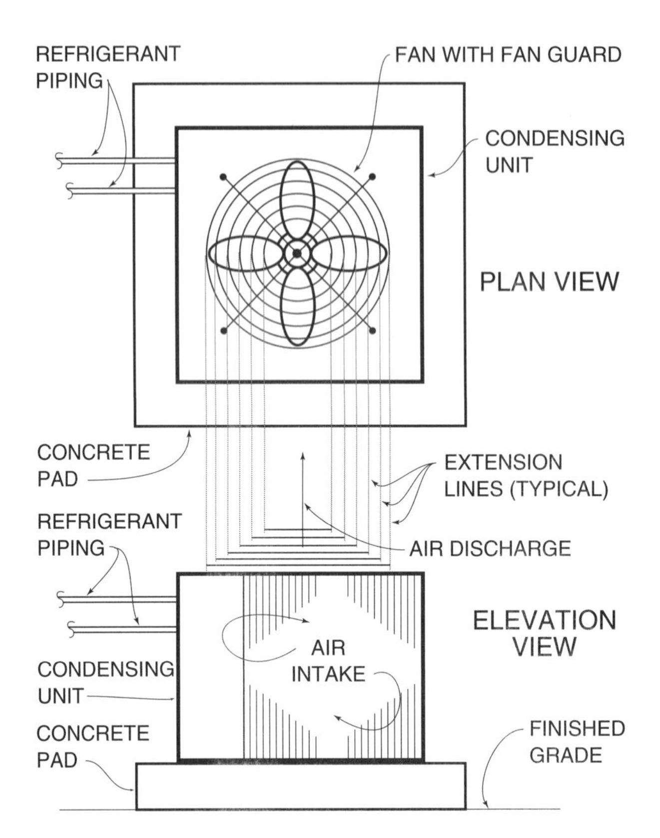

FIGURE 10.12 ■ Plan view and elevation view of condensing unit

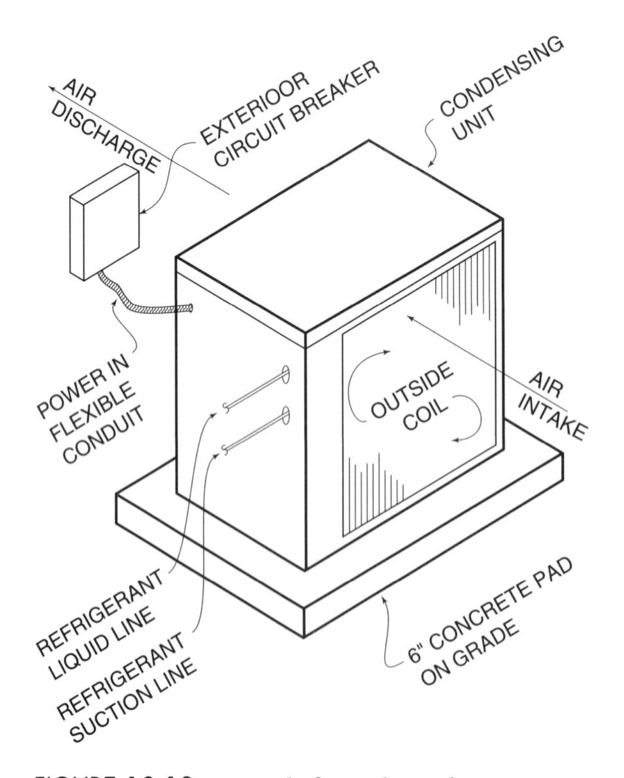

FIGURE 10.13 ■ Detail of outside condensing unit

CHAPTER 11

Creating Construction Drawings

OBJECTIVES

After studying this chapter, you should be able to:

- explain how the sensitized iron salts process is used to create blueprints
- list disadvantages of using the sensitized iron salts process
- explain how ultraviolet light is used in the Diazo process
- outline the major steps in making construction drawings with the electrostatic printing process
- list advantages and disadvantages of the electrostatic printing process
- describe the way CAD machines can be used to make construction drawings
- discuss the use of printers in the creation of construction drawings

11.1 INTRODUCTION

Construction drawings are the plans that are used by the construction crew on the job site. Construction drawings, commonly called "blueprints," are reproductions of working drawings. Chapter 8 discussed how working drawings are created using conventional drafting and Computer-Aided Drafting (CAD). Working drawings are then reproduced by various methods to become construction drawings. The discussion in this chapter will deal with the reproduction process.

11.2 HOW CONSTRUCTION DRAWINGS GOT THE NAME "BLUEPRINT"

As discussed in Chapter 8, it is not practical to use working drawings as construction drawings. Working drawings are made on translucent drawing film that is usually very fragile. These drawings could not withstand use on the job site.

Years ago, working drawings were drawn on a heavy paper or linen. The architect and engineer drew the original designs or drawings and then hired drafters to copy them on heavy paper or linen. (See Chapter 8 for a description of drawing paper or film.) The process was costly, and only a few sets of drawings were produced to construct a building.

The first "blueprint" process for reproducing drawings was discovered around 1840. This process enabled the architect and engineer to make copies of the working drawings produced in their offices. The process consisted of sensitizing a heavy paper with iron salts. This paper underwent a chemical change when exposed to light.

First, the sensitized paper was placed on a pane of glass. Then, the architect's or engineer's working drawing (drawn on translucent paper) was placed over the sensitized paper. A second pane of glass was placed over the working drawing sheet. These pieces of glass were placed in a frame to hold the glass and paper firmly in place (Figure 11.1). The frame was then exposed to sunlight for a specified time. The sunlight caused a chemical reaction on the sensitized paper. Where there were lines on the working drawing, the sunlight was prevented from reacting with the iron salts.

After exposure to sunlight, the sensitized paper was washed with water. The area of the paper that was not protected from sunlight by lines on the working drawing turned blue. The area protected from sunlight by lines on the working drawings turned white.

The sensitized paper was then run through a second bath containing potassium dichromate to fix the color. A final rinse with water finished the process. Then, the sensitized paper was spread out to dry. The blue color of these reproduction drawings gave them the name "blueprints."

This method of reproducing plans was a trying and time-consuming task. If the paper was exposed too long to sunlight, the reproduction bleached out so much that the print was not readable. If not enough exposure to sunlight was made, the print was too blue and unreadable. Even with these drawbacks, the reproducing process was far faster than tracing the working drawings over and over.

There were other disadvantages to the blueprint process. Since the sensitized paper had to be washed with water, washed in a potassium dichromate, and then run through a final water rinse, the print was stretched out of shape when it dried. After drying, the paper was wrinkled and very brittle. Because of the wet baths and then drying, the drawings were stretched to a point that a scale used on the blueprint gave only an estimate of the true measurement. For accurate measurement, one had to rely on the dimensions given on the plan. Very few blueprints are made today using the sensitized iron salts process; however,

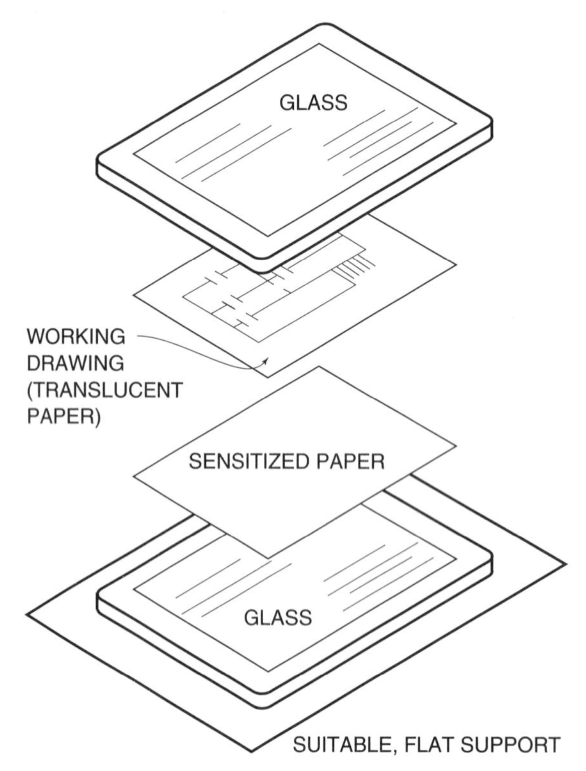

FIGURE 11.1 ■ Original method for exposing prints to light—Blueprint

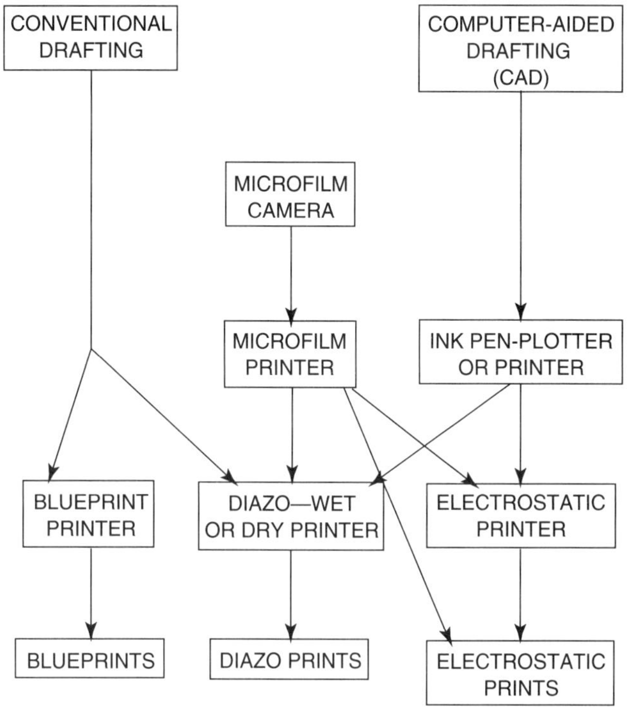

FIGURE 11.2 ■ Paths for making prints today

if blueprints are required, they can be produced by machine. Figure 11.2 shows a block diagram of the various printing processes for producing construction drawings from working drawings. A brief discussion of each of the processes follows.

11.3 THE DIAZO PROCESS FOR MAKING CONSTRUCTION DRAWINGS

The most common method of reproducing working drawings for construction drawings is the *Diazo process*. The Diazo process makes prints with white backgrounds and black lines or white backgrounds with blue lines. As a rule, architects prefer the white background with black lines, and engineers prefer white backgrounds with blue lines.

The Diazo process uses a heavy-duty sensitized paper for printing. This paper is coated with a chemical that reacts to ultraviolet light. A specially designed machine is used to reproduce prints at a fairly fast rate (Figure 11.3).

The Diazo print machine consists of one section that exposes the sensitized paper to ultraviolet light and a second section that develops the print. The exposure section is driven by a variable speed motor that limits the time the paper is exposed to the light. Larger machines have a series of conveyor belts that rotate around a clear cylinder inside the machine. Several fluorescent tube lights that emit ultraviolet light are located inside the clear cylinder. Figure 11.4 is a diagram of the path that a sheet of copy paper follows in a typical Diazo print machine.

Sensitized paper is placed on the work surface of the machine with the chemical (yellow) side up. The working drawing that has been drawn on translucent paper is placed on top of the sensitized paper. The two papers are then pushed into the machine and around the clear cylinder, exposing the sensitized paper to the ultraviolet light. The light passes through the working drawing, and areas not protected by lines on the working

FIGURE 11.3 ■ Printing machine *(Courtesy of BluRay, Inc.)*

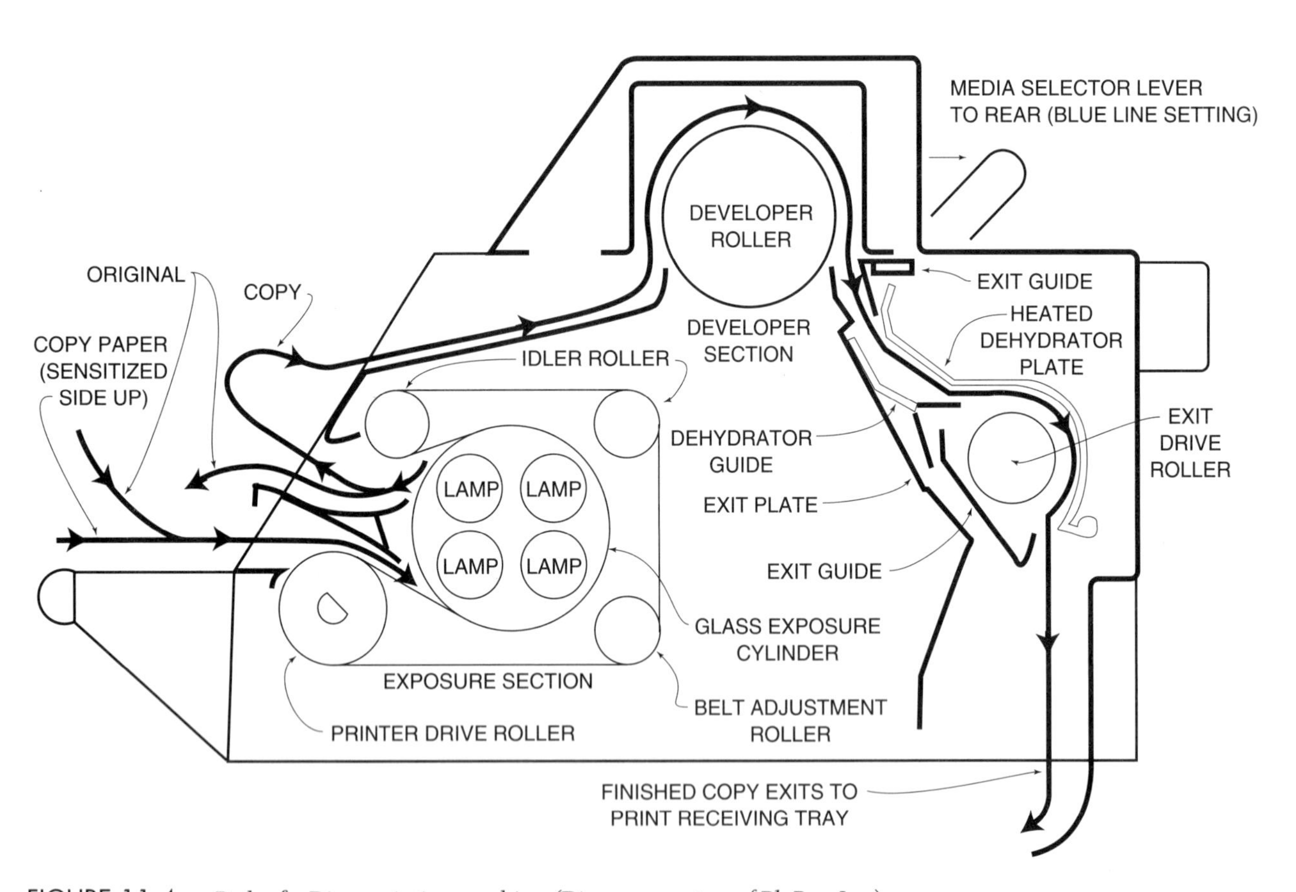

FIGURE 11.4 ■ Path of a Diazo printing machine *(Diagram courtesy of BluRay, Inc.)*

FIGURE 11.5 ■ Table model of a Diazo printing machine *(Courtesy of BluRay, Inc.)*

drawing are exposed to the ultraviolet light. Areas that are protected by lines on the working drawing are not exposed to the ultraviolet light.

When the exposure process is complete for the two pieces of paper, the working drawing is discharged from the machine. The machine operator then inserts the sensitized paper into the developing section of the machine, where it is developed into a finished construction plan.

The Diazo printing machine is manufactured in many sizes and styles. Figure 11.5 is a picture of a table model of a typical Diazo machine often used in architectural and engineering offices. Much larger machines are also available that develop prints at a rate of 16 feet per second or higher. These machines are used in architectural and engineering offices and in commercial printing shops where a large volume of prints is required.

Diazo prints have been proven to withstand heavy use on the job. The process produces a good quality print with crisp lines and a good contrast between the lines and background.

There are, however, two disadvantages to Diazo prints. If these prints are exposed to direct sunlight for a period of time, they will bleach out to a white paper. If the prints are exposed to water, the blue or black lines will run on the background.

11.4 THE ELECTROSTATIC PRINTING PROCESS

The electrostatic printing process, which has only recently become available, is very popular. The printing process is the same as the process used in the office copier. Since the Xerox Company was one of the first to put an office copier machine on the market, the name Xerox is commonly (although erroneously) used to refer to all office copiers.

The electrostatic printing machine can print full-size copies (24' × 36") of the original working drawings prepared by the architect and engineer on microfilm cards containing the image of the original working drawings (Figure 11.6).

In the electrostatic printing process, the working drawing is exposed to light, and the image of the working drawing is projected through a lens to a negatively charged drum inside the machine. The drum is discharged by the projected light, and areas of the drum protected by lines on the working drawing remain negatively charged. Then, the drum is rotated past a roller where black toner particles are attracted to the negatively charged areas on the large drum surface. As the drum continues to rotate, it comes into contact with the print paper, which is positively charged. The toner particles are attracted to the positively charged print paper. The print paper, with the toner particles attached, is then passed through two rollers where heat and pressure fuse the toner to the print paper.

Prints produced by this method are easily read. The background is white and the lines are black. The quality is the same as that obtained from a good office copier. The print paper is usually fairly heavy, of good quality, and it holds up well on the construction site. When this process is used, the print can also be enlarged or reduced with the electrostatic printing machine. The features are not available with the blueprint or the Diazo processes.

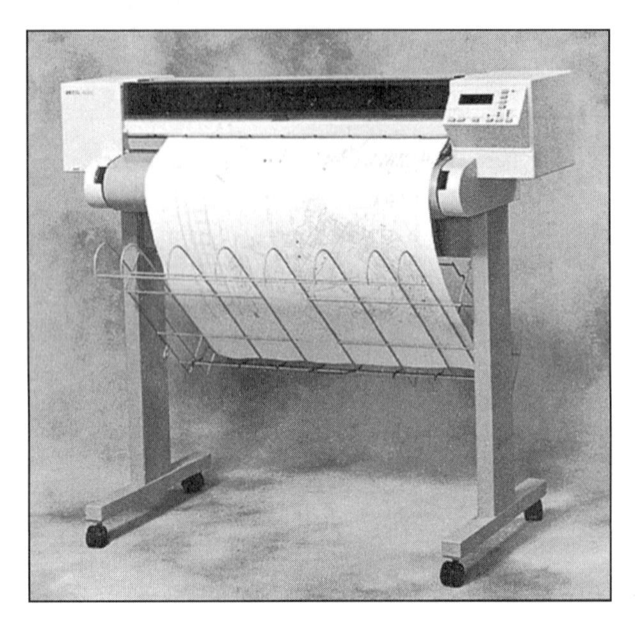

FIGURE 11.6 ■ Electrostatic printing machine
(Courtesy of BluRay, Inc.)

Construction prints produced by the Diazo process and the electrostatic process, as noted for the blueprint process, should not be scaled to determine distances. In the Diazo process, the print paper and the working drawing are rotated over a large cylinder to establish the image on the print paper. Rotation over the cylinder distorts the actual image on the print paper so that the image is no longer the same as the one that was drawn on the working drawing. With the electrostatic printing process, the image is distorted as the lines and image are projected on the negatively charged drum.

Electrostatic prints are gaining in popularity in the industry. In some cases, this type of printing costs more than the Diazo process, and cost is sometimes a factor. The quality of the print is usually very high, with sharp lines and a good contrast between the lines and the background. The paper is about the same quality as that used in the Diazo process, and it withstands heavy use on the job. When exposed to sunlight, the electrostatic print does not fade as quickly as the Diazo process print. When wet, the paper wrinkles, but the lines do not run like the Diazo process print.

All in all, construction drawings produced by both the Diazo process and the electrostatic process are of good quality. The architect and engineer have choices as to the process used for making construction drawings to relay information to the craftspeople on the job.

Finally, some municipal agencies are now requiring black line drawing so that microfilm storage can be used.

11.5 OTHER METHODS OF PRINTING CONSTRUCTION DRAWINGS

Another method of printing construction drawings is gaining in popularity. This method is the printing process used with a Computer-Aided Drafting (CAD) machine. If a one-page drawing is needed on the job to clarify a construction detail, the architect or engineer sometimes does the drawing on the CAD computer and prints the drawing with the pen plotter that is driven by the CAD computer (see Figure 8.5). The drawing is printed on a heavier grade paper than is usually used for working drawings. The paper has a white background with lines that have been drawn in black ink. One disadvantage of a plan created in this manner is that it is not durable enough for heavy use on the job site.

The CAD computer also prints plans with a dot matrix printer that operates similar to a typewriter. Rows of pins strike a carbon ribbon, transferring carbon to the paper and creating lines, letters, and symbols on the drawing. These plans have a white background with black lines, and the quality is similar to that of a typewritten letter. These prints are more durable than those created with a pen plotter. Like pen plotter prints, dot matrix prints are not used to a great extent as construction prints because they will not withstand heavy use on the job site.

Laser printers similar to the office copier are also available for use with the CAD computer. Electronically transmitted information is transmitted from the computer to the printer paper. This transmitted information attracts toner particles to the print paper to form the drawing image on the paper. Toner particles are then fused to the paper with heat and pressure. Prints made by the laser printer have black lines on a white background. These prints are more permanent than the pen plotter prints; however, they are used on a very limited scale as construction prints.

Prints made by CAD and a pen plotter, a dot matrix printer, and a laser printer are usually restricted to the creation of working drawings that can be reproduced by the Diazo printing method or the electrostatic printing method described above. The pen plotter, the dot matrix printer, and the laser printer were designed and are used for drawing on translucent drawing paper or translucent drawing film. (See Chapter 8.)

11.6 MICROFILM AND COMPUTER DISK STORAGE OF DRAWINGS

To save space and for ease in filing record prints, architects and engineers often use a microfilm camera and microfilm printer. The microfilm printer prints the drawing on a standard size drawing sheet or drawing film, which can then be printed on a heavier paper using the Diazo or electrostatic process.

Using CAD also offers an opportunity to save drawings using a minimum amount of space. A CAD machine can copy drawings to a computer disk, which can be stored for future reference or to make future copies of the working drawings. The amount of storage space needed for this system is much less than is needed for the old way of storing full-size working drawings.

SUMMARY

- Construction drawings are reproductions of working drawings.
- Construction drawings must be durable enough to withstand heavy use on the job site.
- "Blueprints" are made by treating a heavy paper with iron salts, exposing it to sunlight, and subjecting it to the developing processes.
- The blueprinting process is more time-consuming and less satisfactory than processes that have been developed more recently.
- The most common process used today to reproduce construction drawings is the Diazo process, which exposes sensitized paper to ultraviolet light and then uses a developing method to complete the process.
- The electrostatic printing machine uses a process like that of the office copier.
- Prints made with the electrostatic printing machine can be enlarged or reduced in size.
- Drawings can be reproduced with a pen plotter, a dot matrix printer, or a laser printer used in conjunction with a CAD machine.
- Microfilm and disk copies save space and are a convenient way to store working drawings in the office.

REVIEW QUESTIONS

1. Define construction drawings.
2. Why are the original working drawings not used on the job site?
3. Discuss the major steps in the blueprinting process.
4. What are three disadvantages of the blueprinting process?

5. What is the most common method used today for reproducing drawings?
6. Outline the steps followed in reproducing prints by the Diazo process.
7. Describe the electrostatic printing process.
8. What are three advantages of the electrostatic printing process?
9. Make a chart on which you compare the advantages and disadvantages of each of the processes for reproducing construction drawings that have been discussed in this chapter.
10. Describe how the CAD machine may be used in conjunction with other machines to reproduce prints.
11. What is the advantage of saving working drawings on microfilm in the architectural or engineering office?

CHAPTER 12

Who Uses Construction Drawings

OBJECTIVES

After studying this chapter, you should be able to:

- discuss the process and order by which various persons and agencies review construction drawings
- tell why construction review agencies have the authority to deny permission for or stop the construction of a building
- define "invited bidding" and "open bidding" of contracts
- define and describe the importance of a "bid bond"
- describe how equipment suppliers determine their bids
- discuss reasons why record drawings are important

12.1 INTRODUCTION

Previous chapters have discussed how working drawings are made, the categories of working drawings, and how construction drawings are made from working drawings. This chapter will focus on the people who use construction drawings.

When working drawings have been completed by the architect or engineer, several sets are printed and distributed as construction drawings to the owner and to various companies and agencies for review and approval. In this chapter, the discussion will progress in the same manner as the usual approval process, ending with record drawings (a form of construction drawings) being turned over to the owner at the end of construction.

12.2 PLAN REVIEW AND APPROVAL BY THE OWNER

When a set of construction drawings has been completed, it is presented to the owner for a thorough review. During this review, the owner presents questions to the architect and engineer. Any changes to be made are directed to the architect and engineer. When changes have been made on the working drawings (see Chapter 8), the corrected construction drawings are printed (see Chapter 11). Afterward, they are again presented to the owner for review and approval.

12.3 REVIEW AND APPROVAL BY BUILDING CONSTRUCTION AGENCIES

After being reviewed by the owner, the construction drawings are submitted to local and state building construction review agencies for review and approval. These agencies have been granted the responsibility and authority by local and state governments to require changes in the plans and, if necessary, to deny permission to build. Building construction review agencies evaluate the drawings to verify that the design, the materials, the methods of construction, the location of the building, and all aspects of the drawings and specifications comply with the requirements and intent of the building codes of the area (see Chapter 13). This review may take several weeks to complete; however, no construction can start before the review has taken place and plans have been approved.

The architect or engineer makes any corrections or changes necessary to secure the approval of the reviewing agencies. Afterward, a building permit can be issued and then construction can start. A set of corrected drawings again goes to the owner for final review. When final approval is received from the owner, the architect and/or engineer is ready to print the construction drawings and proceed with the next stage of the building process.

12.4 CONTRACTORS FOR BIDDING (PRICING)

In most cases, construction plans are issued to several construction companies for competitive bidding. The list of bidding companies is agreed on by the owner and the architect/engineer. This process is called *invited bidding*. Companies on the bid list are the only ones allowed to submit a construction bid. On construction projects that are financed with public money (taxes), the bid list is usually open. An *open bid list* means that any company large enough to do the construction can bid on the job.

A *bid bond* is usually required for construction projects. This bid bond is an insurance policy or available security offered by the bidding company to cover liability for the bid if the bidding company should not be able to complete the work. In that case, the insurance company or security company would hire another contractor to finish the construction project.

Plans are issued to contractors in all areas of construction: for example, building construction contractors, plumbing contractors, mechanical contractors, and electrical contractors.

12.5 EQUIPMENT SUPPLIERS

Equipment pricing is important for determining the amount of a bid in any area of construction. Equipment suppliers need access to construction plans to be able to price the specified equipment. In most cases, these suppliers contact the bidding contractors and use the contractors' plans for "taking off" the equipment for pricing. "Taking off" equipment means listing measurements as well as the number and kinds of equipment needed for the job. (It is not uncommon for each bidding contractor to request and receive several sets of plans for workers who will do the taking off and pricing.)

12.6 CONSTRUCTION PLANS FOR CONSTRUCTION CONTRACTORS

When the bids have been taken and the owner has issued a contract to the successful bidder, the architect/engineer issues the specified number of sets of plans to the contractor. The specified number of plans for construction is usually listed in the specifications (see Chapter 13). Contractors issue the plans to the foremen who will do the work on the site.

During construction, the job is inspected periodically by employees of the Building Inspection Department to see that work is being done properly. These periodic inspections are very important because the Building Inspection Department has the authority to stop construction until incorrect work is corrected. Along with permits, an approved set of drawings should be on the job at all times.

12.7 AS-BUILT DRAWINGS AND RECORD DRAWINGS MADE FROM CONSTRUCTION DRAWINGS

During construction of the building, certain changes are almost always necessary to accommodate the installation of various materials and equipment. It is almost impossible to coordinate every piece of pipe, every section of ductwork, and every piece of conduit so that there are no conflicts. The changes necessary to eliminate these conflicts are made on the job, and the contractor is usually required to keep a set of construction drawings marked up to reflect these changes. These are known as *as-built drawings*.

At the end of the job, the marked-up construction drawings are turned over to the architect/engineer for correction to show changes made on the job. After these corrections have been made, the architect/engineer copies several sets of plans of the finished project for the owner's use. Since these plans give the owner a record of how the building was constructed and the locations of various components of the finished construction, they are also called *record drawings*. The record drawings will be used for reference in case of future renovation of the building, for making repairs, and for identifying piping and valve locations, gas service piping, and other mechanical and plumbing services.

The number of corrected construction drawings turned over to the owner should comply with the agreement made between the owner and the architect/engineer at the beginning of the design.

SUMMARY

- Copies of working drawings issued to the owner, contractors, and review agencies are called construction drawings.
- The owner has the opportunity to have changes made in the construction drawings before and after the review agencies approve them.
- No construction can begin until plans have been approved and a building permit has been issued by agencies established for that purpose.
- Inspectors for the building construction agencies can stop construction if work is not done properly.
- "Invited bidding" means that only those companies on the "bid list" can bid on a job.
- "Open bidding" means that any contractor large enough to do the job may submit a bid.
- Construction drawings are needed by equipment suppliers who are submitting bids on a job.
- All contractors on a job are issued copies of the construction drawings.
- Any changes made during construction are recorded on the construction drawings, which are issued to the owner at the completion of the job as "record drawings."

REVIEW QUESTIONS

1. Briefly discuss the persons, agencies, and contractors who may ask for changes in the construction drawings before the building is constructed.
2. Why does the architect or engineer go over the construction drawings with the owner at several points in the review process?
3. Why is it important that the building construction review agencies react favorably to the construction drawings?
4. What authority does the inspector hired by the construction review agency have?

5. Define:
   A. open bidding
   B. invited bidding
   C. bid bond
   D. "taking off" equipment for pricing
6. How does the architect/engineer usually determine how many sets of construction drawings are to be issued to various contractors?
7. What are record drawings?
8. Why is it important for the owner to have a correct set of record drawings?

CHAPTER 13

Specifications

OBJECTIVES

After studying this chapter, you should be able to:

- explain why specifications are an important part of a set of plans
- tell how specifications are usually organized
- give examples of information found under Division 1—General Requirements
- tell what types of work are described under Division 2—Site Work
- find specific information listed under each of the other divisions on a set of specifications

13.1 INTRODUCTION

Specifications contain written information that is not shown on the plans but is necessary for the completion of the job. Specifications contain information on (1) procedures for starting the job, (2) meetings planned to schedule phases of construction, (3) how payments by the owner are to be made, (4) where equipment is to be stored until needed, (5) insurance that must be carried by the contractors, and (6) other details. On very small jobs where specifications consist of some ten or twelve pages, architects may print them on a plan sheet rather than binding them in a separate book that is included with a set of plans. (See Sheet M-2 of plans at the end of this textbook for an example of specifications printed on a plan sheet.) In most cases, the specifications are written, printed, and bound into a book. Specifications are often referred to as "specs" in the office and on the job.

The length of the specifications varies with the size and type of construction. If the job is small and simple, and if the construction is standard procedure, the specifications can be short. If the job is large and has many special features and necessary procedures, the description of procedures may be lengthy. Specifications written for federal government jobs are known for being long and detailed. The government dictates what is to be included in the specifications, and the specifications become elaborate because each item is described in great detail.

In this chapter, the major divisions of a set of standard specifications will be listed along with a brief discussion of each division. There are sixteen major divisions of specifications; however, some architects and engineers consider plumbing as a separate division rather than grouping it with mechanical (Division 15). In that case, plumbing becomes Division 16 and electrical becomes Division 17.

13.2 DIVISION 1—GENERAL REQUIREMENTS

This division includes general administrative and technical provisions that may not be listed elsewhere and which may apply to more than one division. General requirements include contractual/ legal requirements, a summary of work to be done, an explanation of work to be performed later, a description of meetings to be held (construction coordination meetings, scheduling meetings, and progress meetings, for instance), quality control, and submittal. "Submittal" refers to catalog descriptions or "shop drawings" of equipment, which are submitted by the contractor for approval by the architect and/or engineer.

Division 1 also includes a description of temporary facilities, preparation of the job site for construction, and details about how the job is to be closed out when construction is completed. Listed in the General Requirements are such details as how many sets of construction plans are to be furnished to the construction contractor for the project.

Professional organizations for architects and engineers have drawn up and published a standard section entitled "General Conditions" for use in the General Requirements division. These documents cover major topics similar to those listed above. Special items that pertain to a particular job are included in the Supplemental General Conditions that are written by the architect or engineer. Special laws or building code requirements are listed in the Supplemental General Conditions.

13.3 DIVISION 2—SITE WORK

Division 2 describes work to be done on the site, including soil testing, core drilling, standard penetration tests, and seismic exploration. All these tests must be conducted, and the requirements are expressed in this division.

The demolition of existing structures, materials, and debris is included in Division 2. The capping and removal of existing services is usually mentioned in this division even if the work is also included under the trade division. For example, the capping of an existing sewer line is mentioned in Division 2 and is then described in detail in Division 16 (plumbing).

In addition, the site work division describes the various steps in the clearing of the site. Clearing of the site includes structure moving, clearing and grubbing of shrubs, tree removal, pruning, and tree relocation. Also described are grading and earthwork, grading and drainage, paving and surfacing, and landscaping.

Site improvements such as fences, gates, guardrails, lighting, and irrigation systems (site watering systems) are also included in this division.

13.4 DIVISION 3—CONCRETE

All concrete work is described in this division. The major subdivisions included under Division 3 are concrete form work, expansion and contraction joints, cast-in-place concrete, specially placed concrete, and recast concrete.

13.5 DIVISION 4—MASONRY

Division 4 of the specifications deals with brick, stone, fire brick, glass brick, clay backing tile, and ceramic veneer. Related topics such as joint reinforcement, mortar, anchors and tie systems, masonry joints, control joints, and masonry accessories are also included.

13.6 DIVISION 5—METALS

The metals used in the building are specified under Division 5. If the building has a steel structure, the steel members are specified in this section. Metal roof decking, metal floor decking, and permanent metal forms are described in this section.

Metal fabrications such as stairs, handrails, railings, gratings, castings, and lintels are included in this division. In some cases, ornamental metals are used for these fabrications and are described in detail in this section.

13.7 DIVISION 6—WOOD AND PLASTICS

Materials such as rough carpentry (framing), heavy timber construction, prefabricated structural wood, finished carpentry, and architectural woodwork (cabinets and built-in woodwork) are described in this division.

Plastics are used in many ways in modern construction. This section of the specifications is devoted to such items as joists, studs, columns, beams, hangers, connecting devices, and other miscellaneous plastic items not specifically identified under other headings.

13.8 DIVISION 7—THERMAL AND MOISTURE PROTECTION

Division 7 contains written descriptions of roofing materials, waterproofing materials, flashing and sheet metal trim, insulation, roof accessories, and sealants.

13.9 DIVISION 8—DOORS AND WINDOWS

Metal doors and frames, wood and plastic doors, entrances and storefronts, metal windows, special windows, wood and plastic windows, hardware and specialties, and glazing are the main areas of concern in Division 8.

13.10 DIVISION 9—FINISHES

Lath and plaster, gypsum wallboard, terrazzo, acoustical treatment, ceiling suspension systems, wood flooring finishes, carpeting, special flooring, floor treatment, painting, special coatings, and wall covering are found in this division of the specifications.

13.11 DIVISION 10—SPECIALTIES

Division 10 of the specifications is a catch-all section. If items required for the construction do not fall under the other divisions of the specifications, they are usually found in Division 10.

Typical items in this division are chalkboards and tackboards, louvers and vents, grilles and screens, pest control, fireplaces, flagpoles, lockers, storage shelving, directional signage, and sun control devices.

13.12 DIVISION 11—EQUIPMENT

Division 11 lists equipment that is normally furnished and installed by the general contractor as a part of the construction of the building.

Equipment described in this division includes maintenance equipment, bank and vault equipment, food service equipment, vending equipment, athletic equipment, laundry equipment, library equipment, medical equipment, waste-handling equipment, and loading dock equipment.

13.13 DIVISION 12—FURNISHINGS

Major headings in this division are artwork, window treatment, fabrics, furniture, rugs and mats, and furnishing accessories.

13.14 DIVISION 13—SPECIAL CONSTRUCTION

The description and necessary information for constructing special areas are given in this division. Special construction areas include clean rooms, operating rooms in hospitals, incinerators, instrumentation rooms, nuclear reactors, radiation treatment rooms, sound and vibration rooms, vaults, and swimming pool spaces.

13.15 DIVISION 14—CONVEYING SYSTEMS

Dumbwaiters, elevators, hoists and cranes, lifts, material handling systems, moving stairs and walks, and pneumatic tube systems are typical items described in Division 14.

13.16 DIVISION 15—MECHANICAL

Division 15 covers information on air conditioning, ventilating, and heating in a building. The mechanical division of the specifications generally has major sections consisting of general provisions (general description of the work, procedures, etc.), basic materials and methods, equipment, air distribution, piping, insulation, controls, and instrumentation.

NOTE: Plumbing specifications are also included in the standard mechanical specification division; however, many engineers now write plumbing specifications as a separate division. With plumbing as Division 16, electrical becomes Division 17.

13.17 DIVISION 16—PLUMBING

The plumbing division is similar to the mechanical section. Major sections consist of general provisions (general description of the work, procedures, and so forth), basic materials and methods, equipment, piping systems and services, and insulation. Division 16 also includes fire protection (sprinklers) and special services, such as medical gases, fuel gases, compressed air, and process-piping systems.

13.18 DIVISION 17—ELECTRICAL

The electrical division usually consists of general provisions, basic materials and methods, power generation, power transmission, power service and distribution, lighting, special systems, communications, controls, and instrumentation.

13.19 RESPONSIBILITY AND LIABILITY

The design and calculations for the construction project are the responsibility of the registered architect and registered engineer. (See Chapter 8 for information on the registration process for architects and engineers.) After satisfactorily fulfilling the registration requirements, architects and engineers are granted professional seals to be placed on all plans they are responsible for. Seals are also required on the title page of the specifications. (See Figure 14.3 for an example of the engineer's seal. The architect's seal is similar.)

With registration, the architect and the engineer assume, by law, responsibility for all work done by employees working under their guidance. In most cases, these professionals carry liability insurance to protect them against possible financial ruin if a design failure should occur.

In the specifications, the architect and engineer create a legal document with specific instructions to workers and construction technicians as to how the construction is to be performed. Specifications must be carefully written and followed to protect against poor workmanship, which could cause problems during and after construction.

SUMMARY

- Specifications contain information that is not shown on the plans.
- Except on very small jobs, specifications are usually printed and bound into book form.
- There are usually sixteen divisions of information in a set of specifications (except when plumbing is separated from the mechanical division).
- Administrative, legal, and technical provisions are covered in Division 1 of the specifications.
- Divisions 2 through 17 include specific information pertaining to each area of construction on the job.
- The registered engineer and architect are responsible for the design and calculations on a construction project.

REVIEW QUESTIONS

1. Define specifications.
2. List five types of information that might be found on a typical set of specifications.
3. If a job is too small to necessitate printing the specifications in book form, where are the specifications placed?

4. Explain why specifications for government jobs are usually very long and complicated.
5. List three examples of information found under General Requirements.
6. What are "shop drawings"?
7. What information is listed under "Supplemental General Conditions"?
8. Give five examples of items or details that might be found under Division 2—Site Work.

ADDITIONAL STUDENT EXERCISES: MATCHING SPECIFICATIONS

DIRECTIONS: Match each of the following Divisions of Specifications to the correct item from Answers a through m.

Division 2—Site Work ____________________

Division 3—Concrete ____________________

Division 4—Masonry ____________________

Division 5—Metals ____________________

Division 6—Wood and Plastics ____________________

Division 7—Thermal and Moisture Protection ____________________

Division 8—Doors and Windows ____________________

Division 9—Finishes ____________________

Division 10—Specialties ____________________

Division 11—Equipment ____________________

Division 12—Furnishings ____________________

Division 13—Special Construction ____________________

Division 14—Conveying Systems ____________________

ANSWERS

a. elevators
b. flagpoles
c. tree removal
d. artwork
e. brick work
f. entrances and storefronts
g. painting
h. vending equipment
i. roofing materials
j. recast concrete
k. finished carpentry
l. ornamental handrails
m. swimming pool areas

CHAPTER 14

Title Blocks

OBJECTIVES

After studying this chapter, you should be able to:

- identify the title block on a set of plans and explain the significance of information on it
- explain how sheets are titled and numbered
- tell when and how the revision of drawings takes place
- locate on the title block and briefly discuss the scales used for drawing the plans
- explain the significance of the seal(s) of the architect and engineer
- discuss reasons why there is usually a liability note in the title block

14.1 INTRODUCTION

The title block is an important part of every sheet in a set of drawings. Traditionally, the title block is located in the lower right-hand corner of each sheet of drawings (Figure 14.1). Some architects and engineers have chosen to locate the title block along the right-hand edge of the sheet, across the bottom of the sheet, or even across the top of the sheet. Since there is no standard rule describing the title block or its location, architects and engineers can place it wherever they wish (Figure 14.2). Regardless of the placement of the title block, the same information is shown.

BUILDING FOR JOE JONES
321 EAST 32ND STREET
SMALL CITY, U.S.A.

DRAWN BY: JOHN DOE	JOHN B. HART INC. ARCHITECTS & ENGINEERS 1520 PROFESSIONAL LANE SMALL CITY, U.S.A. TEL: 800-321-1234	DATE:
CHECKED BY: J. B. HART		SCALE: 1/8" = 1'-0"

REVISIONS:

DATE:	DESCRIPTION:	BY:	NO.

SHEET NO. M-3 OF 6

FIGURE 14.1 ■ Typical title block

FIGURE 14.2 ■ Various locations of title blocks

A set of building plans consists of many sheets of drawings, each of which has a title block to identify the job, the title of the drawing, and other important information.

The title block is the same on each sheet (see the set of plans at the end of this textbook). Most architects or engineers devise a standard title block and have it cut and printed on the drafting paper or film used by the firm. With this standard title block form already printed on each sheet, the drafter only has to fill in the information pertaining to that particular sheet. In this chapter, the discussion will focus on the information shown in the title block.

14.2 THE NAME OF THE JOB

In the set of drawings for a construction project, each title block contains the name of the job. In some cases, the owner's name is included with the job name. The job address is also listed. Having the job name and the address of the project on the title block prevents confusion as to the project to which any sheet belongs.

In some cases, the architect or engineer works on more than one project for the same owner at the same time. For example, if the architect or engineer is preparing plans for a company that will have two different office buildings on adjoining sites, the job name will be the same; however, the street address will be different. Listing the street address identifies the drawings if they should become mixed.

14.3 THE ARCHITECT'S AND/OR ENGINEER'S NAME

The architect is usually the prime designer who is responsible for hiring engineers as consultants. In this case, the architect's title block is used on all plans for the project. If the job is an engineering job or the engineer is the prime designer, the title block is the engineer's. It does not matter which title block is used (the architect's or the engineer's), but in most cases, only one type of title block is used for all drawings pertaining to a job.

Some architects and engineers have a small, stick-on title block that is put on the plans. If, for example, all drawings have been placed on drawing paper with the architect's title block, the engineer puts a small, stick-on title block with the name of the engineering firm, address, telephone number, and other pertinent information on sheets prepared by the engineer. This information identifies the firm that has done the engineering design work. If the engineer's title block is used, the architect puts a small, stick-on title block on the architectural portion of the drawings.

14.4 SHEET TITLE

Each sheet in a set of drawings has a sheet title. The title should represent the major item shown on that sheet. For example, the first floor plan may be drawn on the same sheet with several details and sections. Since the main purpose of the sheet is to show the floor plan, the sheet title reads "First Floor Plan."

It is acceptable for the title to contain several items. In the above example, the sheet title could also read "First Floor Plan, Sections, and Details."

14.5 NUMBERING THE SHEETS IN A SET OF DRAWINGS

The sheet number is an important item on the title block. Sheet numbers are usually broken down as follows: Architectural (A), Plumbing (P), Mechanical (M), and Electrical (E). (Mechanical is the term usually used for air-conditioning and heating drawings.) Examples and explanations of sheet numbering are given below:

A-1 of 12: The architectural section has a total of twelve sheets. A-1 is the first of those twelve sheets.

P-1 of 3: The plumbing section has a total of three sheets numbered P-1, P-2, and P-3.

M-1 of 5: The mechanical section has a total of five sheets numbered M-1 through M-5.

E-1 of 8: The electrical section has a total of eight sheets numbered E-1 through E-8.

This method of numbering sheets allows the reader to tell at a glance whether all sheets are in place and, if not, what is missing. (See the set of plans at the end of this textbook for other examples of numbering on construction plans.)

14.6 REVISIONS

Sometimes, it is necessary to make changes on working drawings after they have been issued to contractors as construction plans. The changes are made on working drawings and are identified by the drafter so that the changes can be noted by the contractor (see again Figure 14.1). Each revision is numbered and dated in the title block of the working drawing. The working drawing is then reproduced and reissued to the contractor. The contractor gives the owner the estimated cost for making the revision to construction. If this price is agreed upon by the owner, the owner in turn gives approval, in writing, to the architect or engineer. The architect or engineer instructs the contractor, in writing, to proceed with the revision as shown on the revised plan. At this point, the revised plan becomes the construction plan. If the revision is not approved by the owner, the original construction drawing remains the construction plan for the building.

14.7 DATES ON WORKING DRAWINGS (CONSTRUCTION DRAWINGS)

The architect or engineer decides what date to put on the sheets. All sheets show the same date on the title block regardless of when they were drawn. This date is important because it indicates that all sheets belong to that set of drawings. If the job must be redesigned because of changes requested by the owner, a new date is assigned to the new drawings, and the old drawings are disposed of. The date on each set of drawings identifies the set of drawings and eliminates possible confusion.

14.8 NAMES OR INITIALS FOUND IN THE TITLE BLOCK

Two names or initials appear on the title block. The drafter working on the plan signs or initials it where indicated. Usually a senior designer (or the registered architect or the registered engineer) checks the drawing for accuracy before it is printed and initials or signs each title block as indicated. These two persons are responsible for the accuracy of the plan. Each initials or signs only the sheets for which he or she is responsible.

14.9 THE SCALE USED FOR THE DRAWING

There is a place on each title block where the drafter records the scale used for the drawing. Sometimes, several drawings or details on the same sheet use different scales. If this occurs, a title is listed under each drawing or detail. The scale that was used for that individual drawing or detail is also listed under the drawing or detail.

In the space on the title block where the scale is normally shown, the phrase "as noted" is usually written. When "as noted" is placed in the scale block, the person reading the plans knows to look at each individual drawing for the scale used.

14.10 THE ARCHITECT'S AND ENGINEER'S SEAL

Architects and engineers must be registered professionals in a state to have the right by law to design buildings in that state. (See Chapters 8 and 13 for the major requirements for registration of architects and engineers.)

Registered architects and engineers have seals that are required on all drawings that they are responsible for. Figure 14.3 is an example of an engineer's seal. The architect's seal and the engineer's seal are similar. Both seals are usually placed beside the title block unless there is enough space within the confines of the title block.

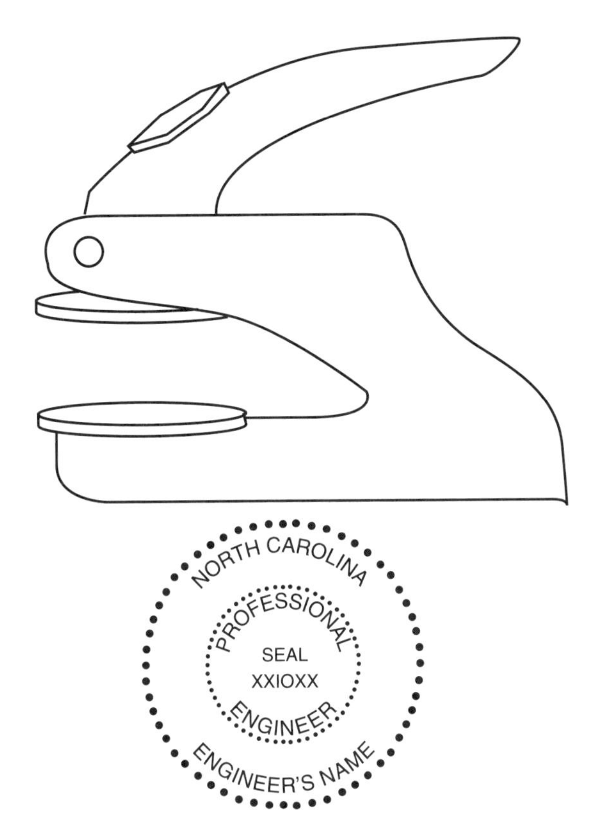

FIGURE 14.3 ■ Simulated typical engineer's seal

14.11 THE LIABILITY NOTE

The architect (or engineer) usually has the following note printed on the drawing paper or film used in the office:

> *"This drawing is the property of the architect. Written consent is required prior to reproducing, altering, or using the contents."*

The architect or engineer who places the professional seal on the drawings indicates that a registered professional is responsible, by law, for the design and drawing of the plan. The liability notation clears the architect or engineer of any responsibility for the design shown on the drawings if this design, or any part of the design, is used on another job or by another architect or engineer or owner. In other words, the architect or engineer has been hired to design the job shown on the plans, and the unauthorized use of any part of the design anywhere other than on the job for which the plans were prepared releases the architect and engineer from liability.

SUMMARY

- A title block is found on each sheet of a set of plans.
- Information about the owner, job name, job address, and job number can be found on each sheet.
- Some sets of drawings use the architect's title block, and other sets of drawings use the engineer's title block.

- Sheet numbers indicate the page number and whether the drawing is part of the architectural, mechanical, or electrical plans.
- Revisions are numbered and dated, reproduced, and reissued to the contractor after the owner's permission to make revisions has been obtained.
- The same date is used on each sheet of a set of drawings regardless of when the actual drawing was done.
- The initials on the title block indicate what persons are responsible for accuracy of the drawing.
- If different scales are used on the same sheet, the scale for each drawing is indicated under that drawing.
- The liability note in the title block protects the architect or engineer in case of the unauthorized use of a set of plans.

REVIEW QUESTIONS

1. Where may the title block be found on a set of drawings?
2. Why is it advantageous to have standard title blocks printed on a company's drawing paper or film?
3. Why does each title block contain the owner's name, job name, and job address?
4. Whose name is in the title block—the engineering firm or the architectural firm? Briefly discuss your answer.
5. How is the title of a sheet determined?
6. Explain how sheets of a set of drawings are numbered for identification.
7. When and how are revisions to drawings made?
8. How is the date in the title block determined?
9. Whose initials are put in the title block?
10. If several scales are used on the same sheet, how is this information shown?
11. What is the purpose of the liability note in the title block?

PART II

FREEHAND SKETCHING AND DRAFTING

CHAPTER 15

Types and Weights of Lines

OBJECTIVES

After studying this chapter, you should be able to:

- identify the most common types of lines drawn by drafters
- distinguish between hard and soft graphite pencils and tell when to use leads of varying degrees of hardness
- distinguish between types of leads used on drafting films
- demonstrate the ability to draw the types of lines discussed in this chapter
- interpret a typical index of standard lines that might be used

15.1 INTRODUCTION

By varying the types of lines on a plan, the drafter helps the reader to distinguish between different kinds of information on the plan. Many types of lines are used on working drawings and in freehand sketching. Object lines, construction lines, extension lines, projection lines, cutting plane lines, dimension lines, hidden lines, center lines, break lines, and border lines will be discussed in this chapter.

The line weight that is selected for plans is very important. The weight of a line describes how dark and how wide the line is drawn. Usually, wider lines are darker than thinner lines. For ink drawings, the ink is usually the same color (conventionally, black), and all lines are the same so far as darkness is concerned. Line width for ink drawings is determined by the choice of pens used by the drafter. It is not uncommon to dilute the ink with water to produce a lighter line. This chapter will include a discussion of other techniques that are used to obtain different shades of lines drawn in ink. Lines drawn with graphite pencils or with pencils designed to draw on plastic film sheets will also be discussed.

15.2 SCHEDULE OF LINES USED IN DRAFTING AND FREEHAND SKETCHING

Figure 15.1 is a schedule with examples of lines that are commonly used in drafting. These lines are drawn with standard drafting pens and black drafting ink. Inking pens are marketed in sets consisting of several pens with different size points. A standard set of pens usually has Numbers 3×0, 00, 1, 2, $2^1/_2$, 3, and 4. Numbers 6×0 and 4×0 are available in the smaller sizes, and Numbers 6 and 7 are available in the larger sizes. Each pen draws a line of a specific width. The pen size is given in Figure 15.1.

Figure 15.2 shows a typical line schedule, or alphabet, for drawing plans. Each line has two characteristics: width and darkness. Some lines are drawn continuously, some are a series of short dashes, and some are a series of short dashes and dots and long dashes and dots. Some are drawn with wide pen points and some are drawn

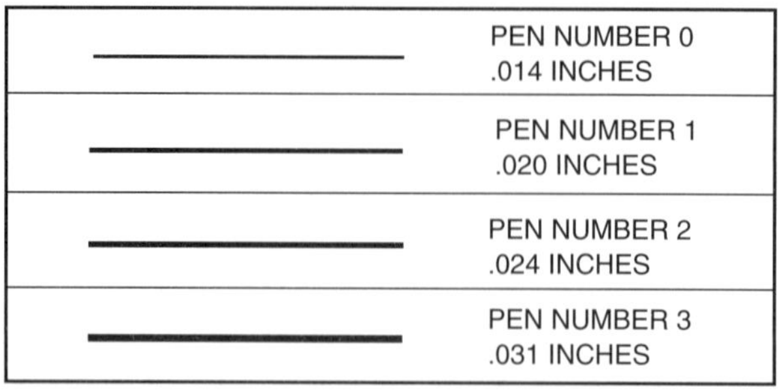

	PEN NUMBER 0 .014 INCHES
	PEN NUMBER 1 .020 INCHES
	PEN NUMBER 2 .024 INCHES
	PEN NUMBER 3 .031 INCHES

FIGURE 15.1 ■ Common line weights

with narrow pen points. Each line is used for a particular purpose. It is very important for the drafter to use the proper lines to relay the information needed and for these lines to be uniform. In other words, all object lines should be the same weight throughout the drawing, and all dimension lines should be uniform throughout the drawing.

15.3 LINES DRAWN WITH PENCILS

Lines drawn with pencils and those drawn with inking pens follow the same general rules. The weight of the lines drawn by a pencil can be controlled by the sharpness of the pencil and by the softness of the lead. If a heavy line is required, the pencil point must be dull so that the resulting line will be heavy. On the other hand, if the pencil point is sharp, the resulting line will be thin. Pencil-drawn lines should be about the same weight and darkness as inked lines (see Figure 15.1). Of course, pencil lines will not be as dark as inked lines; however, when a print is made from the penciled drawing, the print should reflect good, firm lines.

There are standard types of graphite pencils and special leads used with drafting films such as Mylar® or Vellum®. Graphite leads progress from B, which is very soft, up through HB, H, H-2, H-3, H-4, and so on. The higher the number, the harder the lead. The average drafter finds that H and H-2 are in a good range of hardness. This lead is used for drawings and for lettering. Harder leads such as H-4 are sometimes used as guidelines or construction lines. As these lines are drawn lightly on the plan, a sharp pencil point must be maintained so that the lines do not show up on prints. The lines can also be easily erased when the drawing is finished.

Graphite leads and film leads are available in wooden pencils, and singular leads are available for mechanical pencils. Wooden pencils with film lead start with E-0 and progress upward through E-1, E-2, E-3, E-4, and so on. Each higher number represents a harder lead. Some mechanical pencil leads are numbered in a "P" series, with P-1 the softest and P-2, P-3, P-4, and so forth becoming progressively harder. In most cases, the "P" numbers and the "E" numbers are identical in hardness.

Drafters who draw lines with pencils develop techniques to make all the lines on the drawing more uniform and the different weight lines more evident. For example, drafters learn that if the pencil is rotated as the line is being drawn, the line is more uniform from one end to the other.

15.4 OBJECT LINES

Object lines are used to outline the object being drawn. These lines should be drawn heavy (thick) and solid. These lines should be so distinct that there is no question as to the outline of the object being drawn (Figure 15.2).

SCHEDULE (ALPHABET) OF LINES		
NAME OF LINE	EXAMPLE OF LINE	DESCRIPTION OR USE
CONSTRUCTION LINE	THIN	ESTABLISHES SHAPES PRIOR TO DRAWING THE SAME LINES AS OBJECT LINES
OBJECT LINE	THICK	HEAVY LINE USED TO DRAW OBJECTS
BROKEN LINE	MEDIUM	USED TO SHOW PARTS OF OBJECT THAT ARE NOT VISIBLE
CENTER LINE	THIN	SHOWS CENTER OF OBJECT
EXTENSION LINE	EXTENSION LINE THIN	LINES EXTENDED FROM OBJECT SO DIMENSION CAN BE SHOWN
DIMENSION LINES	2' 0" 2' 0" 2' 0" THIN	LINE DRAWN BETWEEN EXTENSION LINES WITH DISTANCES
LEADER	OR THIN	LIGHT LINE (WITH ARROW HEAD DRAWN BETWEEN NOTES AND WHERE THE NOTE APPLIES)
SHORT BREAK LINE	MEDIUM	INDICATES SHORT BREAKS
LONG BREAK LINE	MEDIUM	INDICATES LONG BREAKS
CUTTING PLANE	A 1 OR THICK	SHOWS INTERNAL FEATURES, SECTION SHOWN LOOKING IN DIRECTION OF ARROW
PROPERTY LINE	THICK	SHOWS EXTENT OF PROPERTY
BORDER LINE	THICK	USED TO SHOW BORDER ON PLANS
NEW CONTOUR LINE	60 THIN	LINE INDICATES LAND ELEVATION. NUMBER INDICATES FEET ABOVE SEA LEVEL
EXISTING CONTOUR LINE	60 THIN	LINE INDICATES LAND ELEVATION. NUMBER INDICATES FEET ABOVE SEA LEVEL
PROJECTION LINE	THIN	LINE USED TO EXTEND DIMENSIONS, LOCATION POINTS FROM ONE SPACE TO ANOTHER

FIGURE 15.2 ■ Typical alphabet of lines (Line Schedule)

15.5 HIDDEN LINES

An object on a drawing must be complete and, in most cases, must contain lines to represent all edges and interior sections and surfaces in the object. These edges and interior sections are shown with hidden lines. Hidden lines are usually broken lines that are thinner than object lines (Figure 15.3).

15.6 EXTENSION AND DIMENSION LINES

Extension and dimension lines are used to show dimensions of an object. Extension lines are used to show the limits of the dimension given. These lines are thin, solid lines that are projected from the object at the exact place where the dimension is given. There is a small space between the end of the extension line and the object line. The dimension line is drawn between the two extension lines with arrowheads on each end where the dimension line and the extension line touch (Figure 15.4). Extension and dimension lines are thin, solid lines that should be the same weight.

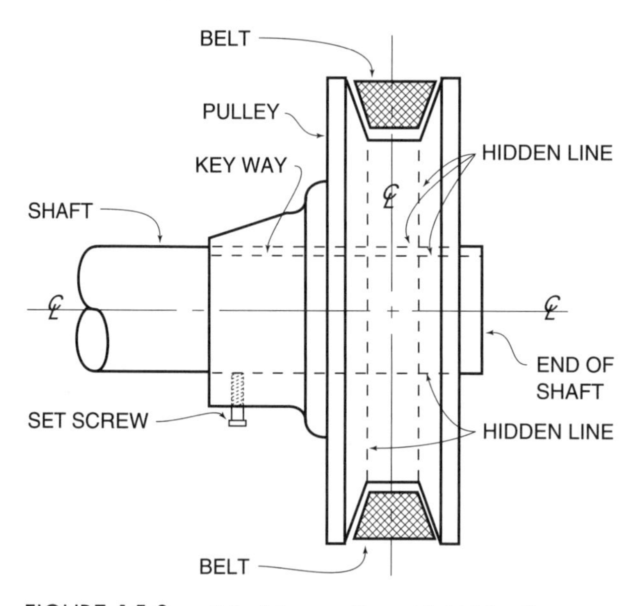

FIGURE 15.3 ■ Belt-driven pulley with hidden lines

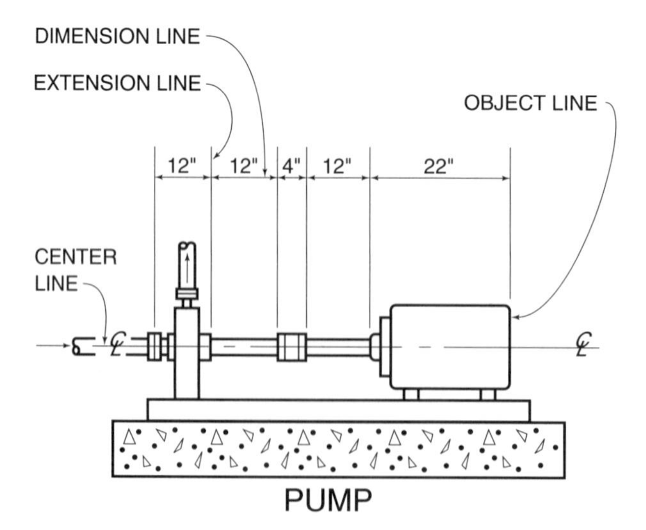

FIGURE 15.4 ■ Extension and dimension lines

Extension and dimension lines are very important because the size of the object is reflected by these dimensions. A mistake in a dimension line can create many problems in construction.

15.7 CONSTRUCTION LINES

Construction lines are used to help develop the drawing. Construction lines are thin lines that are drawn lightly on the plan (Figure 15.2). If object lines are put on the drawing while the object is being developed, mistakes are hard to erase. If construction lines are used instead, they are easily erased and changed. When the object has been fully developed, object lines can be drawn over the construction lines for the finished drawing. Unused construction lines are erased when the drawing has been completed. See Figure 15.5 for an example of an object being developed—first with construction lines and then with object lines applied over the construction lines later.

15.8 PROJECTION LINES

Projection lines are like construction lines, although these fine lines may be drawn either heavy or light depending on how the lines are to be used. In orthographic projection (Chapter 16), projection lines are used to project dimensions from one view to the other (Figure 15.6). Sometimes, these lines are left on the drawing, and sometimes, the lines are erased when the drawings have been completed. In either case, projection

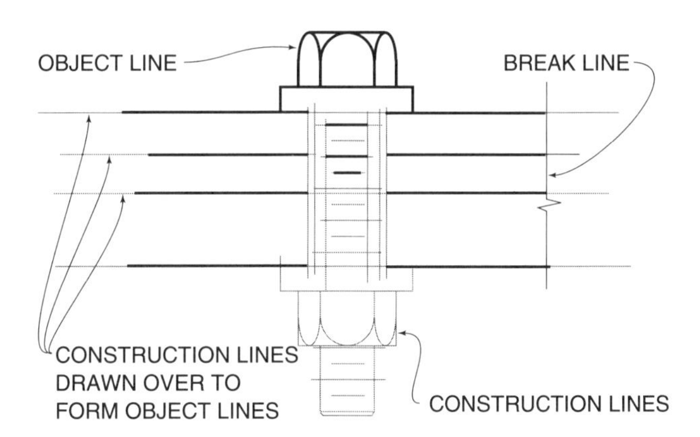

FIGURE 15.5 ■ Object and construction lines

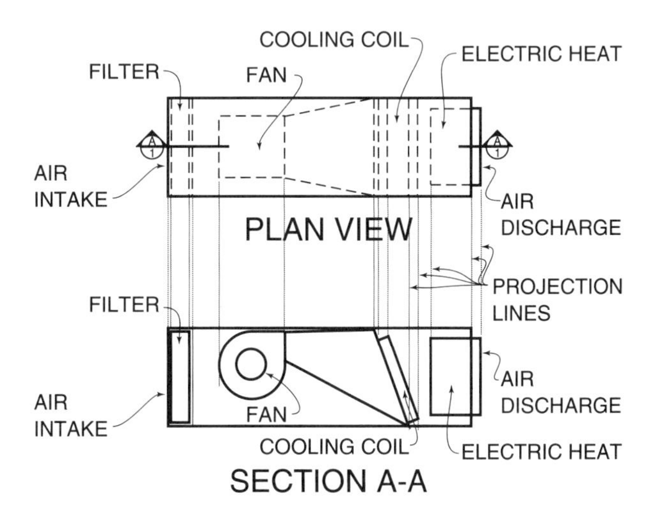

FIGURE 15.6 ■ Using projection lines

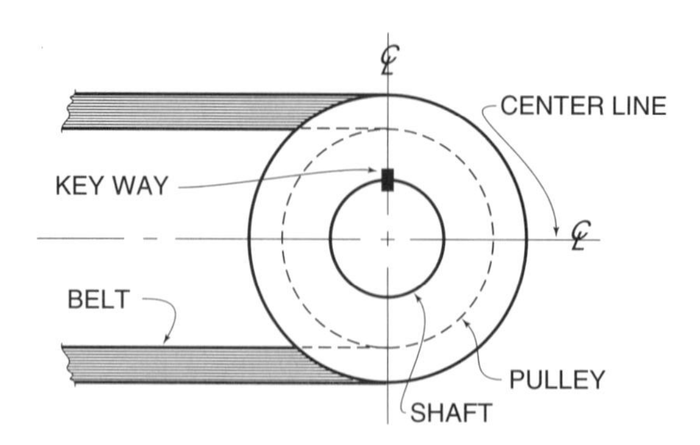

FIGURE 15.7 ■ Pulley and shaft showing center lines

lines are fine lines used in drafting to aid the drafter who needs to project dimensions from one area of a drawing to another.

15.9 CENTER LINES

It is important to show the center line of some objects or drawings so that dimensions can be referenced to the center of the drawing. The center line is usually shown as a series of long lines and short lines (Figure 15.2). The center line is a sharp, thin line that is drawn heavy enough to print along with the object line.

Two center lines are usually drawn at 90 degrees to each other so that the exact center is indicated where the two lines cross (Figure 15.7).

15.10 CUTTING PLANE LINES

Cutting plane lines can be shown with several types of symbols. One of the most common is a heavy line with a long dash and two short dashes drawn in series. The line is drawn through the object and extended for a short distance outside the object line. At this point, a short line is drawn showing the direction in which the cutting plane is cut, and an arrowhead is placed on the end of the short line (Figure 15.2). The cutting plane is numbered to identify the section if it is located somewhere else on the plan.

A second method of identifying a cutting plane is to draw a heavy line consisting of long dashes through the object. The line is extended to the outside of the object, and a short line with arrowheads is added similar to the cutting plane described above. The cutting plane is identified with a number or letter so that the section can be identified when it is drawn on the sheet. Refer to Figure 10.7 in Chapter 10 for additional examples of the use of cutting plane lines.

One of the most popular cutting plane symbols used in architectural and mechanical drawings is a heavy line drawn from a point inside the object to a point outside the object on both sides of the object or where the cutting plane ends. A continuous line between these two locations is not necessary. A circle is drawn at each end of the cutting plane, and the circle is cut in half with a horizontal line. An arrowhead is created around the circle to show the direction the section is shown. The top half of the circle contains the section number. The bottom half of the circle gives the sheet number where the section is shown (see Figures 15.2 and 15.6).

15.11 BREAK LINES

It is sometimes necessary to increase the scale of details of various parts of a building. With the larger scale, it is not practical or necessary to draw the entire portion of the building. When this is the case, it is desirable to show what is needed and omit the unnecessary portion of the building. Break lines are used to show that there is a continuation of the detail; however, the continued drawing is not necessary. Figure 15.2 demonstrates two types of lines used as break lines. The wavy line is drawn freehand, and the ruled line and zigzags are drawn with drafting instruments. As a general rule, the wavy freehand line is used when the break is short, and the ruled line with zigzags is used when the break is long (Figure 15.5).

15.12 BORDER LINES

Border lines are used to frame the drawing sheet. Horizontal border lines are drawn across the top and bottom of the sheet, and vertical border lines are drawn on the left side and right side of the sheet. The border lines set the size of the drawing and are containment lines for the drawing (or drawings) on that sheet. Border lines are usually very heavy, wide lines that are two or three times the weight of object lines (Figure 15.2).

15.13 SCHEDULE OF LINES USED IN OFFICES

Many types of lines are used by drafters in various offices. An office often encourages uniformity by using a line schedule to show that certain lines stand for specific things on drawings made in that office. A typical alphabet, or schedule, of standard lines for one office is included in Figure 15.2. As a rule, standard lines are identified and approved by the office manager for use by drafters in that office.

SUMMARY

- Line weight is determined by the size of the pen used by the drafter.
- Line weights drawn by pencil are controlled by the sharpness of the pencil and quality of the lead.
- Object lines are heavy solid lines that outline the object being drawn.
- Interior sections are shown by thin, broken lines called hidden lines.
- Extension and dimension lines show the dimensions of an object.
- Construction lines are light, thin lines that are usually erased when the drawing is completed.
- Projection lines project dimensions from one view to another.
- Center lines show the exact center of an object.
- Various symbols indicate cutting plane lines.
- Break lines are used to indicate that only a portion of an object is being drawn.
- Border lines frame the drawing sheet on the top, on the bottom, on the left side, and on the right side.
- Some offices have their own schedule or alphabet of standard lines to be used in that office.

REVIEW QUESTIONS

1. Why does a drafter need more than one inking pen to draw a plan?
2. How does the drafter control the weight of lines drawn by pencil?
3. What is the most common lead used by drafters using standard graphite pencils?
4. Why should the drafter learn to rotate the pencil as lines are being drawn?
5. Why is it especially important that dimension and extension lines be accurate?
6. What is the purpose of construction lines, and why are they usually erased after the drawing has been completed?
7. Demonstrate one method of drawing cutting plane lines by using an example from the textbook.
8. How does the use of break lines save time for the drafter?
9. Why do many offices have an index of lines for that office?

CHAPTER 16

Orthographic Projection

OBJECTIVES

After studying this chapter, you should be able to:

- explain why orthographic projection is used for drawings
- define a viewing plan
- describe the three basic views used in orthographic projection
- describe two other views—other than the three basic views—that are often used in drawing plans
- describe the various types of lines used in orthographic projection and give their functions
- distinguish between the views used in drawing plans
- explain how the various views of building plans correspond to the five views used in orthographic projection

16.1 INTRODUCTION

In this unit, orthographic projection will be discussed and illustrated. In order to relay information clearly from designer to fabricator or craftsperson, it is important to show graphically the shape of an object, the dimensions of the object, and the relationship of one view to other views of the object. Orthographic projection is one of the most common methods for drawing plans of an object to relay important information about it.

Orthographic projection is often referred to as making three-view drawings of an object. In most cases, three views are enough to relay the needed information to the fabricator or craftsperson. The three basic views are the *plan view*, the *front view*, and the *end view*.

Although the three basic views are usually sufficient, a total of five views can be developed by using projection lines from one view to the other. The five views that can be developed are the plan view, the front view, the right side or end view, the left side or end view, and the rear view. The plan view can be a floor plan or roof plan. The front view is usually called the front elevation, the right and left side views are called the right and left elevations, and the rear view is called the rear elevation.

16.2 THE VIEWING PLANE

Understanding orthographic projection requires an understanding of the viewing plane. A plane is an imaginary, flat, transparent surface similar to a pane of glass. When this plane is placed parallel to the surface of an object, the surface of the object parallel to the plane is shown in its true size and shape. If the surface of the object is not parallel to the plane, the surface is shown as it is seen according to size and shape (Figure 16.1).

When surfaces are not parallel to the plane, the orthographic projection view shows these surfaces smaller in width or height (whichever the case may be). See Figure 16.2 for examples of object surfaces that are not parallel to the viewing plane.

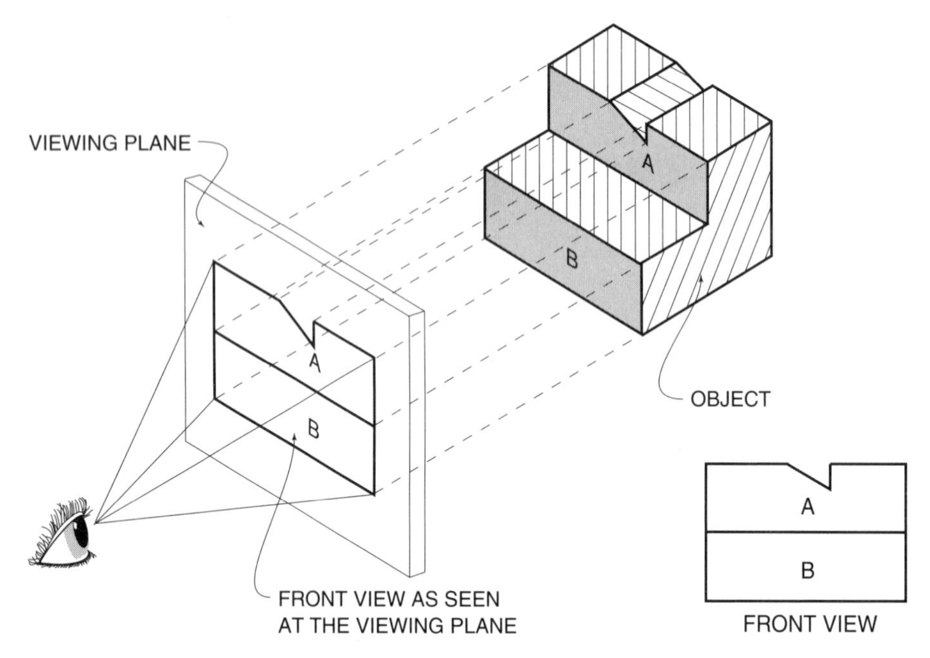

FIGURE 16.1 ■ Object as seen through a viewing plane—front view

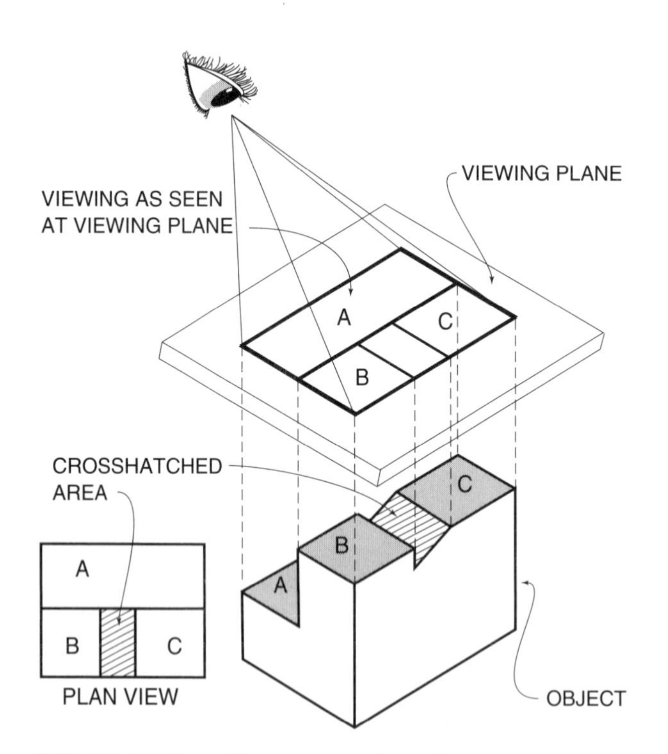

FIGURE 16.2 ■ Object as seen through a viewing plane—plan view

For the plan view, the viewing plane is parallel to the top of the object (Figure 16.2). The shaded portion represents the surface of the object that is parallel to the viewing plane. The crosshatched portion represents the surface of the object that is not parallel to the viewing plane, and this surface is not shown actual size. (Surfaces that are not parallel to the viewing plane are shown as they appear rather than actual size.)

The front view of the object is shown in Figure 16.1 along with the viewing plane. The viewing plane is vertical, and surfaces of the object that are parallel to the viewing plane are shown shaded. Surfaces that are not parallel to the viewing plane are shown crosshatched. As in the plan view, surfaces that are not parallel to the viewing plane are shown as they are seen and do not reflect true shape and size.

The right side view is shown in Figure 16.3. The viewing plane is shown in the vertical position, and the surfaces that are parallel to the viewing plane are shown shaded. Surfaces that are not parallel to the viewing plane are shown crosshatched. Again, surfaces that are not parallel to the viewing plane are shown as they are seen and do not reflect true shape and size.

The left side view can be shown in the same manner as the right side view, and the rear view (back view) can be shown in the same manner as the front view, with the viewing plane in the vertical position at the rear of the object.

16.3 COMBINING VIEWING PLANES

If viewing planes were placed on all four sides of an object and on top of the object, the object would be encased in a transparent box. If the image of the object, as seen through the various viewing planes, were projected to each side of the box, the plan view, the front view, the rear view, the right side view, and the left side view would be visible on the viewing planes. In Figure 16.4, the plan view is projected straight down. The other views are aligned with the plan view. When all the various views are considered, the true lengths, widths, and heights are apparent.

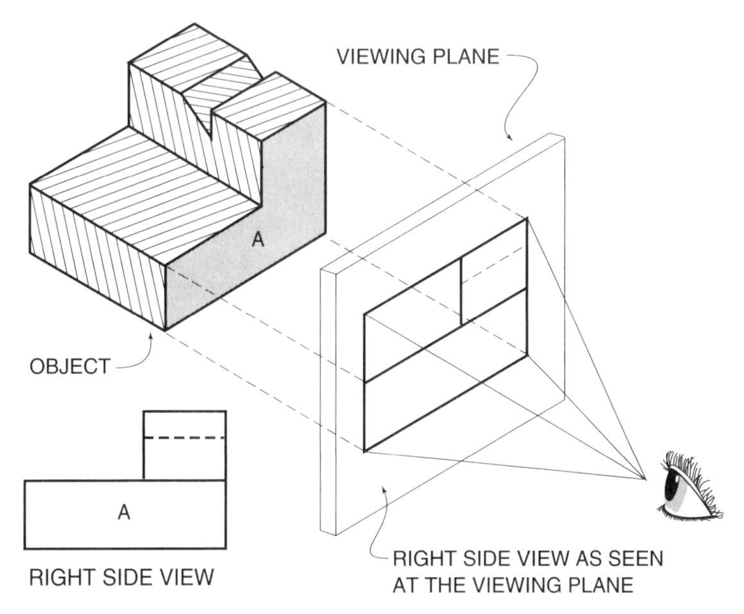

FIGURE 16.3 ■ Object as seen through a viewing plane—right side view

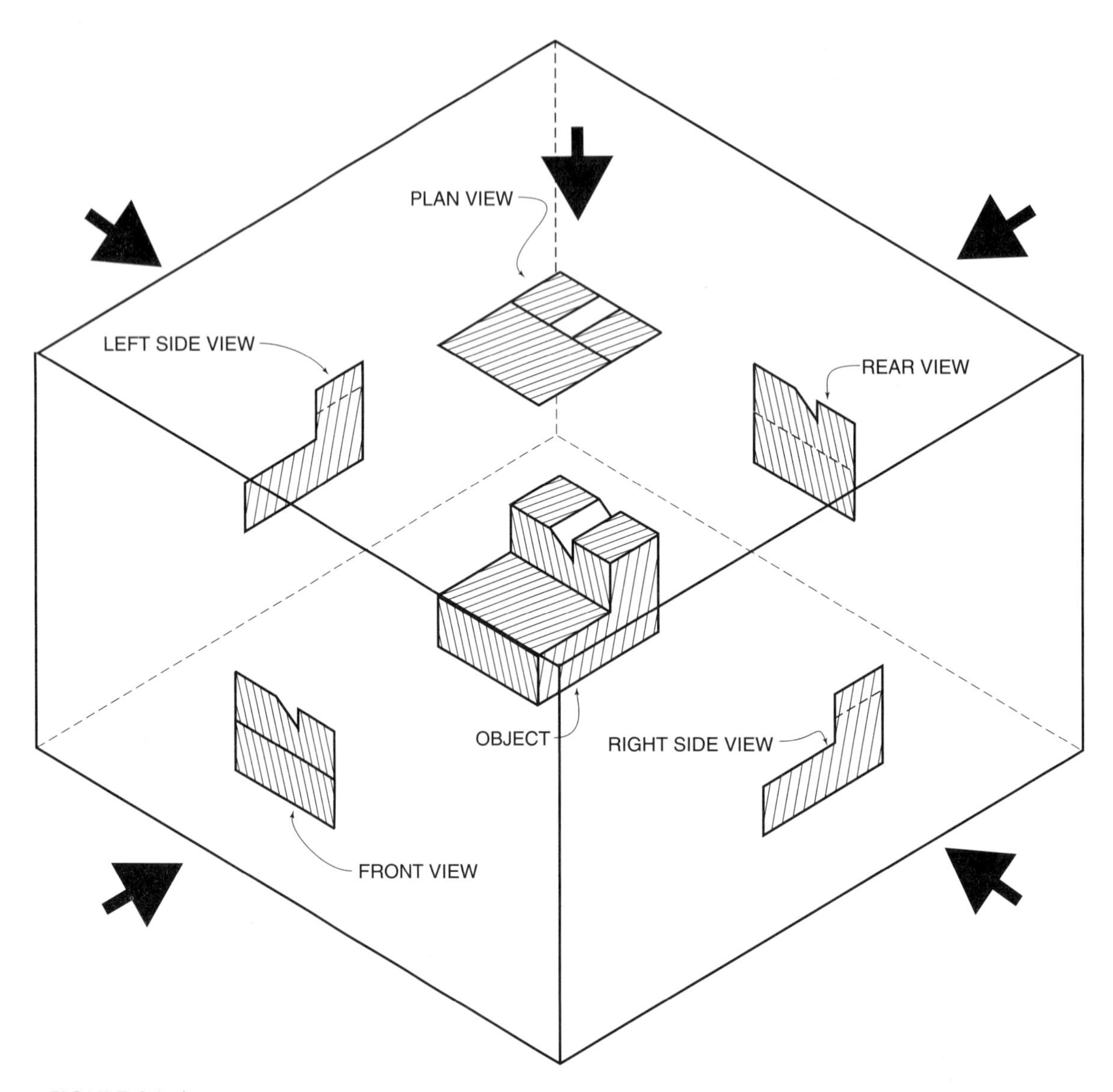

FIGURE 16.4 ■ Plan view, front view, right side view, and rear view with five viewing planes

16.4 TYPES OF LINES USED IN ORTHOGRAPHIC PROJECTION

Several types of lines are used in orthographic projection drawings. Briefly described in this unit will be object lines, construction lines, hidden lines, miter lines, projection lines, extension lines, and dimension lines. A more detailed discussion of types of lines used in drafting was included in Chapter 15.

Object lines are drawn for the purpose of outlining the object. Object lines are moderately heavy and dark because they show the shape of the object (Figure 16.5). A 2H pencil is normally used to draw the object lines. Refer to Chapter 15 for a more detailed description of types of pencils and drawing pens, line weights, and drawing techniques.

Construction lines (often called projection lines) are usually drawn with a harder pencil (H4 or H6). These lines are thin and are usually not as dark as object lines. Construction lines are used to transfer distances from one view to the other in orthographic projection (see Figure 16.5). Construction lines are usually removed from the finished drawing.

Hidden lines are broken lines that are used to outline surfaces that cannot be seen in the other views because the surfaces are located behind another surface. Hidden lines are usually drawn with the same thickness and weight used for regular object lines. The H2 pencil is usually used to draw hidden lines (see Figure 16.5).

Miter lines are construction lines drawn at 45 degrees to the horizontal of the drawing. Miter lines change the direction of distances when construction lines are drawn from one view to the other (see Figure 16.5). When projection lines are extended from the object in one view to the miter line and then are drawn up or down from the miter line, the true size or dimension of that part of the object is transposed to the other view.

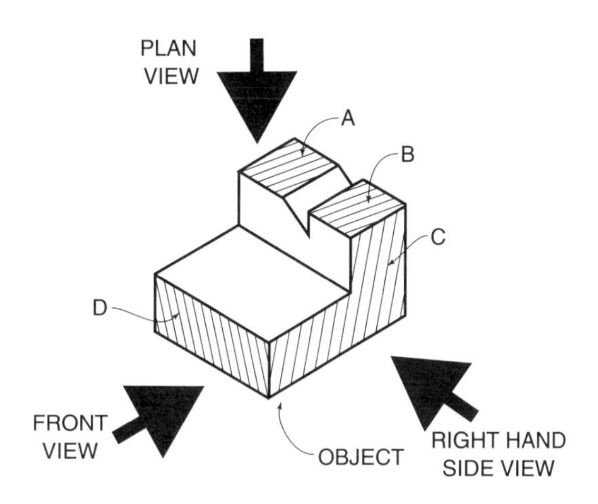

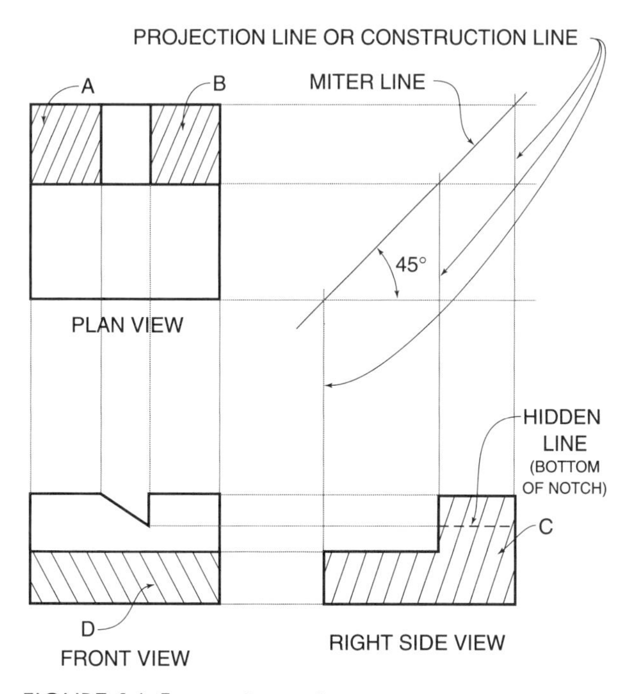

FIGURE 16.5 ■ Orthographic projection

Projection lines are a type of construction line, and they are usually removed from the finished drawing. Projection lines, as described in the preceding paragraph on miter lines, are drawn from the object to the miter line. By projecting these lines up or down on the drawing, actual distances or dimensions can be transferred from one view of an object to another. This procedure aids the drafter, who does not have to measure each dimension on each view of the object (Figure 16.5).

Extension lines are construction lines that are drawn a little heavier than projection lines and are left on the finished drawing. These lines are extended out from the object to reflect the distance between various points on the object. Extension lines do not touch the object, and they are used to indicate where measurements of the object start and stop (see Figure 16.8).

Dimension lines are construction lines drawn like the extension lines described above. These lines are drawn between extension lines and have arrowheads on each end of the line. This is the usual method for showing dimensions of the object. The dimension or measurement is written on the dimension line. Dimensions are usually given in feet and inches. In some cases, dimensions are given in decimal form or in metric form.

16.5 USING ORTHOGRAPHIC PROJECTION TO CONSTRUCT THE THREE BASIC VIEWS

In Figures 16.6, 16.7, and 16.8, the object is shown isometrically in the upper left-hand corner of the drawing. The orthographic method of drafting is used to show the three basic views (plan view, front view, and right side view) of the object.

Figure 16.6 is an example of the plan view of a building. Projection lines have been extended to the right and to the bottom from the major points on the plan view.

The front view can be developed using projection lines from the plan view and the known dimensions of the height of the object (Figure 16.7). Projection lines are then extended to the right from the main points of the front view. A miter line is drawn through the projection lines extended to the right from the plan view. The projection lines are extended down from the miter line to a point where they intersect the projection lines from the front view (Figure 16.7). The projection lines extended down from the miter line now intersect the projection lines from the front view. By darkening the projection lines between these intersections, the right side view can be drawn without having to do any measuring (Figure 16.8).

Surfaces that cannot be seen from the imaginary viewing plane are shown with broken lines (hidden lines) as illustrated in Figure 16.8. The blueprint reader can tell that there is something there, even if the hidden detail cannot be seen. By looking at all three drawings, the reader can determine what the hidden lines are showing.

The location of the miter line drawn through the horizontal projection lines determines the location of the right side view (see Figures 16.6 through 16.8). Before drawing the miter line, the drafter must be sure that the right side view will be located where one view can be easily related to the other.

When the drawing is complete, the drafter can add extension lines and dimension lines as shown on the plan view and the front view to give exact dimensions of the object (Figure 16.8). One dimension of Figure 16.8 is shown on the right side view; however, this dimension could also be given on the front view.

16.6 SELECTING THE PROPER VIEWS

When an orthographic projection drawing is constructed, it is important that the views that are selected show the most detail of the object being drawn. Once the front view and the right side view are selected, the plan view can be drawn, and the front view and right side view can be developed. It is sometimes necessary to draw the front view, the right side view, the left side view, and the rear view to show all the necessary details of the object. If the front view and the right side view do not do an adequate job of showing necessary details, then the left side and rear views must be shown.

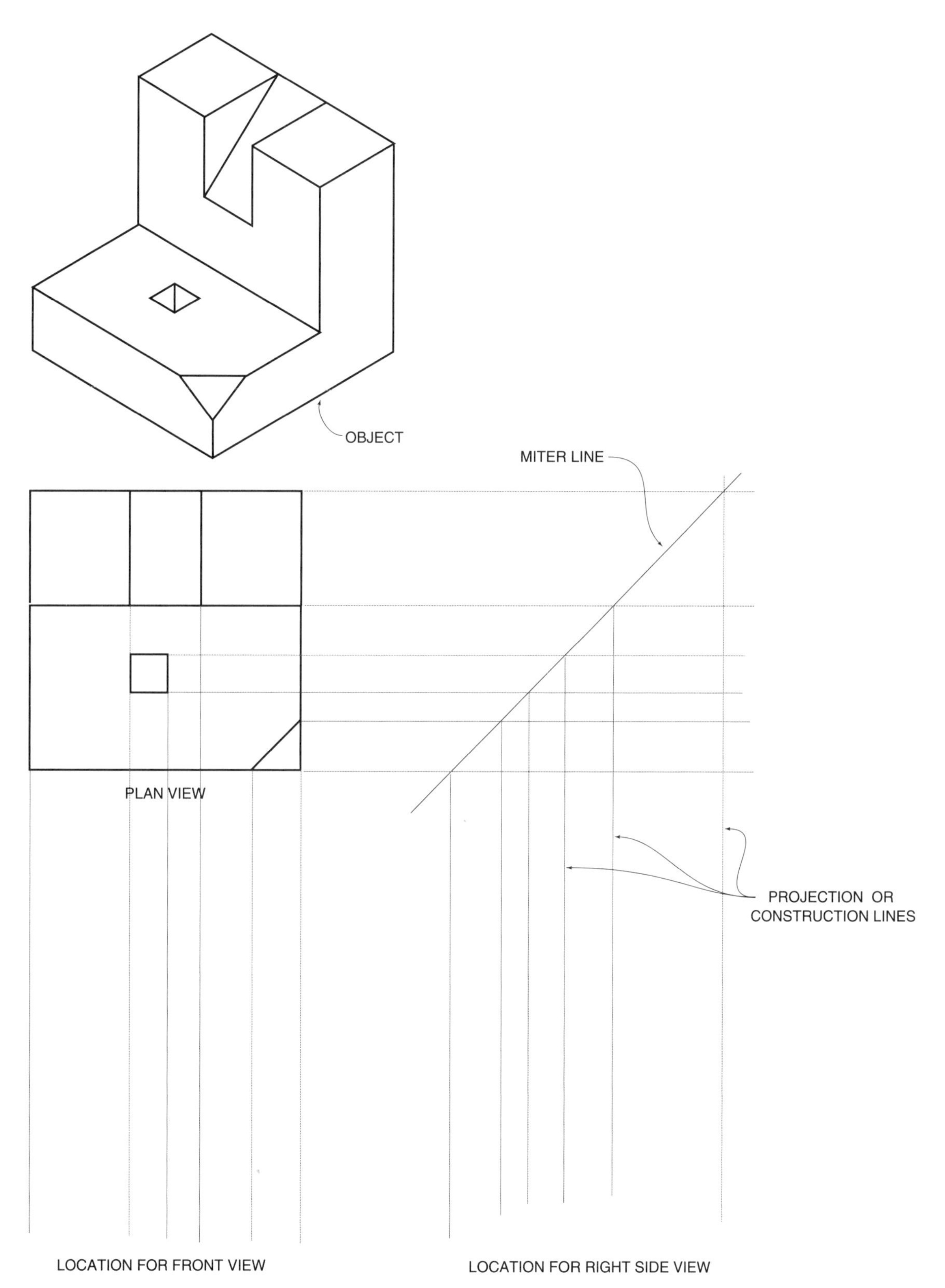

FIGURE 16.6 ■ Orthographic projection showing miter line and projection lines from plan view

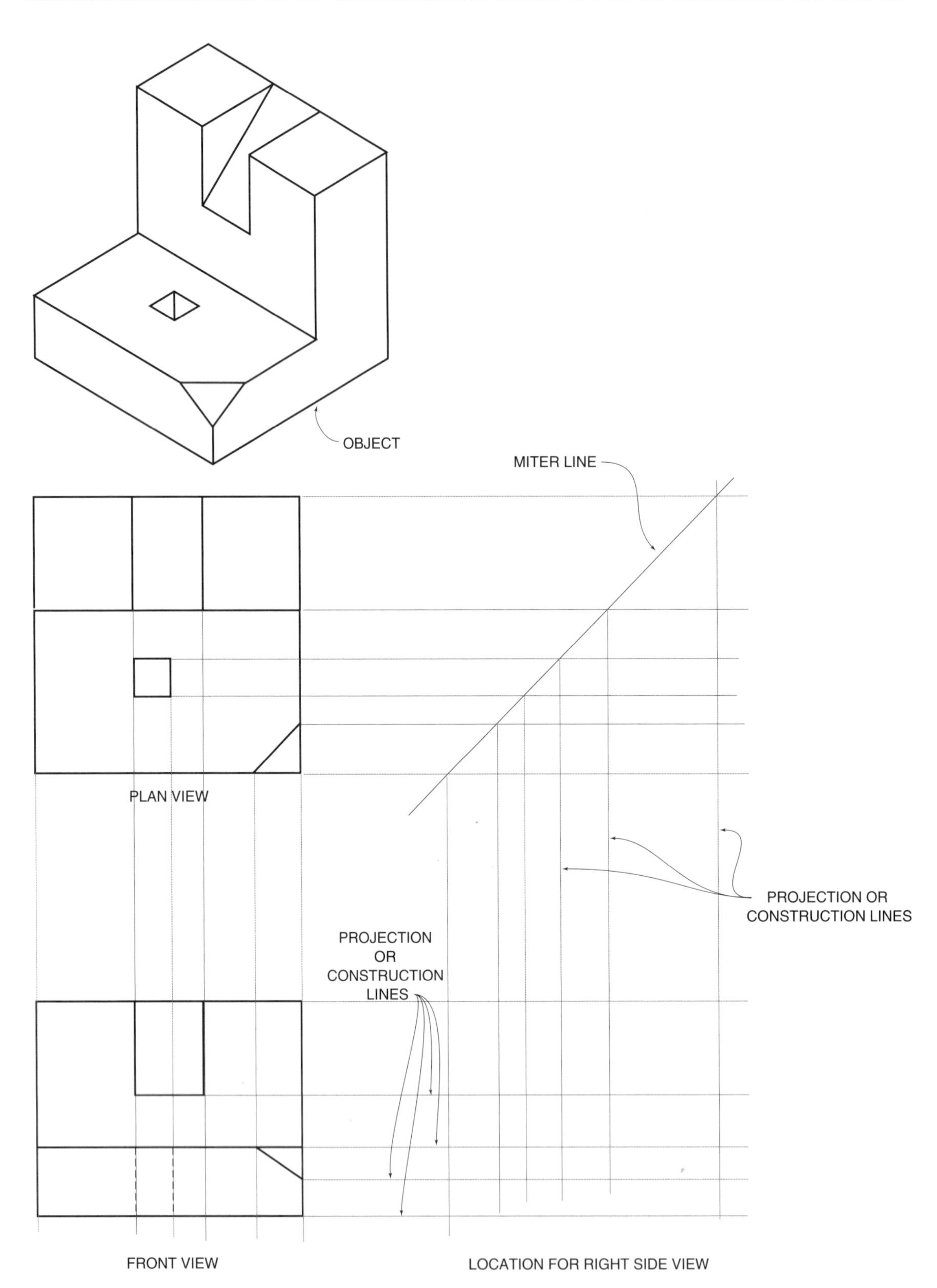

FIGURE 16.7 ■ Orthographic projection showing miter line and projection lines from plan view and front view

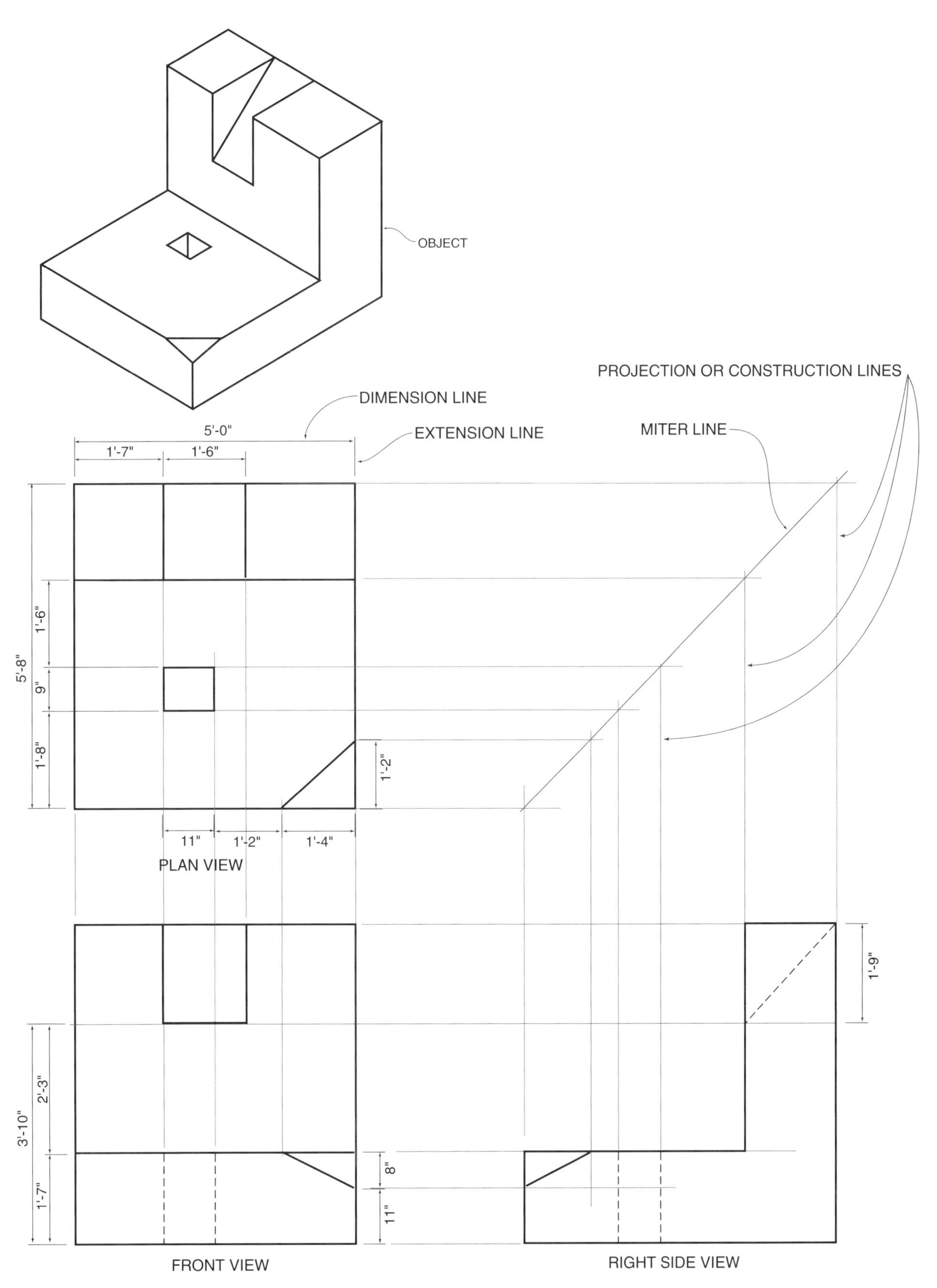

FIGURE 16.8 ■ Orthographic projection showing plan view, front view, and right side view with projection or construction lines and extension and dimension lines

It is always important to locate the drawings on a sheet so that the entire drawing can be shown and so that each view relates to the other views.

16.7 WORKING DRAWINGS AND ORTHOGRAPHIC PROJECTION

Architectural and engineering plans for a building use the orthographic projection procedure on the working drawings. A comparison of building plans with the orthographic projection procedure is as follows:

1. The floor plans and the roof plan represent the plan view of the building.
2. The front elevation, the right side elevation, the left side elevation, and the rear elevation plans represent the front view, the right side view, the left side view, and the back view of the building.

SUMMARY

- Orthographic projection refers to the making of three-view drawings of an object.
- The three basic views of an object are the plan view, the front view, and the end view (right side view).
- The five views that can be created in orthographic projection are the plan view, the front view, the right side or end view, the left side or end view, and the rear view.
- A viewing plane is an imaginary, flat, transparent surface through which the drafter views an object.
- Surfaces that are parallel to the viewing plane are shown in true shape and size.
- Surfaces not parallel to the viewing plane are not shown in their true width or height.
- Lines used in orthographic projection include object lines, construction lines, hidden lines, miter lines, projection lines, extension lines, and dimension lines.
- Orthographic projection views should be carefully selected to show the most important details.

REVIEW QUESTIONS

1. Briefly explain why orthographic projection is used on plans by architects and engineers.
2. Describe the three basic views of an object.
3. Define a viewing plane.
4. An object is drawn in its true size and shape if:

   A. the surface of the object is parallel to the viewing plane.

   B. the surface of the object is not parallel to the viewing plane.
5. Define each type of line and tell how it is used:

   A. Object line

   B. Construction line

   C. Hidden line

   D. Miter line

   E. Projection line

   F. Extension line

   G. Dimension line
6. Floor plans of a building represent the (plan view, front view) of a building.
7. Building plans of the front elevation represent the ________________________ view in orthographic projection.
8. The rear elevation is the same as the ________________________.

ADDITIONAL STUDENT EXERCISES: COMPLETING ORTHOGRAPHIC PROJECTION VIEWS

For Problems 1 through 5 (Figures 16.9–16.13), the student should draw the missing lines needed to complete the three basic orthographic projection views of the indicated isometric object.

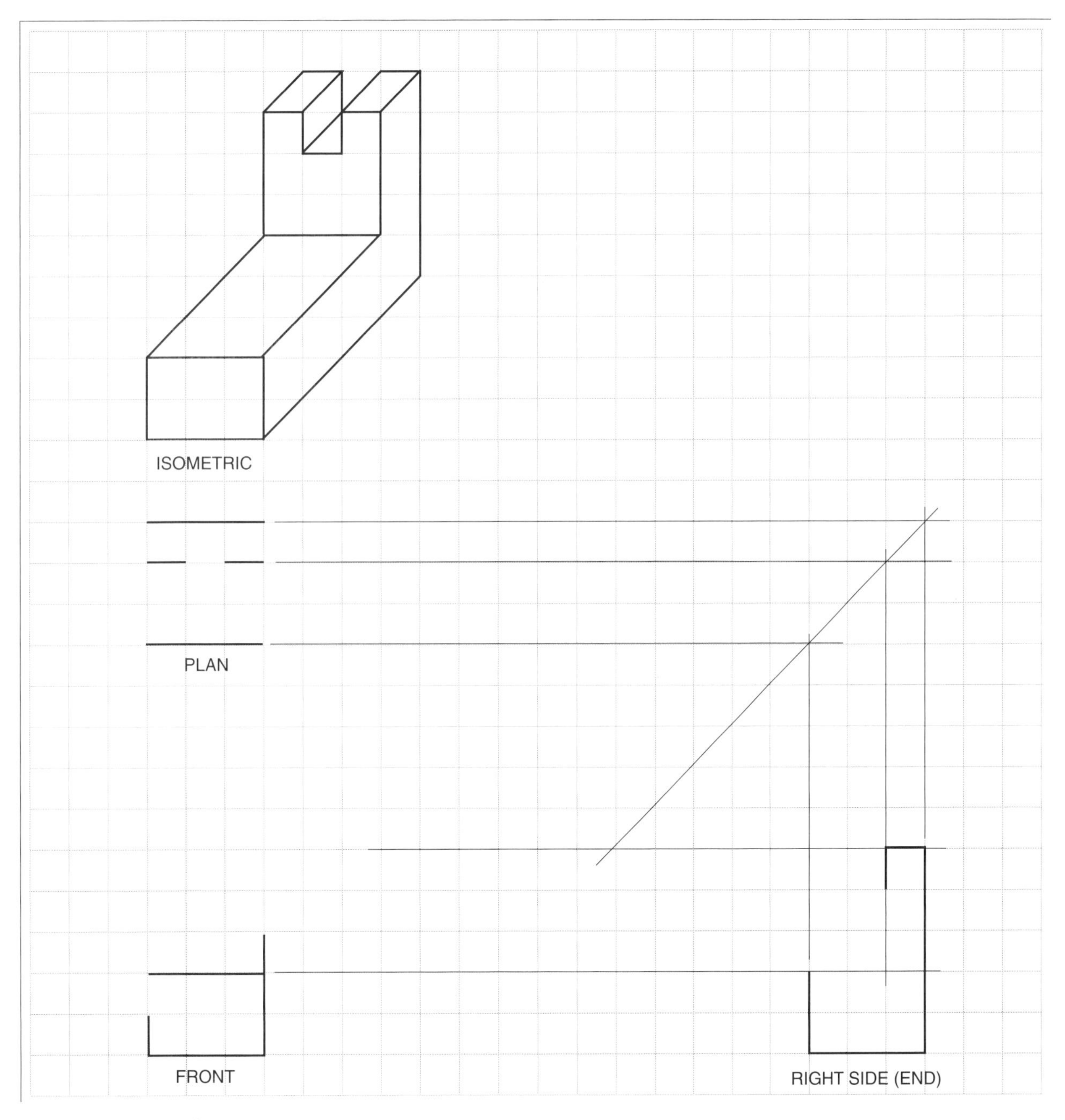

FIGURE 16.9 ■ Exercise 1

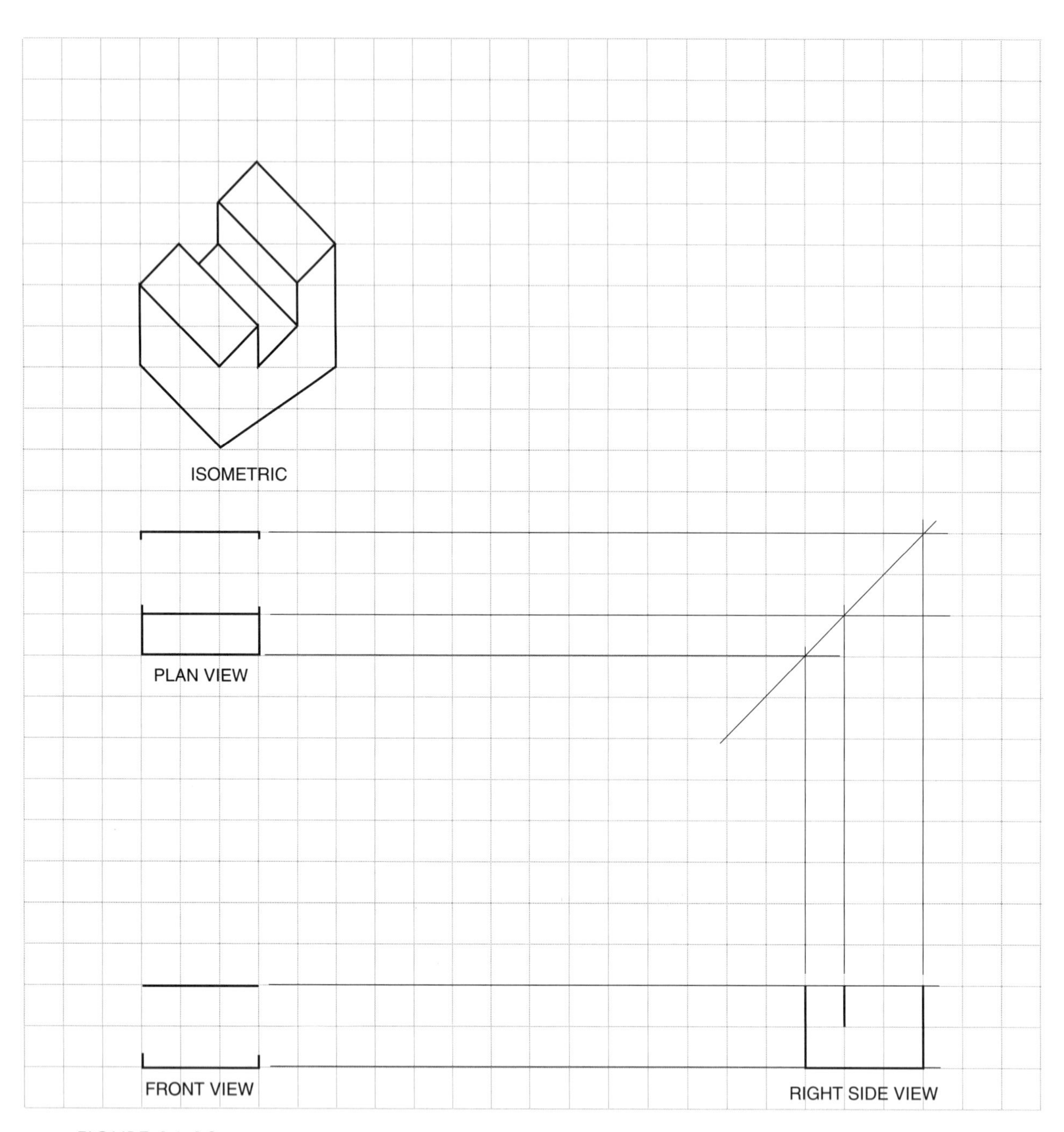

FIGURE 16.10 ■ Exercise 2

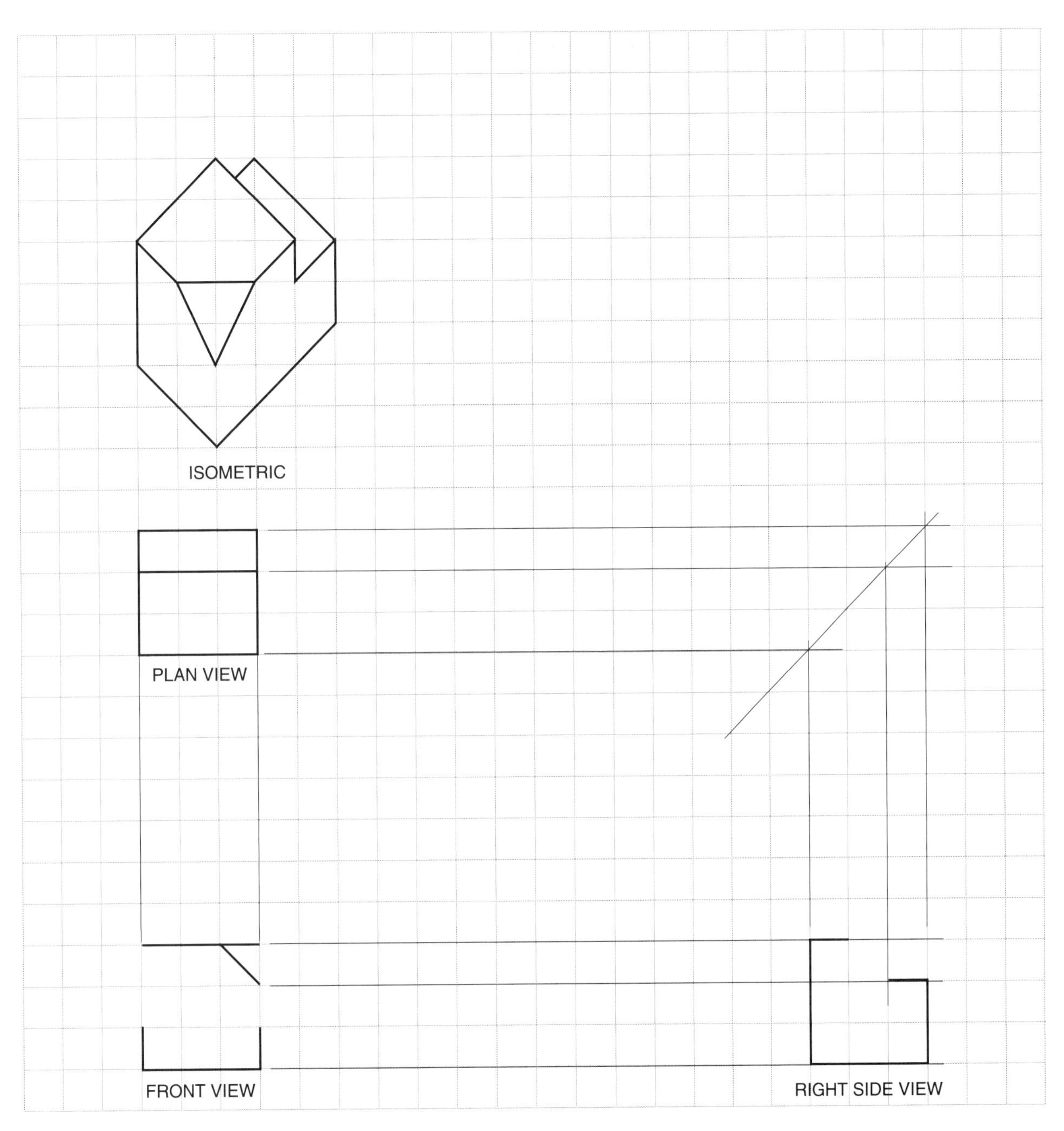

FIGURE 16.11 ■ Exercise 3

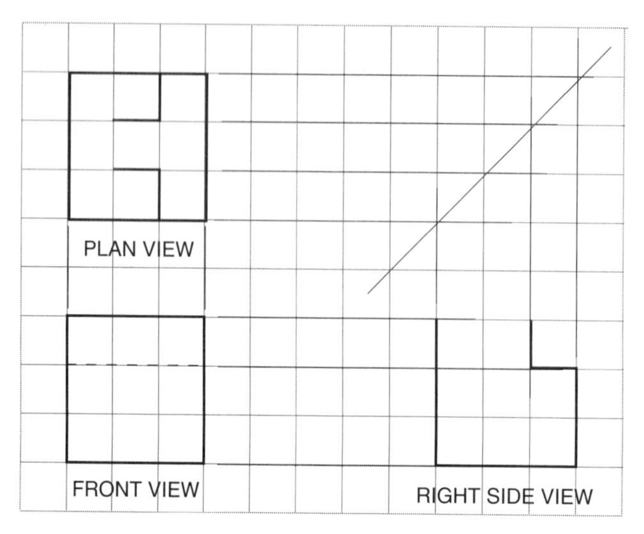

FIGURE 16.12 ■ Exercise 4

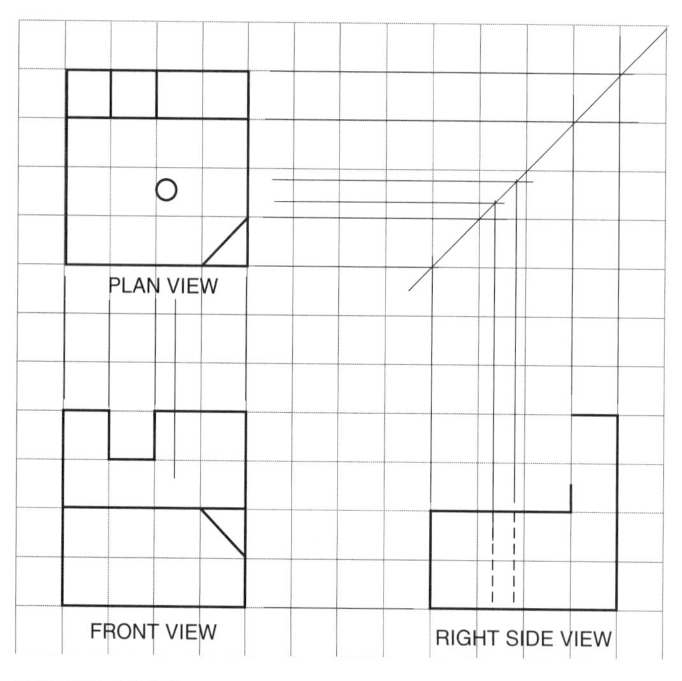

FIGURE 16.13 ■ Exercise 5

CHAPTER 17

Oblique Drawings

OBJECTIVES

After studying this chapter, you should be able to:

- describe three types of drawings that show three-dimensional views of objects
- explain how construction lines can be drawn to show the three basic views of an object
- demonstrate the ability to create an oblique drawing of an object
- explain how variations of oblique drawings are made

17.1 INTRODUCTION

Information can sometimes be relayed better when the object being drawn takes on the feeling of a three-dimensional rather than a two-dimensional drawing. A two-dimensional drawing shows width and height, whereas a three-dimensional object has width, height, and depth. An oblique drawing is one type of pictorial drawing that relays this three-dimensional feeling. The oblique drawing is very helpful for showing details of air-conditioning equipment.

Normally, objects are viewed in three dimensions. The plan view, front view, and side view are seen simultaneously. The three basic views of an object are discussed in Chapter 16.

There are several ways to draw an object so that it appears to be three-dimensional. One method is to make isometric drawings. (Isometric drawings are discussed in Chapter 18.) A second method is to make perspective drawings with diminishing points. (Perspective drawings are not included in this book because they are not usually used in the heating and air-conditioning fields.) Oblique drawings are probably the easiest of the drawings that give a three-dimensional impression. Oblique drawings show two or more surfaces at one time on the same drawing. The front view appears in its true size and shape, but the other sides are not shown in their true size and shape because they are drawn at an angle.

17.2 HOW OBLIQUE DRAWINGS ARE CREATED

The orthographic projection method of drafting is discussed in Chapter 16. Using orthographic projection, the drafter draws an object showing the three basic views (the plan view, the front view, and the right-side view). The drawings are not drawn to scale.

Oblique drawings consist of a front view or an end view of an object drawn in the same manner as in orthographic projection. This view is drawn with the surface parallel to the imaginary vertical viewing plane. (See Chapter 16 for an explanation of the viewing plane.) Figure 17.1 shows the basic steps in making an oblique drawing. Step 1 demonstrates how the front view is started, and Step 2 shows how the front view is completed.

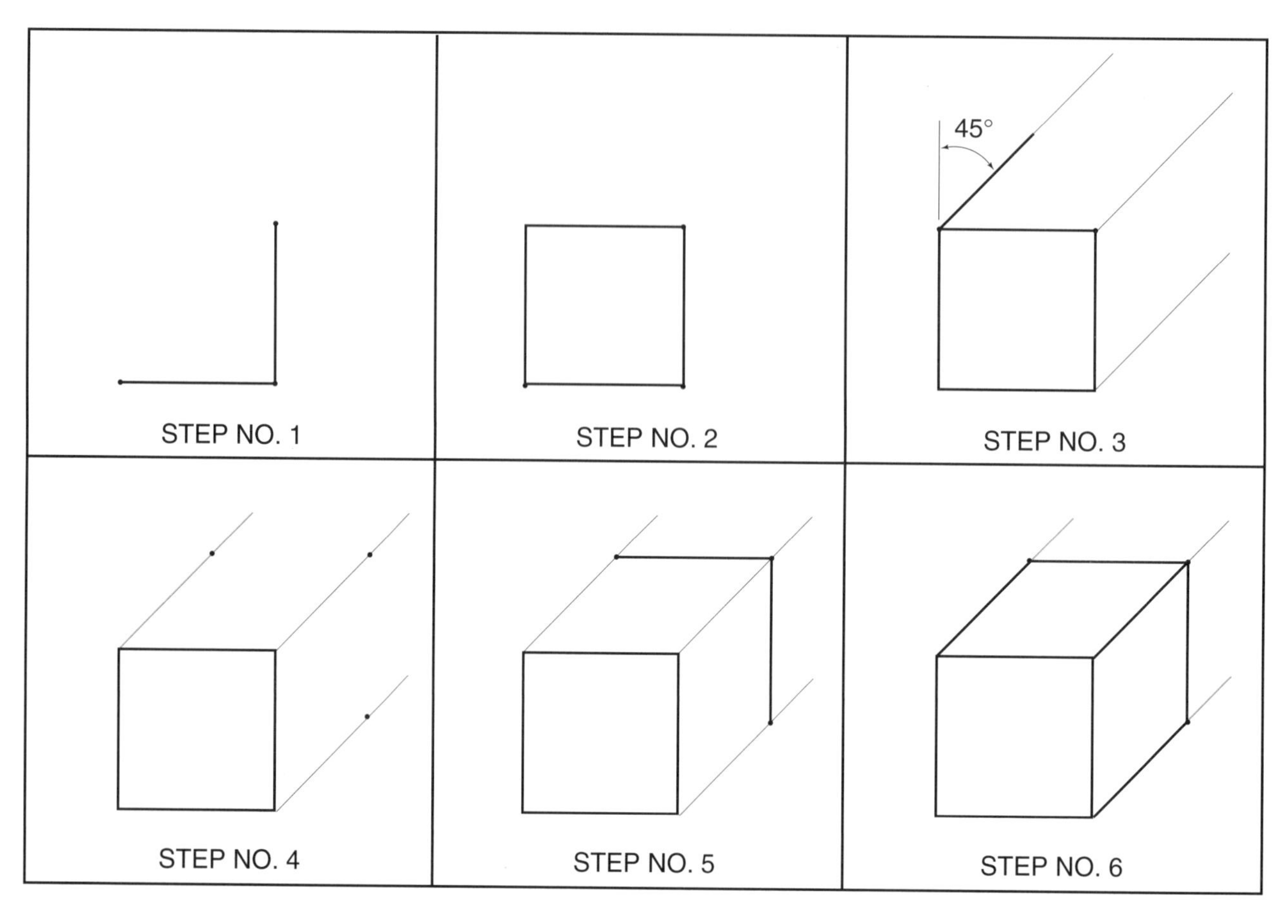

FIGURE 17.1 ■ Developing an oblique drawing to the right

When the front view or side view has been established, the drafter projects construction lines to the upper right or upper left from each corner or change in direction of planes of the surfaces of the object. These construction lines are projected back from the front or side view of the object at 45 degrees from the horizontal of the drawing (see Step 3). When all the projected construction lines have been drawn, the drafter locates points along these lines to establish distances from the front or side view, thus revealing the back outline of the object (see Step 4). The points should be placed in such a manner that the drawing will be in proportion. In other words, the distance along the projected construction lines should be representative of the distance between the front of the object and the back of the object.

When these points have been established and the distances verified, the drafter connects the points with vertical and horizontal lines to form the outline of the back of the object (see Step 5). Step 6 shows the finished drawing of the object. Angular lines can be drawn where needed to show inclined planes in the object.

NOTE: In this discussion, consider the back (or rear) view as the side that is opposite the front or end view (the view that was first drawn).

Intermediate points can now be placed on the projected construction lines where needed so that object lines can be drawn between these points to indicate changes in direction of planes on the object.

The only views of the object that are not shown in these two drawings are the back view and the bottom view. If these views are required, the drafter can rotate the object and use the oblique method of drafting to show these sides or views.

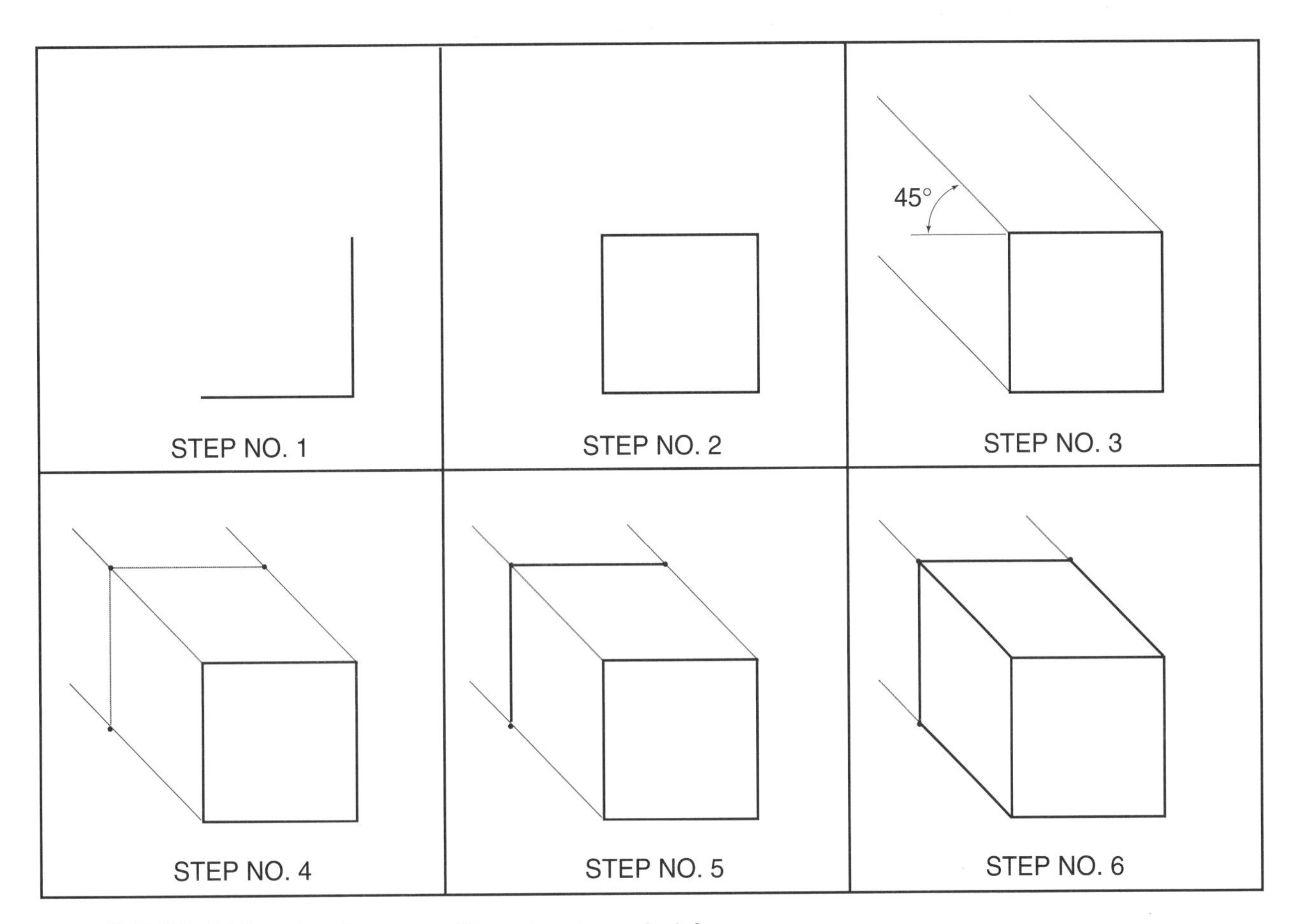

FIGURE 17.2 ■ Developing an oblique drawing to the left

Figure 17.2 shows the drawing process for extending the construction lines to the left. The steps are the same as those described in Figure 17.1 except that the construction lines are extended to the left instead of the right.

17.3 USING THE ARCHITECT'S SCALE FOR OBLIQUE DRAWINGS

The architect's scale is a valuable tool in freehand sketching because it helps establish distances and helps keep the drawing in proportion. If a scale drawing is needed, the architect's scale is used for the entire drawing.

The oblique drawing is not intended to be a scale drawing. Even though many of the dimensions are drawn to scale, the oblique drawing is not a scale drawing because some of the surfaces appear distorted if they are at an angle to the viewing plane. The purpose of an oblique drawing is to give a pictorial view or three-dimensional concept of an object.

17.4 LINES DRAWN ON OBLIQUE DRAWINGS

Lines drawn on freehand sketches and on instrument drawings should be drawn as construction lines in the beginning. After the object has been constructed with these construction lines, object lines can be drawn over the construction lines to show the object clearly. When this step has been completed, the construction lines should be removed to make the drawing neater.

When the oblique drawing is completed, the object will have a three-dimensional appearance. In the drawing, the object is shown as it actually looks from three sides: the front view (if this is the view that the drafter started with), the side view, and the plan view. (See Step 6 of Figure 17.1.) The oblique drawing is helpful because it reveals the three basic viewing planes, and it is drawn on a single plain piece of paper.

17.5 VARIATIONS IN OBLIQUE DRAWINGS

In some cases, drawings may appear to be distorted. This can be corrected by adjusting the lines as necessary to obtain realism (Figure 17.3). This adjustment can be made by shortening or lengthening lines.

Oblique drawings can be drawn at angles other than 45 degrees. In some cases, it is more desirable to project construction lines at 30 degrees or 60 degrees. Figure 17.4 shows an oblique drawing of a heating coil with construction lines drawn at 45 degrees. Figure 17.5 shows the same heating coil with construction lines drawn at 30 degrees. Figure 17.6 shows the same coil with the construction lines drawn at 60 degrees. The angle chosen for the construction lines makes a slight difference in the appearance of the three drawings.

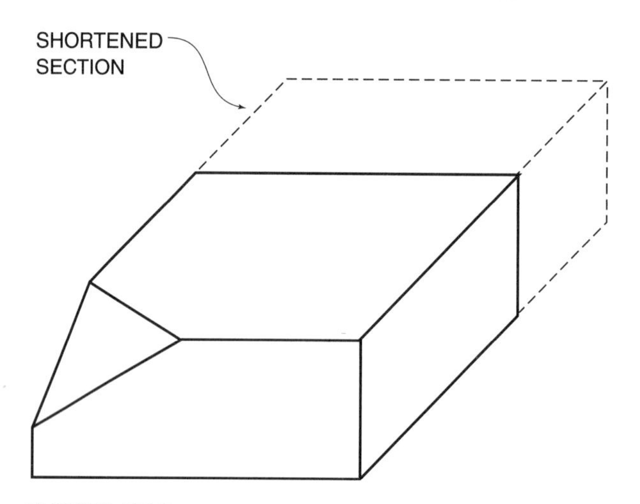

FIGURE 17.3 ■ Adjusting a drawing to obtain realism

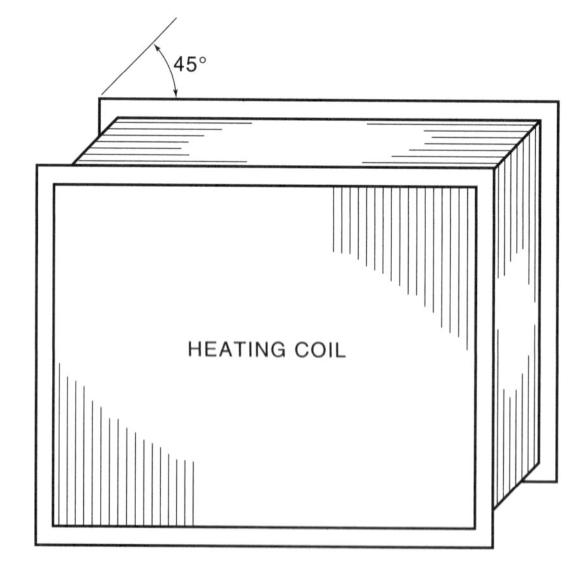

FIGURE 17.4 ■ Drawing of a heating coil with projection at 45 degrees

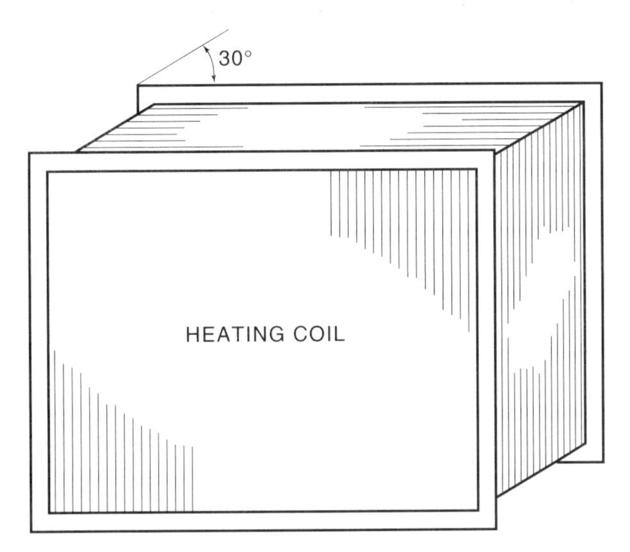

FIGURE 17.5 ■ Drawing of a heating coil with projection at 30 degrees

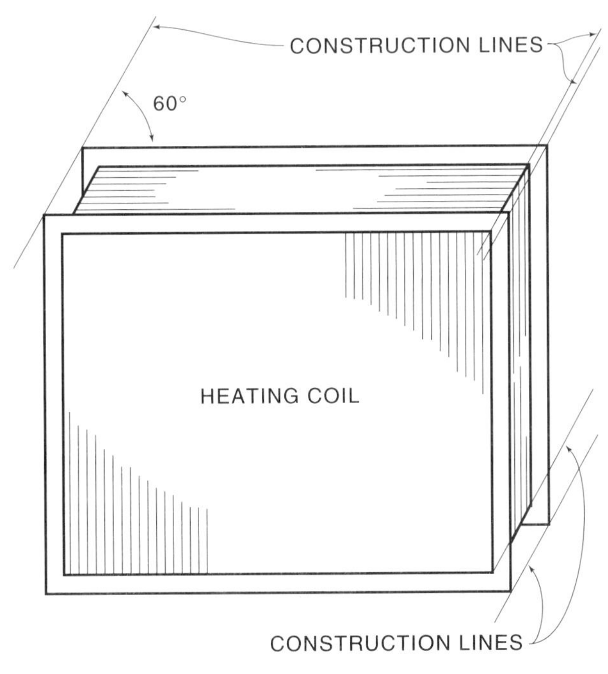

FIGURE 17.6 ■ Oblique drawing of a heating coil (with construction lines) with projection at 60 degrees

SUMMARY

- An oblique drawing is a type of pictorial drawing that gives a three-dimensional effect.
- Isometric drawings and perspective drawings also give a three-dimensional effect.
- Oblique drawings are not scale drawings.
- Construction lines project back from the corners of the front view in order to establish the back outline of the object.
- An oblique drawing reveals the three basic viewing planes of the object (front view, side view, and top view).
- By rotating the object, the drafter can also show the other side view, the back view, and the bottom view.

REVIEW QUESTIONS

1. Describe an oblique drawing.
2. List three types of drawings that give a three-dimensional effect.
3. What are the three basic views of an object?
4. After the front or side view has been drawn, lines are projected back from the corners of the object at how many degrees?
5. How is the back outline established?
6. Is an oblique drawing a scale drawing?
7. How is it possible to use an oblique drawing to show the bottom view of an object?

ADDITIONAL STUDENT EXERCISES: CONSTRUCTING OBLIQUE DRAWINGS

1. Draw an oblique drawing from the orthographic drawings in Figure 17.7.
2. Draw an oblique drawing from the orthographic drawings in Figure 17.8.
3. Using the elevation views shown in Figure 17.9, project the necessary lines to show the duct fitting as oblique drawings.

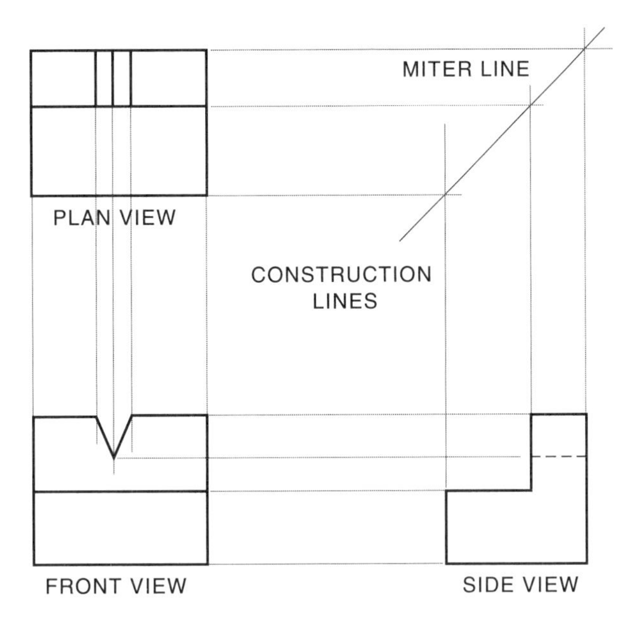

FIGURE 17.7 ■ Exercise 1

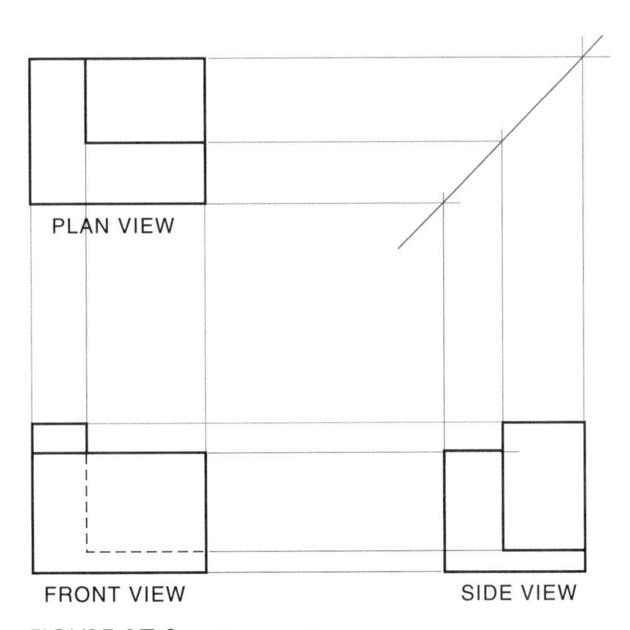

FIGURE 17.8 ■ Exercise 2

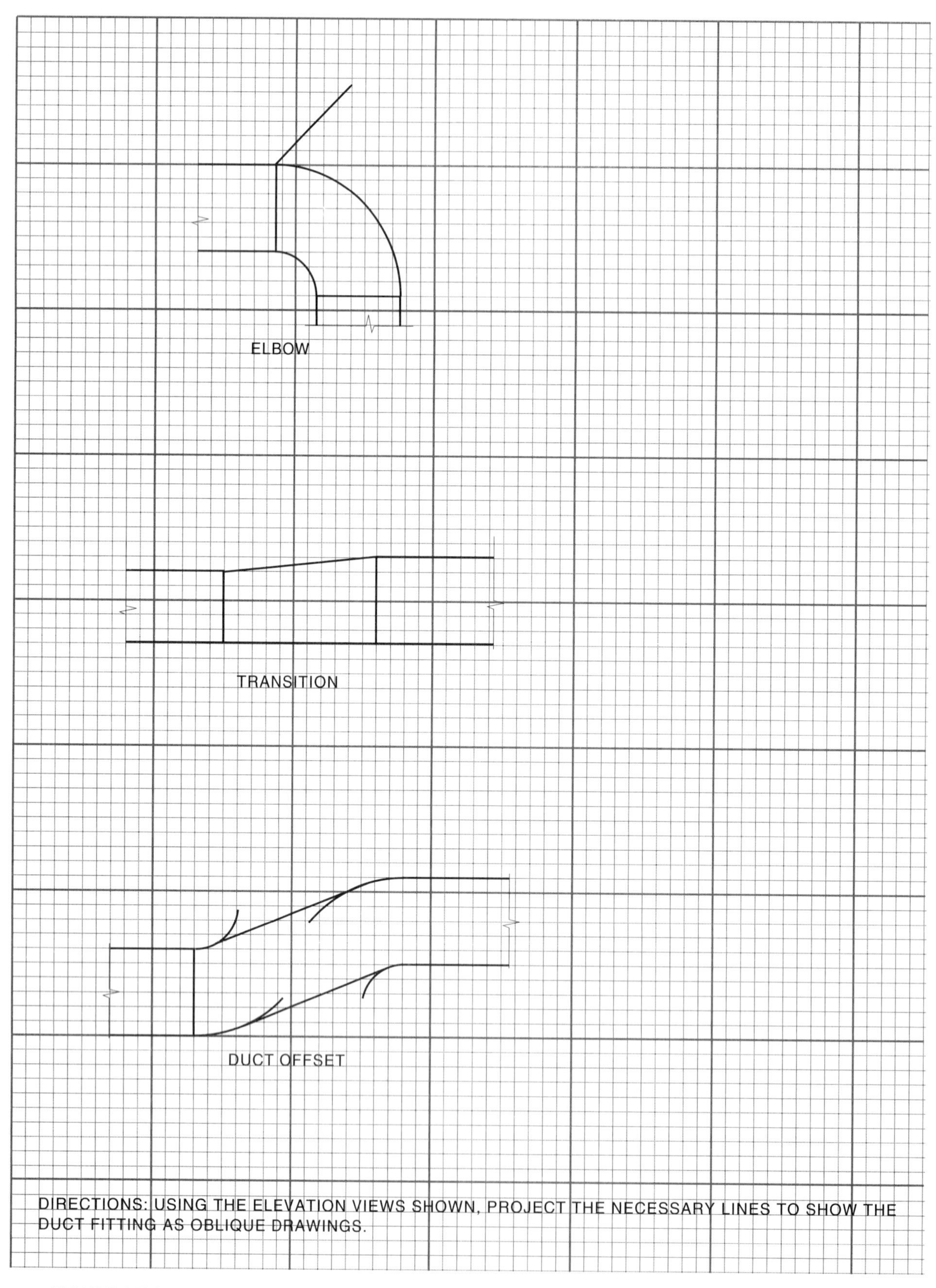

FIGURE 17.9 ■ Exercise 3

CHAPTER 18

Isometric Drawings

OBJECTIVES

After studying this chapter, you should be able to:

- identify the steps used to create the three basic views of an object isometrically
- demonstrate the ability to construct isometric drawings of simple objects
- discuss reasons why isometric drawings are the most commonly used pictorial drawings in construction projects
- compare isometric drawings drawn at 30 degrees from the horizontal with those drawn at 60 degrees from the horizontal
- explain the advantages of using isometric drawings for piping diagrams

18.1 INTRODUCTION

Chapter 17 discusses the use of pictorial drawings to relay information from the drafter/designer to the craftsman or construction worker. The isometric drawing method is also used to give the reader of the drawing a feeling of seeing objects in three dimensions.

As stated in Chapter 17, the *easiest* of the pictorial drafting methods is the oblique drawing method; however, the isometric drawing method is more commonly used because it is *easier to read*. If several objects are shown in the same drawing, there appears to be more space between the objects in an isometric drawing.

A third type of pictorial representation is the perspective drawing with diminishing points. Since this drawing method is very seldom used in the HVAC field, perspective drawings are not included in this book.

In this chapter, we will discuss the development of an isometric in steps. As stated in Chapter 16, the three basic views can be the plan (top), front, and left (or right) side views. This is also true with isometric drawings. The sides are determined by the drafter. The drafter can rotate the object to reveal the most information on the front and right-hand side views.

Figures 18.1 through 18.9 are step-by-step drawings that illustrate how a drawing is developed. Figure 18.10 shows the final (finished) drawing of an object.

18.2 HOW ISOMETRIC DRAWINGS ARE DRAWN

Isometric drawings show three basic views of an object. These three views are the plan view, the front view, and the right side (end) view. The three basic views in the isometric drawing are like the three basic views in orthographic projection (see Chapter 16).

The first step in creating an isometric drawing is to establish the front right-hand corner of the object. The drafter must remember to provide space to develop the front view to the left of this right-hand corner point, the right side (end of the object) from this corner point, and the plan view from this corner point.

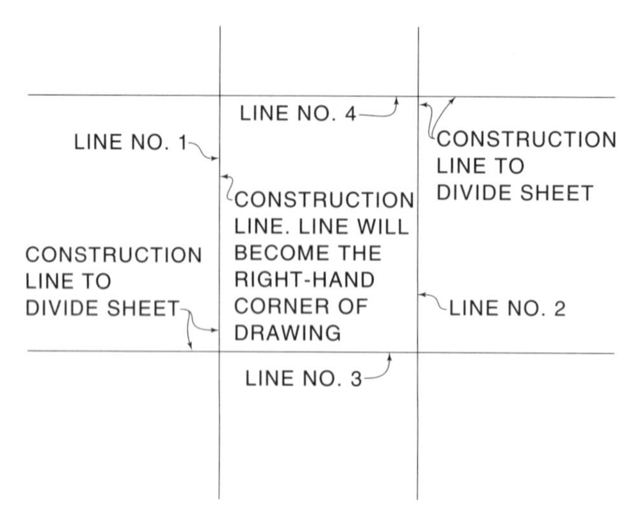

FIGURE 18.1 ■ Dividing the space into three parts

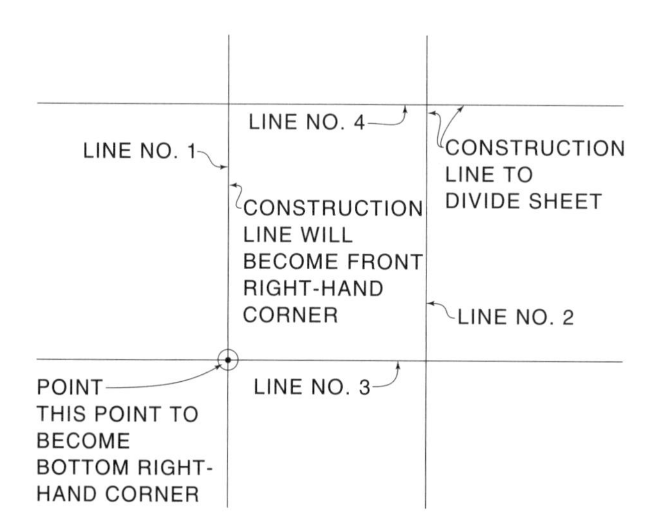

FIGURE 18.2 ■ Locating the front, bottom corner of drawing

To establish the right-hand corner, the drafter should visualize the drafting sheet divided into nine equal parts; i.e., three parts vertically and three parts horizontally (Figure 18.1). In the example, lines are numbered 1 through 4. Next, a construction line should be drawn along vertical Line 1. Line 1 will be the right-hand corner of the isometric drawing. The intersection of Line 1 and Line 3 is the front, bottom corner of the drawing. A point or dot should be placed at this intersection (Figure 18.2).

The drafter should draw a construction line from the point representing the bottom right-hand corner on the right-hand corner line. This construction line should intersect the point on the right-hand corner and project in an upward direction to the right at a 30-degree angle from the horizontal (Figure 18.3). This line will be the base of the front view.

The next step for the drafter is to draw a construction line to the left at a 60-degree angle from the horizontal. This construction line must intersect the right-hand corner line at the right-hand, bottom corner point and project up and to the left (at 60 degrees) from this right-hand, bottom corner point (see Figure 18.3). This 60-degree line will become the base for the front view (Figure 18.4).

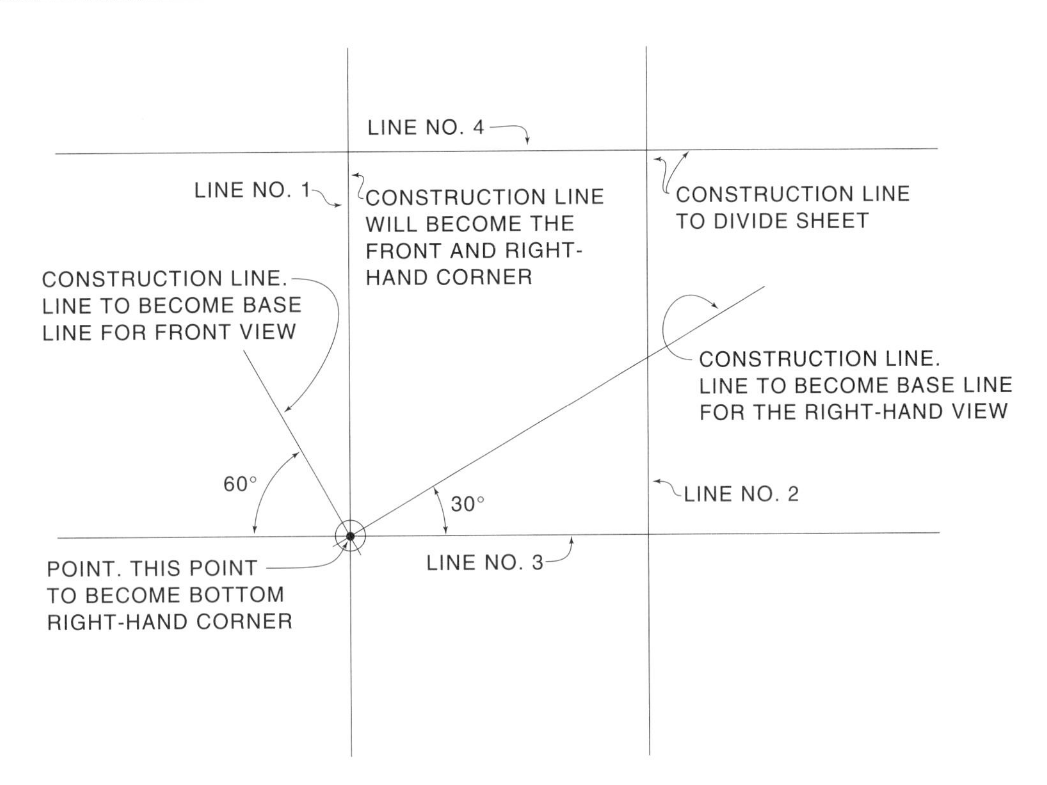

FIGURE 18.3 ■ Establishing baselines for the front view and the right side view

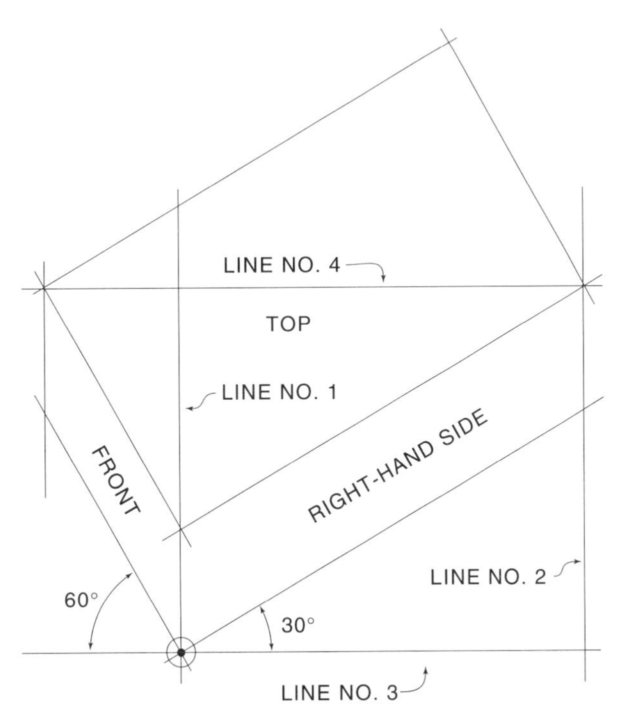

FIGURE 18.4 ■ Establishing the front corner and the cube of an object

With the front, right-hand corner established, and the baselines for the front view and the right side (end) view drawn, the drafter is ready to establish the height of the object, the width of the front view, and the length of the right side (end) view.

It should be noted at this point that the isometric drawing is not a drawn-to-scale drawing. It is, however, important to keep everything in proportion on the drawing. For this reason, the architect's scale is used to establish distances but not for making the drawing *to scale*.

The drafter must use common sense to keep the drawing in proportion. For example, if the object being drawn is two times as long as it is wide, the drawing should reflect this difference. If the height of the object is half the width of the object, the drawing should reflect this difference. The architect's scale is a very useful tool to help establish these proportions.

After the length of the front view, the length of the right side view, and the height of the object have been established, the four corners of the isometric view can be established and a rectangular solid or a cube can be established with construction lines that set the outside limits of the object. The drafter can now use the architect's scale to establish distances on the object where surfaces change directions and the object takes shape.

One important thing to remember is that *parallel lines are parallel to each other, and parallel surfaces are parallel to each other*. Figure 18.5 shows the object being developed. Known points are being identified and object lines are being drawn between these points. It is suggested that as a rule, one view of the object should be worked on at a time; however, in some cases, working on two views at the same time helps with the proportions of the object. Most drafters work to develop the plan view portion of the object first. This requires the

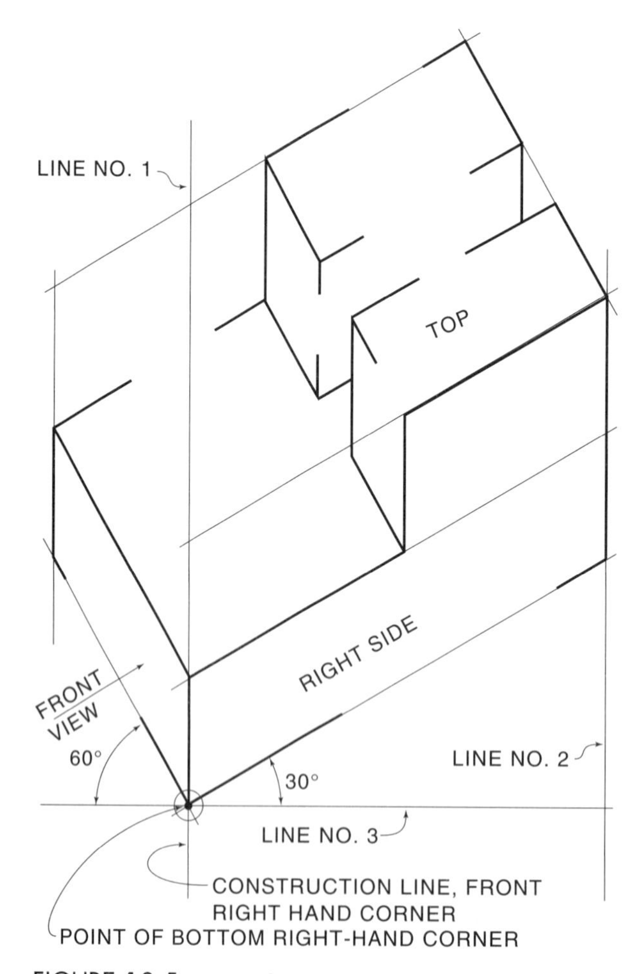

FIGURE 18.5 ■ Developing the object

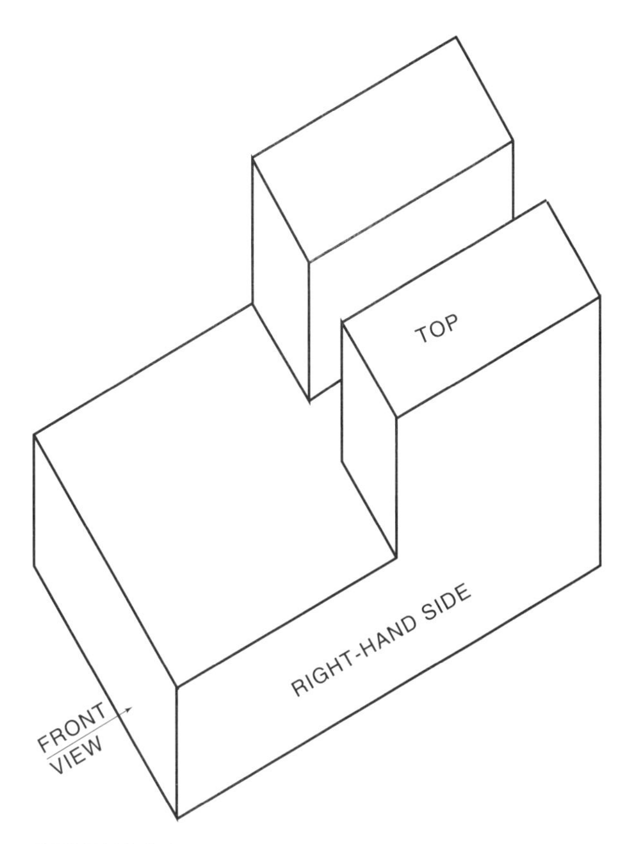

FIGURE 18.6 ■ Finished drawing

development of the top profile of the rear view. When the front corner and the top view have been developed, the other two views are developed from these known items.

The finished drawing of the example is shown in Figure 18.6. This drawing was developed with 30 degrees/60 degrees = base angles.

18.3 A VARIATION OF ISOMETRIC DRAWINGS

In Section 18.2, the baselines were drawn from the right-hand corner (dot or point) at 30 degrees from the horizontal to the right and 60 degrees from the horizontal to the left (Figure 18.3). These two construction lines become the baselines for the front and right side views of the objects.

Some isometric drawings can be developed with more detail if both of these construction lines are drawn at 30 degrees from the horizontal. In Figure 18.7, the construction lines are drawn from the right-hand corner point at 30 degrees from the horizontal. The procedure for dividing the drawing paper into six parts and the location of the right-hand corner point is the same as described for Figures 18.1 through 18.3 except for the angle of the construction baselines.

The construction lines establishing the outside limits of the object are shown in Figure 18.8. The rear outline of the object is developed, and the plan view, front view, and the right side view are developed in the same manner as described at the beginning of this chapter (Figure 18.9). The finished drawing of the object is drawn with the baseline for the front view and the right side view at 30 degrees from the horizontal (Figure 18.10).

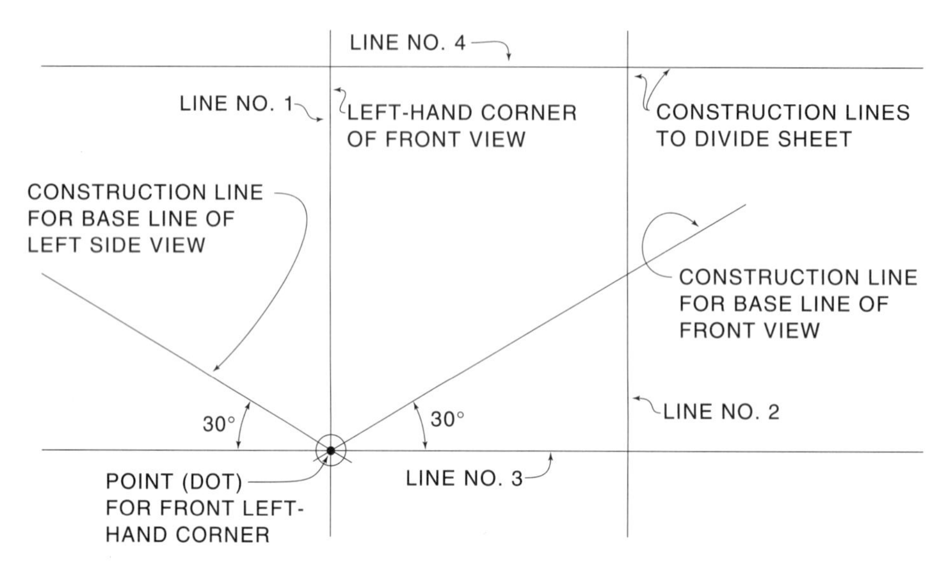

FIGURE 18.7 ■ Establishing baselines for front and right-hand side views

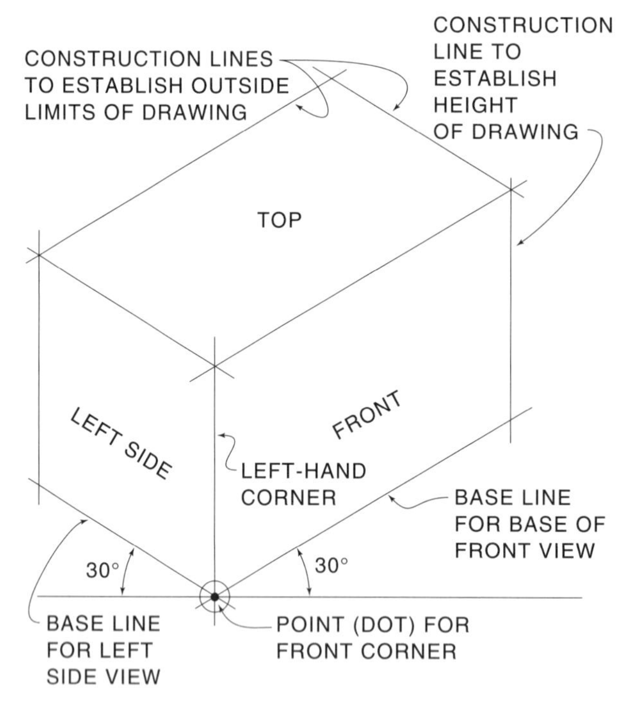

FIGURE 18.8 ■ Establishing the outside limits for an object

When the two isometric drawings are compared (Figures 18.6 and 18.10), the right side view is about the same for each plan. The front view is larger in Figure 18.10 than in Figure 18.6 because the 30-degree baseline makes the front view longer than in Figure 18.6, where the baseline is shown at 60 degrees from the horizontal. The plan view (top) is larger in Figure 18.10 because of the 30 degrees from the horizontal for the baseline in the front view.

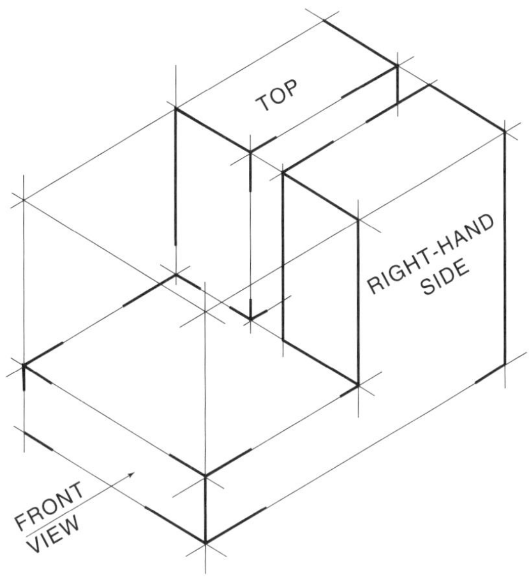

FIGURE 18.9 ■ Developing the views of the object

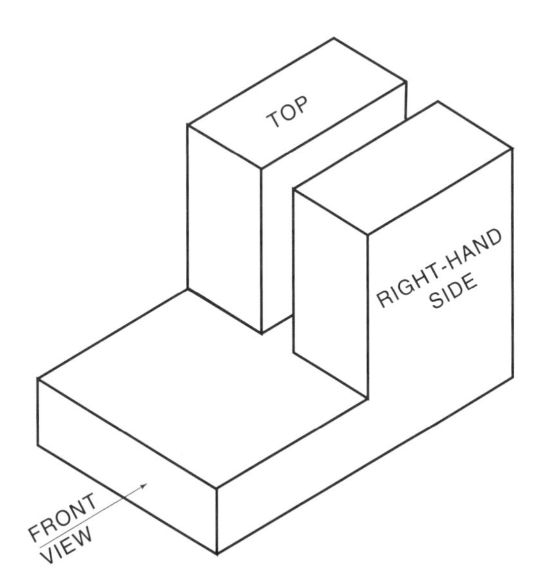

FIGURE 18.10 ■ Finished drawing

The drafter must decide which angles to use when drawing an isometric plan. The best drawing is the drawing that will relay the most information to the worker or technician in the clearest fashion.

It may be necessary to rough out the base drawing up to the point of establishing the outside limits of the object (Figures 18.4 and 18.8) before making the decision as to which base angles should be used from the right-hand corner point.

18.4 ISOMETRIC PIPING DIAGRAM

An isometric drawing of a piping diagram is shown in Figure 18.11. This drawing shows a pump with the piping connections and the fittings, valves, and equipment that should be installed in the piping system. Figure 18.11 was drawn using 30 degrees/30 degrees baselines from the horizontal.

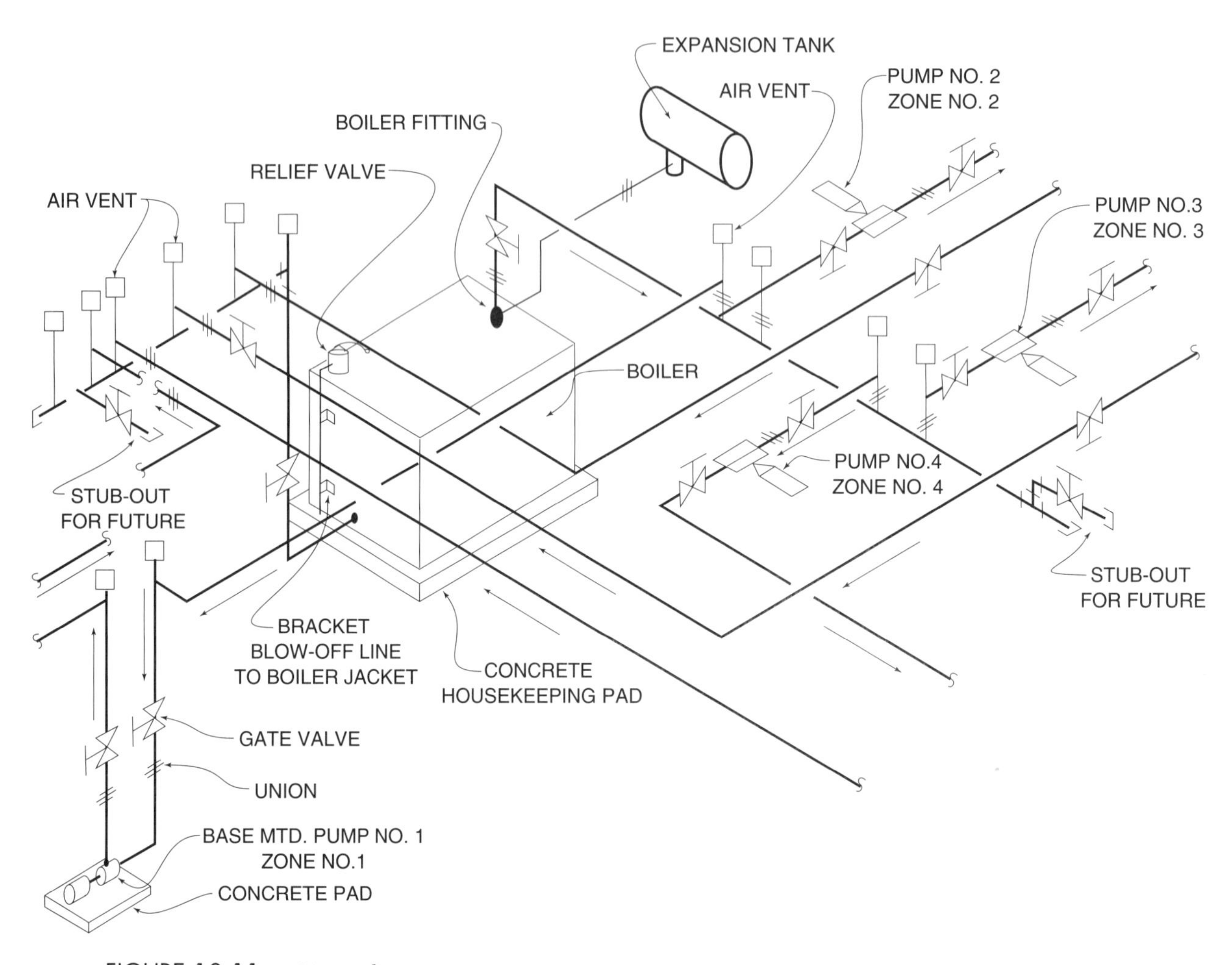

FIGURE 18.11 ■ Piping diagram

Isometric drawings are used to great advantage in detailing piping diagrams. When isometric drawings are drawn in proportion, they show the relationships of the different components in the piping system, and they provide space to add notes, symbols, dimensions, and instructions as to how the system is to be installed.

18.5 WHERE ISOMETRIC DRAWINGS ARE USED

In the HVAC field, designers and drafters use isometric drawings for many different purposes. Some examples of the various ways that isometric drawings are used are described and illustrated on page 136. As mentioned earlier, Figure 18.11 is a piping diagram showing pipe fittings, special equipment, and where the equipment is to be installed. Figure 18.12 is a detail of a base for equipment. Figure 18.13 shows a fan mounted on an equipment base with ductwork connected to the fan. Figure 18.14 shows an offset in ductwork to avoid a pipe. Figure 18.15 shows piping connections to a heating coil located in an air-handling unit. Related controls and fittings are shown in the piping system. In many cases, the isometric drawing is the only place in a set of drawings where information is given to the installer or the craftsperson detailing how the different components are to be installed.

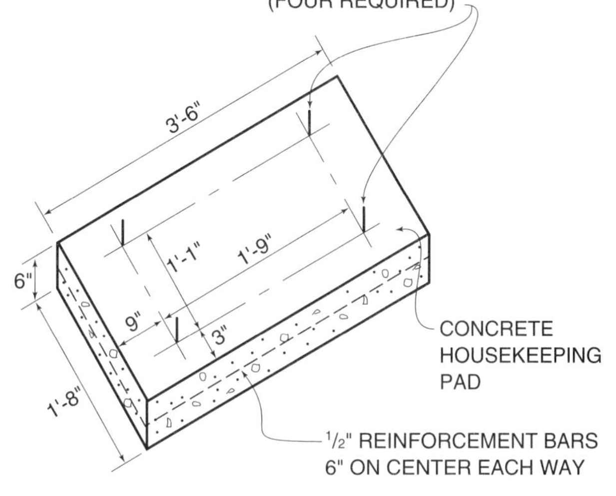

FIGURE 18.12 ■ Detail of an equipment base

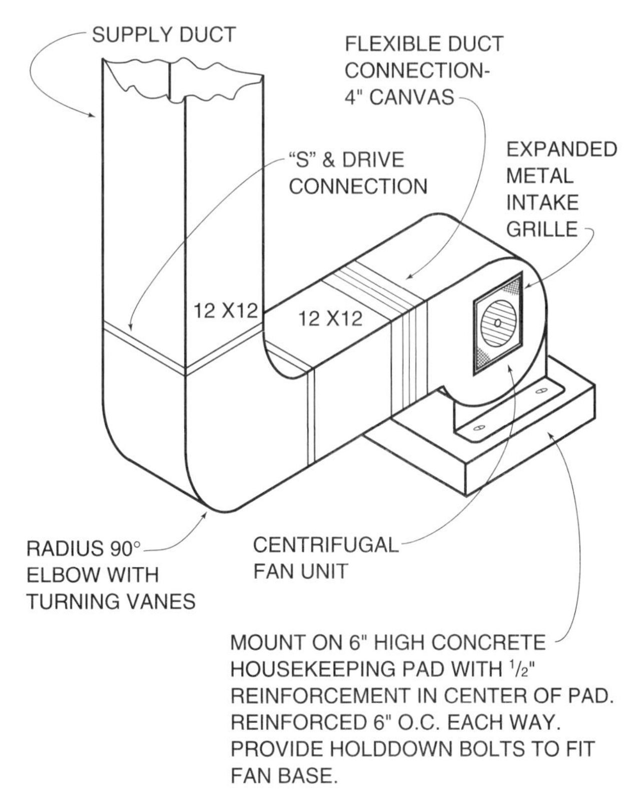

FIGURE 18.13 ■ Detail of ductwork and fan on equipment base

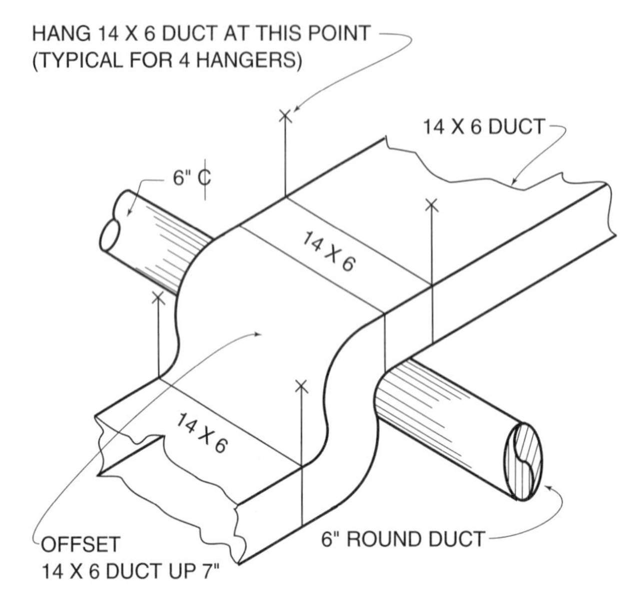

FIGURE 18.14 ■ Detail of ductwork offset over pipe

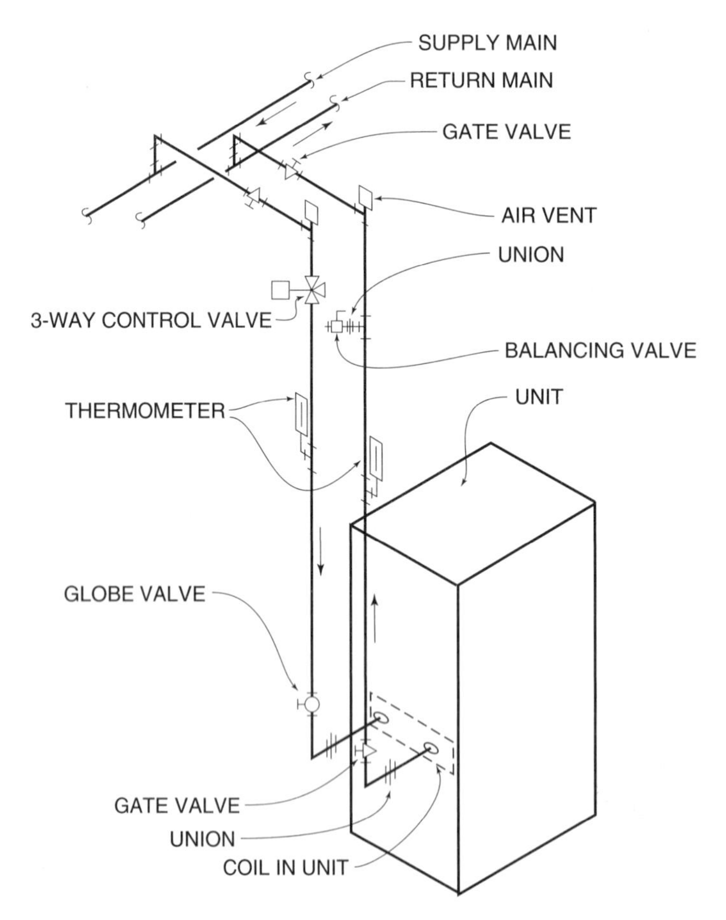

FIGURE 18.15 ■ Piping connections to heating coil in an air-handling unit

SUMMARY

- Although the oblique method of drawing is easier to *draw*, the isometric method of drawing is easier to *read*.
- Isometric drawings show the three basic views of an object (plan view, front view, and right-hand view).
- Although not drawn to scale, isometric drawings should be drawn in proportion.
- Parallel lines must be drawn parallel to each other, and parallel surfaces must be drawn parallel to each other.
- Often, an isometric drawing is vital to show the craftsperson how different components are to be installed.
- The drafter can choose whether construction lines are more effective when drawn at 30 degrees or 60 degrees from the horizontal.

REVIEW QUESTIONS

1. What are three methods of drawing pictorial representations of objects?
2. Which of the three methods listed above is the easiest to draw?
3. Which of the three methods listed above is the easiest to read?
4. What are the three basic views of an object that are shown in isometric drawings?
5. Is an isometric drawing done to scale?
6. What is the reason for using an architect's scale in creating an isometric drawing?
7. Parallel lines should be drawn ______________________________, and parallel surfaces should be drawn ______________________________.
8. Give at least two reasons why isometric drawings can be advantageously used in detailing piping diagrams.

ADDITIONAL STUDENT EXERCISES: FREEHAND SKETCHING ISOMETRIC VIEWS

1. Using graph paper, freehand-sketch isometric views of the objects shown in Figure 18.16 by using orthographic projection.
2. Draw isometric views of the orthographic objects shown in Figure 18.17 by using orthographic projection.
3. Freehand-sketch an isometric drawing of the air-cooled condensing unit shown in Figure 18.18 by using orthographic projection. The unit is mounted on a concrete pad and has a round, vertical air discharge *with* a round fan guard.
4. Freehand-sketch an isometric drawing of the gas-fired furnace shown in Figure 18.19 by using orthographic projection.
5. Refer to Figure 18.11 to answer the following questions:
   A. Are unions provided for quick pump disconnect?
   B. How many zones are being served by the boiler?
   C. Is a base-mounted pump shown?
   D. Does the boiler have a housekeeping pad?
   E. Have provisions been made for future zones?

F. How many safety valves are shown?
G. Is the piping shown located underfloor?
H. Have provisions been made to remove air from the system?
I. How many gate valves are shown on the drawing?
J. Is an expansion tank shown?

6. Refer to Figure 18.12 to answer the following questions:
   A. What is the overall dimension of the equipment base?
   B. What is the size of the reinforcement bars in the pad?
   C. Are there two rows of reinforcement bars?
   D. How thick should the pad be?
   E. What is the size of the anchor bolts?
   F. How far are the bolts located from the edge of the pad? (Two answers exist.)

7. Refer to Figure 18.13 to answer the following questions:
   A. What kind of connection is used in the ductwork?
   B. Does the fan have vibration isolation on the duct connection?
   C. Does the fan have a housekeeping pad?
   D. Is the discharge ductwork insulated?
   E. What is the thickness of the housekeeping pad?
   F. Does the concrete pad have reinforcement?
   G. What is the size of the reinforcement, and how is it spaced?
   H. What is the intake opening to the fan covered with?
   I. Does the fan have hold-down bolts?

8. Refer to Figure 18.14 to answer the following questions:
   A. Why is the offset needed?
   B. Does the ductwork have hangers?
   C. Are the types of duct joints shown?

9. Refer to Figure 18.15 to answer the following questions:
   A. Is the heating coil indicated?
   B. How many gate valves are shown?
   C. How many thermometers are shown?
   D. Is there a globe valve in the system?
   E. Is there a balancing valve shown?
   F. How many unions are shown?
   G. Is the direction of flow indicated?

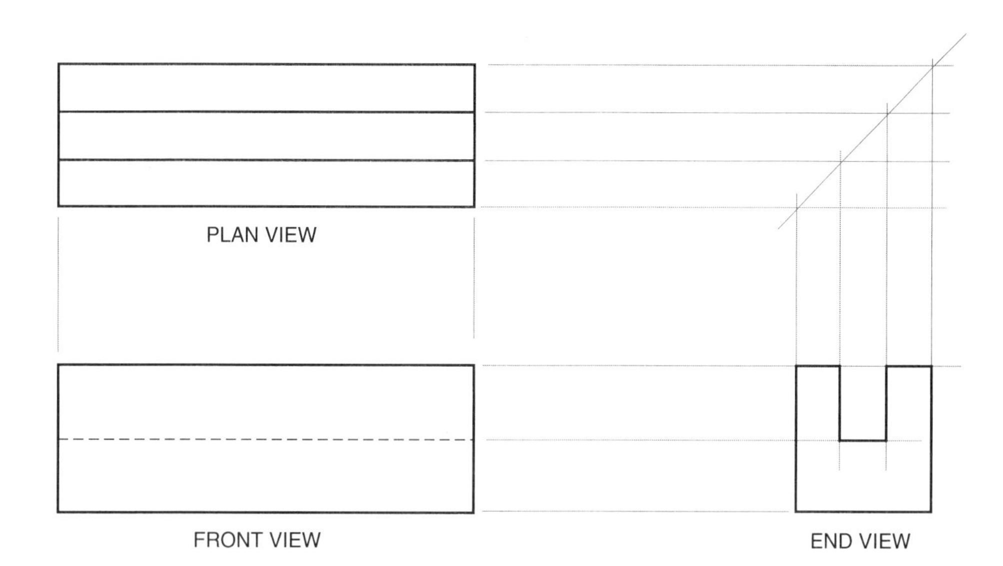

DIRECTIONS: DRAW ISOMETRIC VIEW OF ORTHOGRAPHIC FIGURE ABOVE.

FIGURE 18.16 ■ Exercise 1

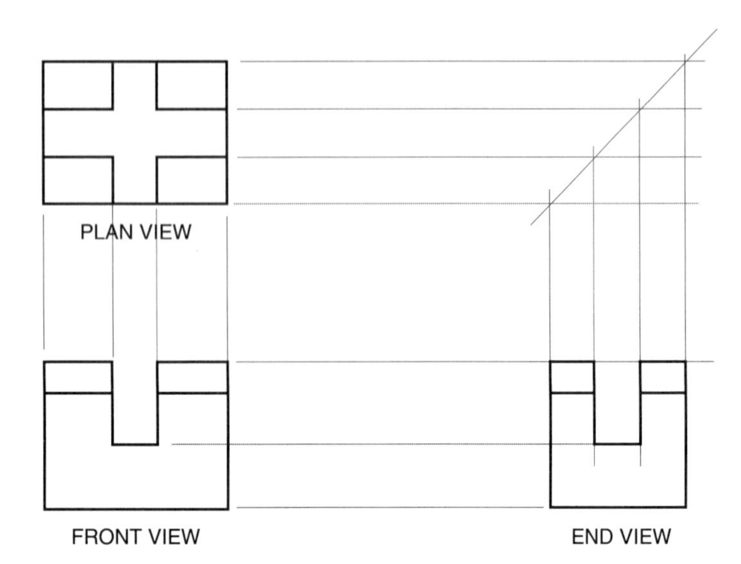

DIRECTIONS: DRAW ISOMETRIC VIEW OF ORTHOGRAPHIC OBJECT ABOVE.

FIGURE 18.17 ■ Exercise 2

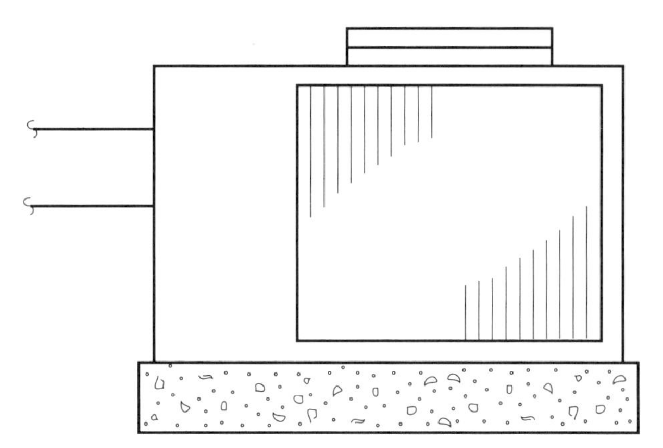

DIRECTIONS: FREEHAND DRAW AN ISOMETRIC DRAWING OF THE AIR-COOLED CONDENSING UNIT SHOWN ABOVE. THE UNIT IS MOUNTED ON A CONCRETE PAD AND HAS A ROUND, VERTICAL AIR DISCHARGE WITH A ROUND FAN GUARD.

FIGURE 18.18 ■ Exercise 3

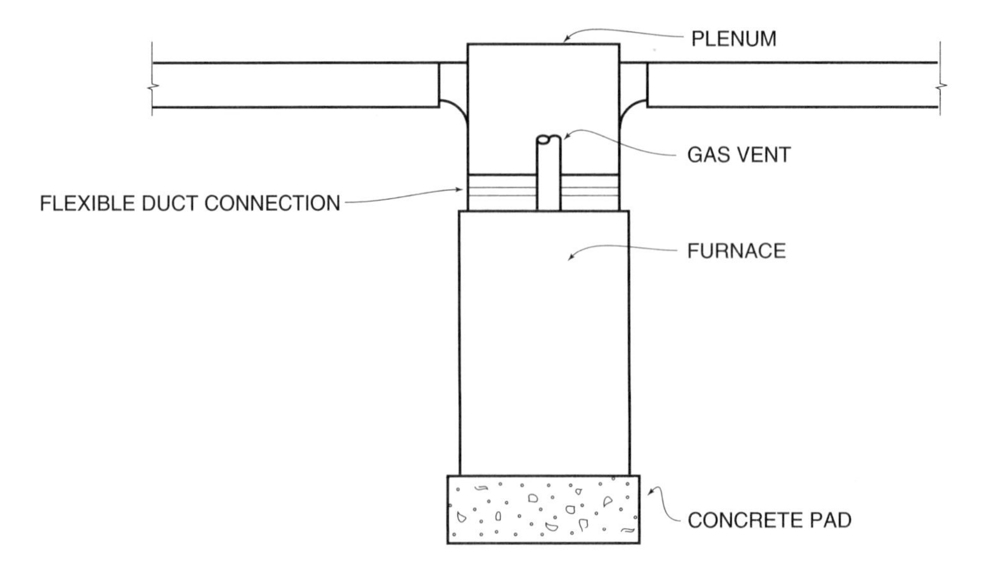

DIRECTIONS: FREEHAND DRAW AN ISOMETRIC DRAWING OF THE GAS-FIRED FURNACE USING ORTHOGRAPHIC PROJECTION.

FIGURE 18.19 ■ Exercise 4

CHAPTER 19

Lettering and Notations

OBJECTIVES

After studying this chapter, you should be able to:

- discuss the use of uppercase and lowercase letters
- demonstrate the ability to draw vertical-style letters and numerals
- demonstrate the ability to draw sloped-style letters and numerals
- describe and illustrate the proper spacing between letters, words, and sentences
- discuss factors that are considered in locating notes on plans
- define a leader and demonstrate its use on drawings

19.1 INTRODUCTION

It is sometimes said that the lettering on a drawing makes or breaks the whole drawing. While accuracy and clarity of the drawing itself are very important, the lettering and notations affect the overall appearance of the drawing and relay much necessary information. The beginning drafter soon learns that a beautiful drawing can become a mediocre or poor drawing if the lettering of notes and titles is of an inadequate quality.

This chapter will discuss several types of lettering commonly used on drawings and will give examples of vertical and sloped lettering. Achieving high quality lettering in manual drafting requires repeated and consistent practice. When computer drafting is being used, the lettering can be selected and formed on the plan in an almost perfect manner.

Proper notation is also covered in this chapter. The wording and location of notes will be discussed along with how the notes should relate to the drawing.

19.2 UPPERCASE LETTERING

The two basic types of letters used on working drawings are uppercase letters and lowercase letters. Uppercase letters are used most often. These letters can be formed in vertical (Figure 19.1) or sloped style (Figure 19.2). Engineers often use sloped letters, and architects seem to prefer vertical letters.

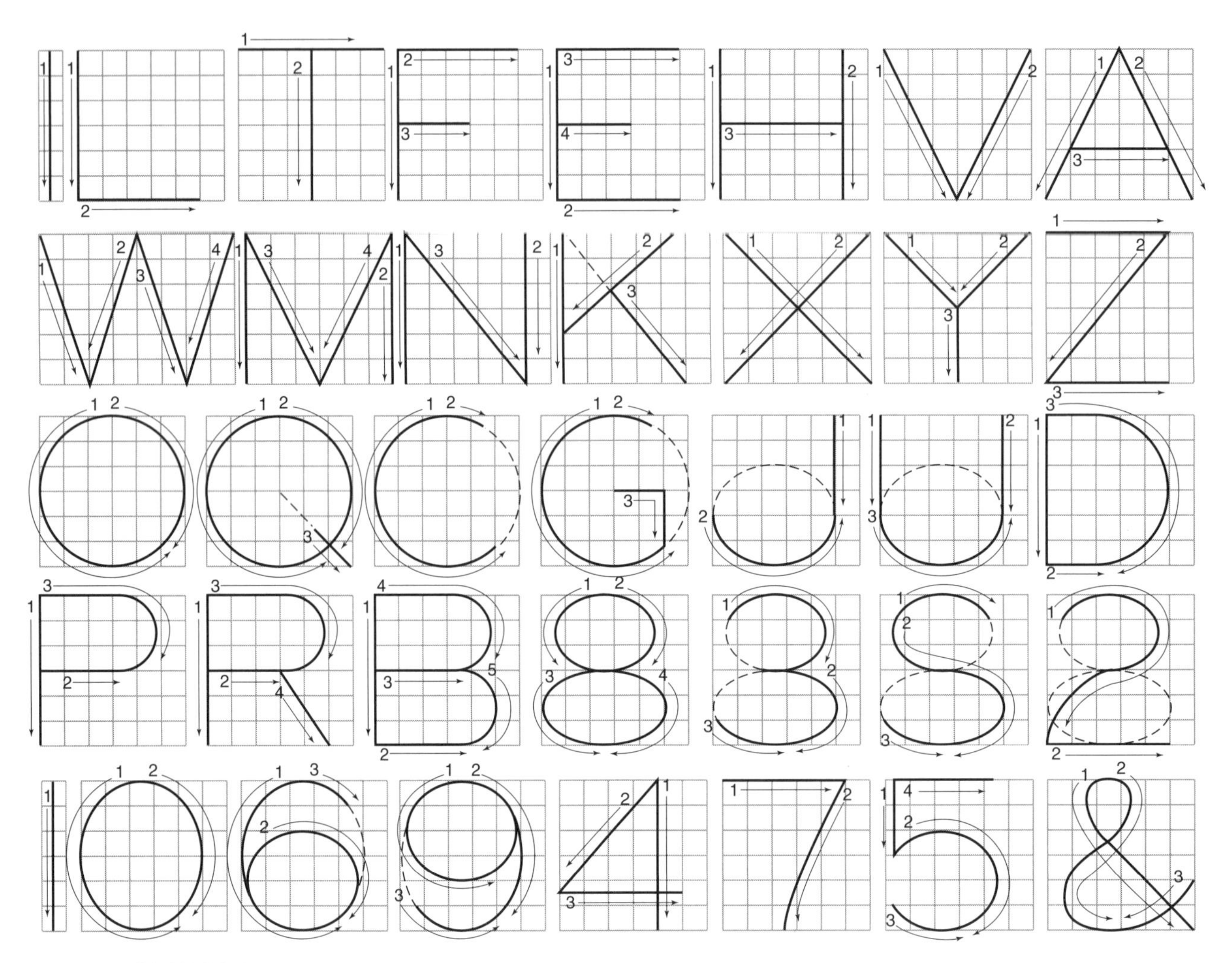

FIGURE 19.1 ■ Vertical uppercase letters

In Figures 19.1 and 19.2, the small arrows indicate how letters are to be formed. For example, to form the "F," the first stroke of the pen or pencil should be the vertical portion of the "F," followed by the second stroke to form the top of the letter, with the last stroke forming the middle line. The drafter must visualize circles while forming the curved portions of the letters B, C, D, G, J, O, R, S, and U. Practice is the secret to good lettering.

Sloped letters are formed in the same manner as vertical letters except that they are sloped some 20 degrees off vertical (Figure 19.2).

Many architects and engineers follow the basic system for forming letters; however, many add their own frill to the letters to give them a special appearance. This practice is acceptable if the owner approves.

19.3 LOWERCASE LETTERING

The suggested method for forming lowercase letters is shown in Figure 19.3. The small arrows on Figure 19.3 indicate the suggested strokes for forming the letters. More circular strokes are required for forming lowercase letters than for uppercase letters. These circular strokes require practice to improve the quality of the letters. Lowercase letters can be vertical or sloped style (see Figure 19.3).

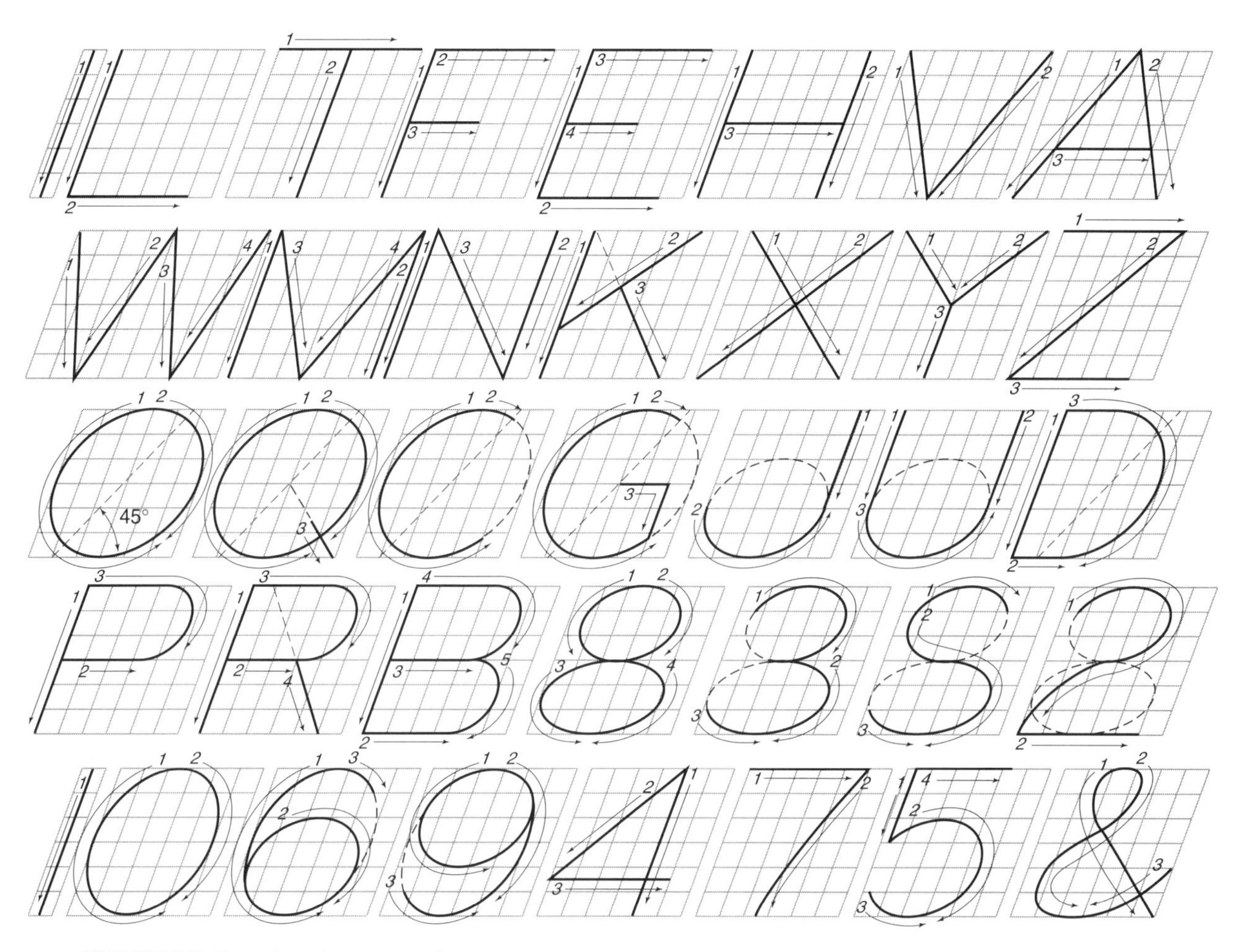

FIGURE 19.2 ■ Sloped uppercase letters

Some owners prefer a combination of uppercase and lowercase letters. The federal government usually requires combination-style lettering. The combination style has the first letter of a sentence in uppercase and the remaining letters in the sentence in lowercase. If sloped lettering is being used, lowercase letters are sloped the same as uppercase letters.

19.4 LETTERING NUMERALS

Numerals (symbols for numbers) should be formed as shown in the same style as letters. If letters are vertical, numerals should be vertical. If letters are sloped, numerals should be sloped. The suggested method for forming slant numerals is illustrated in Figure 19.4. Numerals may also require several partial circles for proper formation. Here again, practice is required to form numerals well.

Fractions may become a problem when they are combined within a line of lettering. Fractions can be formed two ways (see Figure 19.5 for examples). The first method is to establish the dividing line in the center of the lettering line and place the fractional letters above and below this line. Figure 19.5A demonstrates this method. The numerals should be the same size as the letters between the lettering guidelines. Another way is to write the numerator of the fraction between the guidelines, place a sloped line after the numerator, and place the denominator after the sloped line (Figure 19.5B). The two numerals should be drawn the same size as the lettering. Figures 19.5A and 19.5B show sloped numerals. Vertical numerals are formed in the same manner.

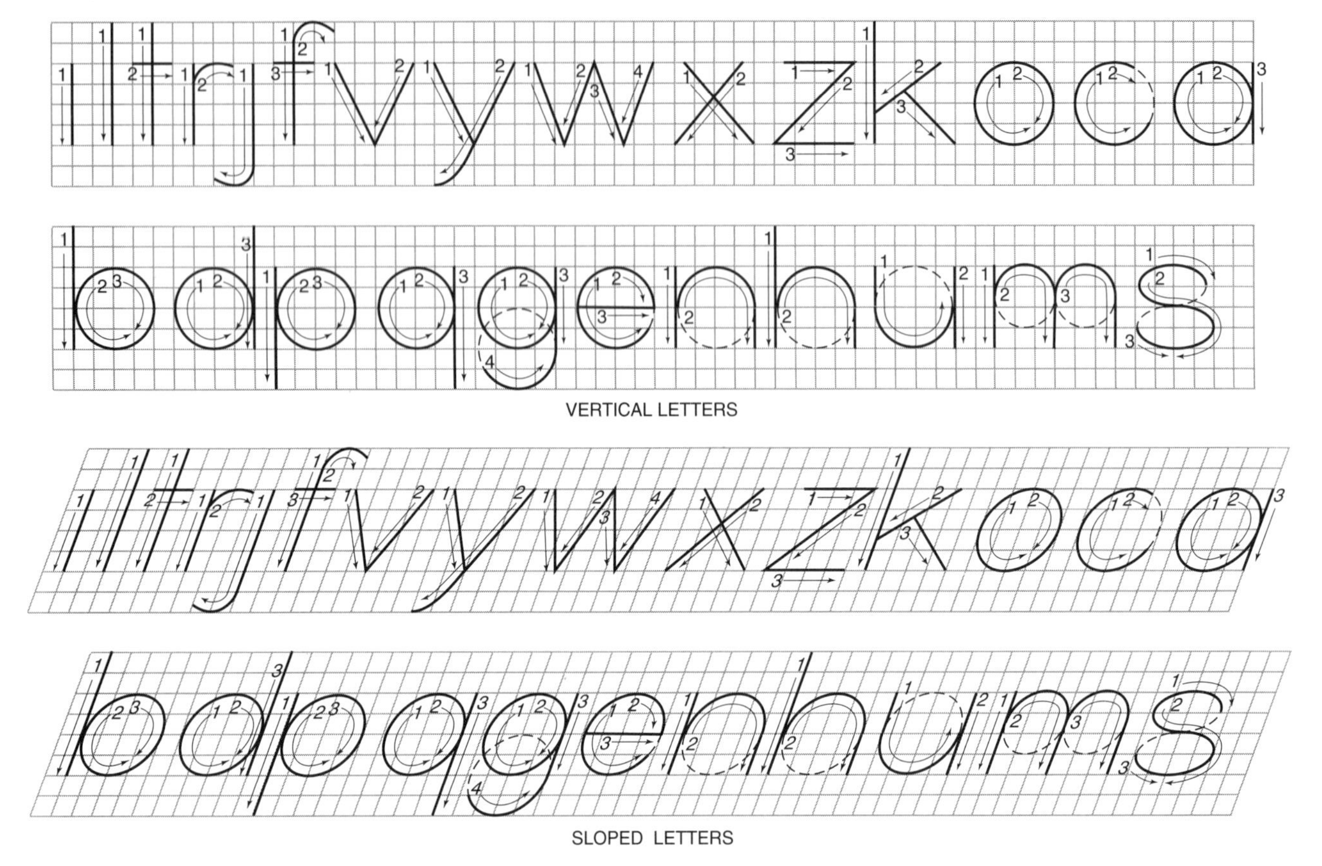

FIGURE 19.3 ■ Lowercase letters

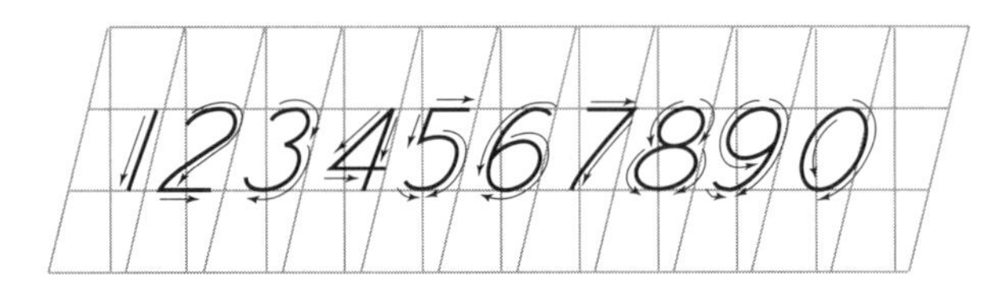

FIGURE 19.4 ■ Forming slant numerals

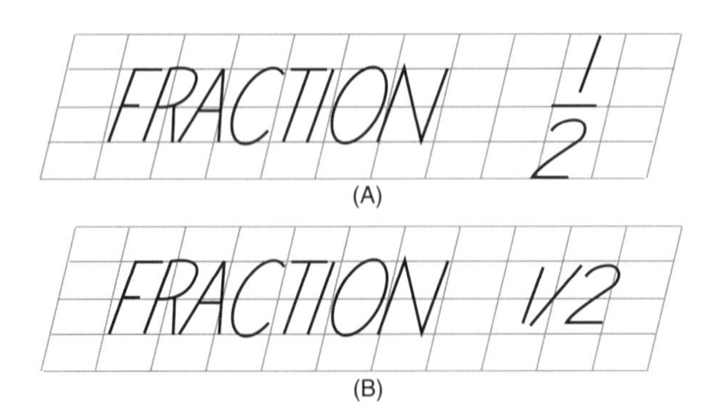

FIGURE 19.5 ■ Lettering fractions

19.5 WORDING NOTATIONS FOR BLUEPRINTS

The architect, engineer, and drafter are responsible for relaying design information to a worker or installer. Plans show where the work is to be performed and what materials are to be used, but additional instructions are often needed. This information is covered by notes. To save space and keep the plan from becoming cluttered, notes must be written in precise language.

It is also important to locate the notation in a logical place on the drawing. The note should be close to the area where it applies. Notes are written instructions that tell the types of materials, the sequence of installation, how different members fit together, equipment usage, how to operate the equipment, and other pertinent information.

To link the note to the object it refers to, a thin line called a leader is extended from the beginning or the end of the notation to the place where the note applies. The leader usually starts near the notation and extends to the reference object, terminating with an arrowhead. The leader is usually drawn straight out from the note and then turned at an angle to the reference object and arrowhead. Notes should always be located so that the leader lines do not cross. A typical note with leader is shown in Figure 19.6. Some architects and engineers draw a freehand leader. Freehand leaders are used in the set of plans at the end of this textbook.

In some cases, notes are too long to place in the location where they are needed. In that case, the drafter will insert brief instructions, such as "See Note No. 1." These long notes are then lettered on a blank space on the plan so that the floor plan is not cluttered.

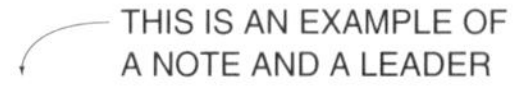

THIS IS AN EXAMPLE OF
A NOTE AND A LEADER

FIGURE 19.6 ■ Typical note with a leader

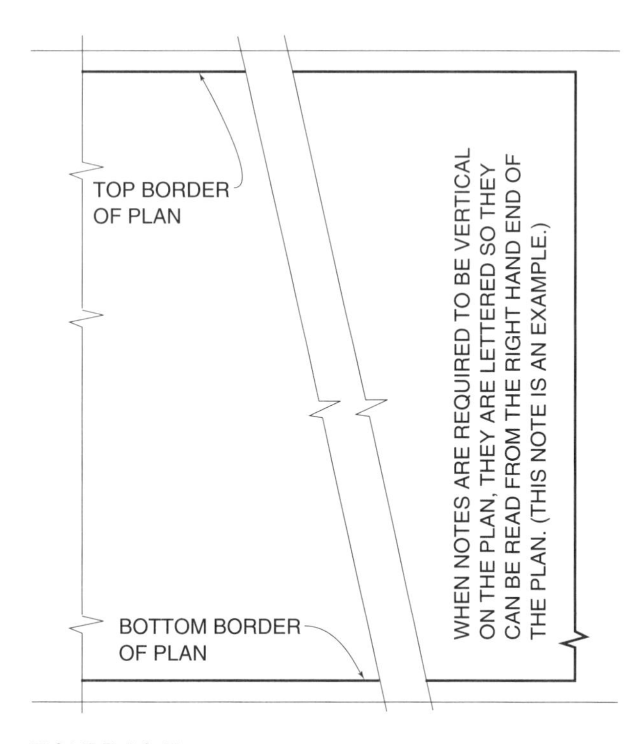

FIGURE 19.7 ■ A vertical note on a plan

Most notes are lettered horizontally, but sometimes, it is advantageous to letter the note vertically. When vertical notes are used, they should be readable from the right-hand side of the plan (Figure 19.7).

SUMMARY

- Two basic types of letterings are all uppercase letters and a combination of uppercase and lowercase letters.
- Vertical letters are formed from top to bottom, and sloped letters are tilted about 20 degrees off vertical.
- The same style should be used for letters and numerals if they appear in the same drawings.
- The drafter should use guidelines to regulate the height of letters and should practice spacing letters in a neat and attractive way.
- If possible, notes should be located close to the area to which they refer, with a leader connecting the note to the proper area.

REVIEW QUESTIONS

1. Following the guidelines in this chapter, print the letters of the alphabet in uppercase letters, vertical style.
2. Following the guidelines in this chapter, print the letters of the alphabet in lowercase letters, vertical style.
3. Following the guidelines in this chapter, print the letters of the alphabet in uppercase letters, sloped style.
4. Following the guidelines in this chapter, print the letters of the alphabet in lowercase letters, sloped style.
5. Print the numerals 1 through 20 in vertical and sloped style.
6. Define a leader.
7. Print a paragraph from this chapter to demonstrate acceptable spacing between letters, words, and sentences.
8. Print, in an acceptable style, three rules for the proper notation on plans.

ADDITIONAL STUDENT EXERCISES: PRACTICE FORMING LETTERS

1. Using the grids provided in Figure 19.8, practice forming vertical letters.
2. Practice forming sloped letters using the grids provided in Figure 19.9.

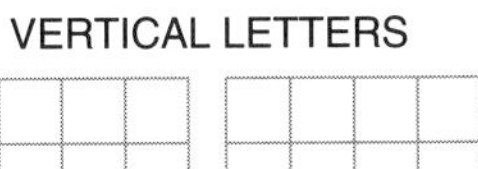

VERTICAL LETTERS

FIGURE 19.8 ■ Exercise 1

SLOPED LETTERS

FIGURE 19.9 ■ Exercise 2

CHAPTER 20

Organizing a Drawing Sheet

OBJECTIVES

After studying this chapter, you should be able to:

- describe how a basic schedule of plans is set up
- discuss the information put on planning sheets
- explain the purpose of a completion log
- tell how a drafter determines the placement of a drawing or drawings on a sheet
- tell why it is important to draw neat, uncluttered plans

20.1 INTRODUCTION

The designer and drafter are responsible for relaying information to the construction worker or technician in a thorough, neat, and orderly manner. To accomplish this task, they must first plan what is to be shown on the sheet and then decide how to arrange this information in a logical manner for the worker or technician. This chapter will give suggestions for the organization of a drawing sheet.

Architectural and engineering offices usually have their own procedures for planning each sheet of drawings that they produce. Some of these procedures will be discussed in this chapter.

20.2 CREATING A BASIC SCHEDULE OF PLANS

At the beginning of a job, the architect and engineer usually call a meeting of the designers and drafters who will be working on the project. The purpose of this meeting is to plan the drawings needed to relay the necessary information to the construction workers or technicians. The number of sheets and the titles of these sheets are listed. The general grouping and numbering of the plans are considered. For example, the site plan or plans come first, followed by the floor plans, elevations, sections, and details. The schedules are the last group of plans.

The size of the job determines the number of sheets of drawings that will be required. The larger the project, the more planning is required to ensure that all essential information is included on the drawings. Larger jobs also require more coordination among different sheets of drawings.

20.3 A SYSTEM FOR ORGANIZING A JOB

A popular system for organizing plans for a project uses an 8½" × 11" sheet of paper to represent each sheet of drawing that is required. The sheet number is placed at the lower right-hand corner of each 8½" × 11" sheet. The sheet number corresponds to the sheet number in the regular drawings. All drawings, details, elevations, sections, and schedules are listed on these sheets. At the end of the planning sessions, the architect or engineer knows *approximately* how many sheets are required to design and draw the plans for the project, but the *exact* number of sheets will not be known until they are actually drawn. During the design, additional sheets may have to be added to relay all the required information.

If a sheet is added to the drawings, an 8½" × 11" sheet should be added to the original planning booklet. Sheet numbers (and sometimes sheet titles) should be changed to reflect the addition of the sheet. The planning booklet containing the 8½" × 11" planning sheets should always be updated. This way, the design team will always know what sheets are required as well as what information is required on each one.

In larger offices, each designer may be assigned a certain sheet to complete. With a project-planning booklet, each designer knows what information is needed on that particular sheet and can assemble the information to make the design.

Coordination with other designers and drafters should take place on a regular basis to ensure that the overall systems will work, that the design work does not overlap, and that important pieces of design work are not being omitted.

A completion log is usually kept on the 8½" × 11" sheets. When this log is updated, each member of the team can look at the planning book to see what sheets have been finished or what percentage of each sheet has been finished. This log helps keep the project on schedule because additional manpower can be added if the job is lagging behind schedule.

20.4 ORGANIZING INDIVIDUAL DRAWING SHEETS

Each drawing sheet must have a title for identification. Once this title has been established, all the information shown on the sheet should pertain to this title. For example, if the sheet title is "First Floor Plan," the first floor plan with all the information related to the first floor plan should be shown on the sheet. Often, the floor plan does not fill the full sheet. It is permissible to draw details, sections, and elevations pertaining to the floor plan on the floor plan sheet. In this case, the sheet title should be changed to include the details, sections, and elevations.

In the example in the preceding paragraph, the floor plan is the most important drawing on the sheet; therefore, it should be shown in a prominent place, usually the center of the sheet. Some architects and engineers prefer to center the floor plan on the sheet from top to bottom but shifted to the right-hand border rather than the center of the sheet from side to side. This placement allows more usable space to the left of the floor plan for details, sections, and elevations.

The most important thing to remember in planning a drawing sheet is that the sheet must be organized and easy to read. Cluttered and unorganized sheets are hard to read and may cause costly mistakes.

SUMMARY

- The architect, the engineer, designers, and drafters on a job hold a planning meeting before any drawing is done.
- For a large job, more sheets of plans and more coordination are needed.
- Many firms make a planning booklet with one 8½" × 11" sheet corresponding to each proposed sheet of drawings.
- The planning booklet must be updated to reflect the number of sheets of drawings and what information is to be put on each sheet.

- A completion log keeps workers informed as to what percentage of the drawings on each sheet has been finished at any given time.
- The most important drawing on a sheet should be the most prominent item on the page.
- If more than one drawing is put on a sheet, the title for that sheet should reflect the contents of the sheet.
- Each sheet should be neat, uncluttered, well organized, and easy to read.

REVIEW QUESTIONS

1. What takes place at the planning meeting before any drawing is done?
2. Who attends this planning meeting?
3. If a planning booklet is prepared as discussed in this chapter, what information goes on each $8^1/_2$" × 11" sheet?
4. What is the purpose of a completion log?
5. How can a person know whether more than one type of drawing has been put on the same sheet?
6. Why is it important to have each sheet uncluttered, neat, and well organized?

ADDITIONAL STUDENT EXERCISES: CONSTRUCTING PROJECT-PLANNING BOOKLETS

1. Review the construction plans for "A Small Bank and Trust Company" included at the end of this textbook. Construct a project-planning booklet for the design of the building. Create a sheet for each sheet of drawings, and describe the drawings that have been drawn on each sheet. Add or omit any items that may improve the quality of the set of plans. Details and sections may be moved from one sheet to another. Give at least one reason for making such changes.
2. After reviewing Chapter 11 and the plans at the end of this textbook, you should be familiar with the various sections of a set of plans and what is generally required for each section. Visualize a set of plans for a residence of approximately 2000 square feet that includes three bedrooms, two bathrooms, a family room, a kitchen, a dining room, and a living room. Construct a project-planning booklet for this residence.

CHAPTER 21

Shading and Crosshatching

OBJECTIVES

After studying this chapter, you should be able to:

- recognize and define crosshatching
- recognize and define shading
- discuss when and why shading, crosshatching, and poché are used on drawings
- explain how shading is used to heighten the three-dimensional effect on drawings
- discuss the use of commercially prepared stick-on shading to achieve a uniform appearance on drawings

21.1 INTRODUCTION

When a drawing has been completed, it is sometimes hard to distinguish the object line separating two objects or two different materials. To help with this problem, crosshatching or shading can be applied to the object to help identify the dividing line. Poché is a type of shading that makes the dividing line between objects more apparent. This procedure is used often throughout a set of building plans to identify different materials used in construction. Shading makes the three-dimensional effect much more pronounced in oblique and isometric drawings. In this chapter, the topics of crosshatching and shading will be discussed and demonstrated.

21.2 CROSSHATCHING

Crosshatching is accomplished by using a 45-degree triangle or a 30/60-degree triangle to draw parallel lines, equally spaced, over a selected area (Figure 21.1). A 45-degree angle is commonly selected for manual crosshatching. A drafting machine can also be used for crosshatching. The drafting machine must be set at the desired angle, and parallel lines, equally spaced, are drawn over a selected area. The crosshatching lines are drawn from one object line to the other object line.

The quality of the lines should be the same as the extension and dimension lines. These lines are usually thin lines; however, they must be heavy enough to print (see Chapter 15). In some cases, crosshatching lines are drawn the same weight as object lines. The choice of line weights for crosshatching is usually left up to the drafter.

For crosshatching, the parallel lines can be drawn in any direction and at any angle. If two kinds of materials (or two different pieces) fit together, crosshatching can be used to help separate them (Figure 21.2). The drafter should turn the crosshatching lines in one direction from the common object line and then turn the crosshatching lines in another direction from the same object line (Figure 21.3).

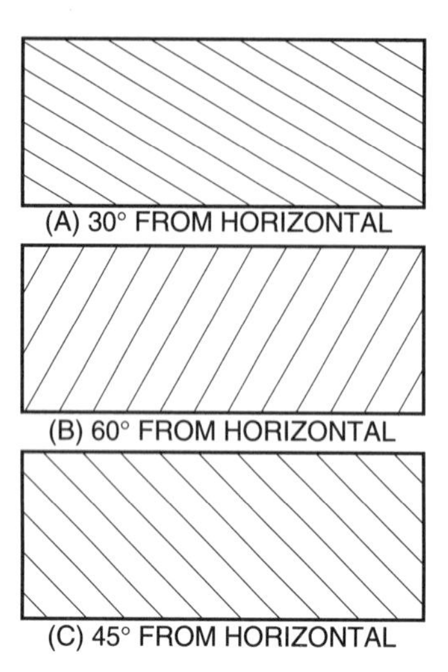

FIGURE 21.1 ■ Crosshatching at different angles

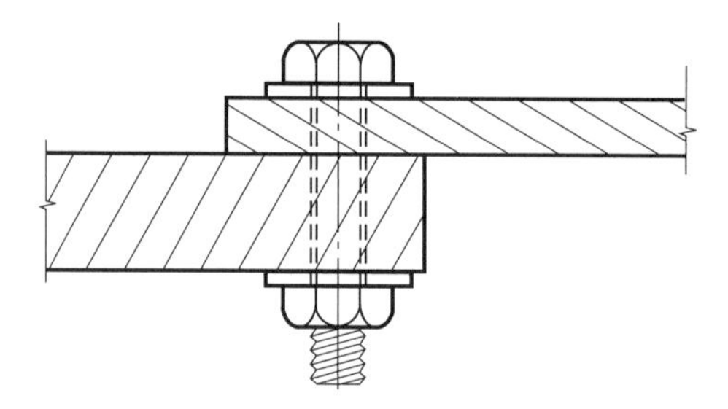

FIGURE 21.2 ■ Two different metals indicated by crosshatching

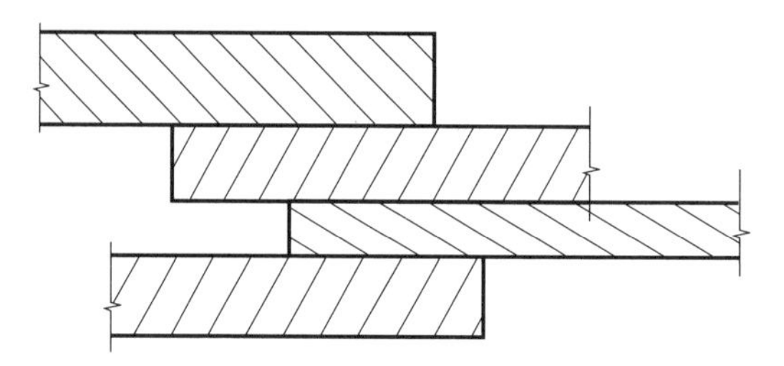

FIGURE 21.3 ■ Crosshatching of different materials

21.3 POCHÉ

Poché (pronounced "poh-shey") can be defined as the process of shading an object with a soft graphite pencil and rubbing the graphite with a finger or a soft tissue to create a uniform shading effect. This procedure has been used for many years to distinguish between different materials on building drawings. One example of the use of poché is the treatment of partition walls on an architectural floor plan (Figure 21.4). When the walls have poché, they are easily seen because of the contrast between the walls and the floor.

When soft graphite is rubbed, it is sometimes spread outside the object lines. If this happens, the drafter must clean the area outside the poché with an eraser.

Poché can be used in conjunction with other drafting methods. When the architect draws concrete block walls, the symbol for concrete block is put on the plans. Poché is added to the symbol for concrete block because it makes the walls look more realistic (Figure 21.5). Concrete walls are often shown with the concrete symbol, and poché is added to give the walls more contrast.

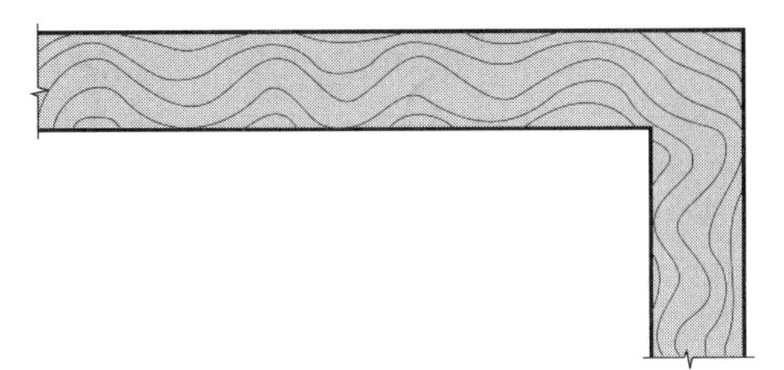

FIGURE 21.4 ■ Partition wall with poché

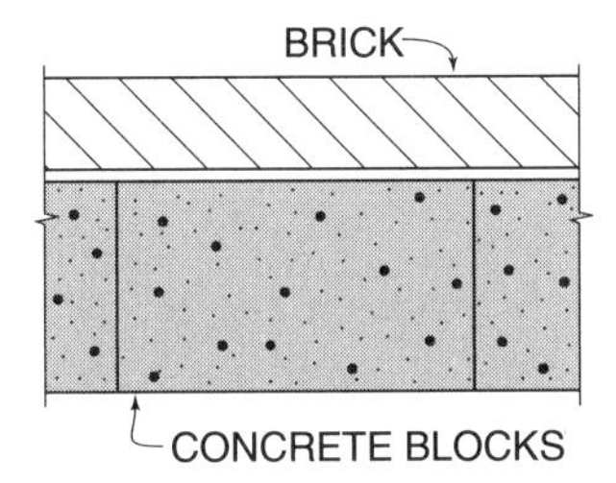

FIGURE 21.5 ■ Brick and concrete block wall with concrete blocks shown with poché

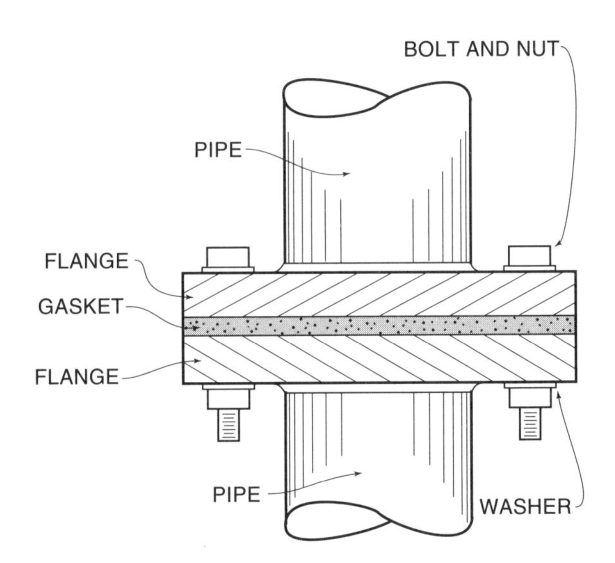

FIGURE 21.6 ■ Pipe flanges with gasket (gasket shown pochéd)

21.4 COMBINING CROSSHATCHING AND POCHÉ

It is often advantageous to use several different methods to distinguish between different materials or parts that are separated by only one object line. To make this distinction clear, crosshatching can be used on one material or part. Crosshatching can also be used for several different materials as long as the parallel lines are drawn at different angles. To break the monotony, poché can be used in conjunction with the crosshatching or it can be used between the sections of crosshatching (Figure 21.6).

21.5 SHADING DRAWINGS

In Chapters 17 and 18, oblique and isometric drawings are described. Oblique and isometric drawings are very important in the air-conditioning and heating fields because they are pictorial representations that give a three-dimensional effect. With shading, the three-dimensional effect is greatly improved.

21.6 PROCEDURES FOR SHADING OBLIQUE AND ISOMETRIC DRAWINGS

A drafter who has finished an oblique or isometric drawing can improve the drawing by shading the object in the drawing. First, the drafter must imagine a floodlight being shone on the object (Figure 21.7). When this light strikes a surface of the object, the surface becomes brighter and shadows are cast on surfaces that are not struck by the light. These shadows appear darker than the lighted surfaces.

Keeping the imaginary light in mind, the drafter should mark the shaded areas on the drawing. The shaded areas are the areas where shadows would result if a light were shown on the object. These markings should be very light so that they can be erased. This imaginary lighting should always strike the object at an angle (Figure 21.8). A 45-degree angle is commonly used; however, other angles are sometimes used with good results. The drafter may try several angles for the imaginary light.

When all shaded areas on the object have been marked, the drafter decides what symbol (crosshatching or poché) is to be used to indicate the shaded area. Special care should always be taken to make the shading uniform.

Commercially prepared stick-on shading can be used for the shading. This stick-on shading is a transparent sheet with printed dots on one side. The other side of the transparent sheet has colorless glue that adheres to the drafting paper. When the sheet is cut to fit the shaded area, the backing is peeled off and the prepared shading is applied as needed on the drawing. When the drawing is run through the printing machine, the dots of the stick-on sheet show up the same as if they were drawn by hand. This procedure gives the finished drawing uniform shading or shadowing that is neat and easily read.

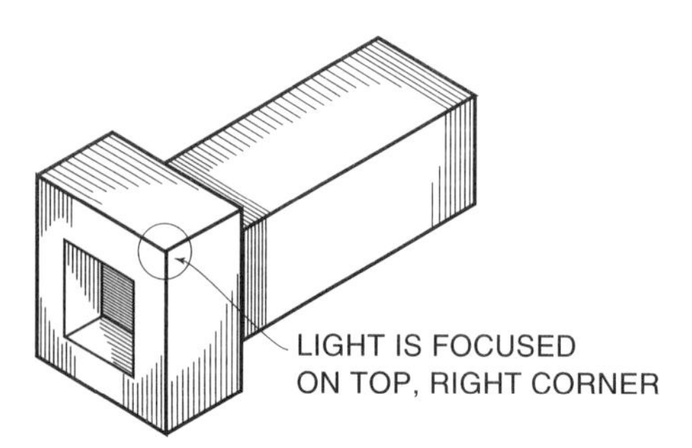

FIGURE 21.7 ■ Pipe flanges with gasket (gasket shown pochéd)

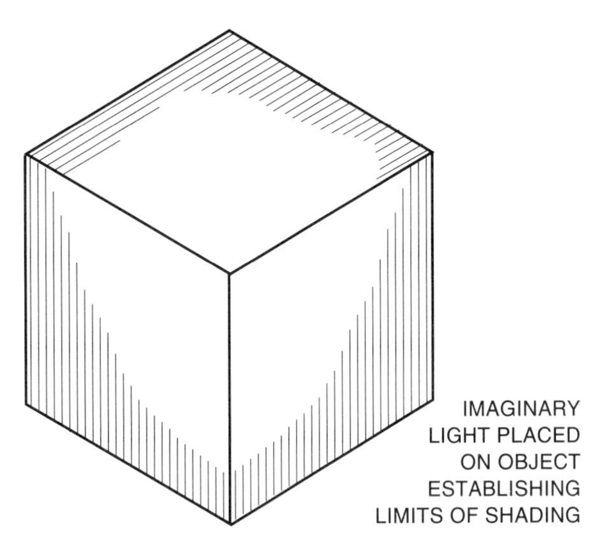

FIGURE 21.8 ■ Marking limits for shading

SUMMARY

- Crosshatching, shading, and poché are used to make the dividing line between objects on a drawing more apparent.
- Crosshatching lines are thin lines drawn close together at an angle (typically a 45-degree angle) from object line to object line.
- Poché is the procedure by which an object is shaded with a soft graphite pencil and then rubbed with a tissue or the finger to achieve uniform shading.
- Shading improves the three-dimensional effect on isometric and oblique drawings.
- To shade accurately, the drafter must duplicate the effect created by a floodlight shining on one surface of an object and casting shadows on other surfaces.
- A uniform shading effect can be achieved by the use of commercially prepared stick-on paper.

REVIEW QUESTIONS

1. When should a drafter use crosshatching and shading?
2. Why is shading important on oblique and isometric drawings?
3. How should the triangle be used to make crosshatching?
4. How can a drafting machine be used for crosshatching?
5. What criteria are used for determining the weight of crosshatching lines?
6. Why are crosshatching and poché sometimes used together?
7. Define poché, and give an example of when it is used.
8. Describe the procedure by which a floodlight defines areas to be shaded.
9. What is the advantage of using commercially prepared stick-on paper for shading?

ADDITIONAL STUDENT EXERCISES: SHADING AND CROSSHATCHING

1. Using the examples in Figures 21.7 and 21.8, shade the vertical tank in Figure 21.9.
2. Using the examples in Figures 21.7 and 21.8, shade the unit depicted in Figure 21.10.
3. Using the examples in Figures 21.7 and 21.8, shade the horizontal tank shown in Figure 21.11.
4. Referring to Figure 21.12, choose the light source spot where light strikes the gas furnace and then crosshatch it to indicate shading.
5. Determine the light source spot where the light strikes and then shade the object shown in Figure 21.13 using the poché method.

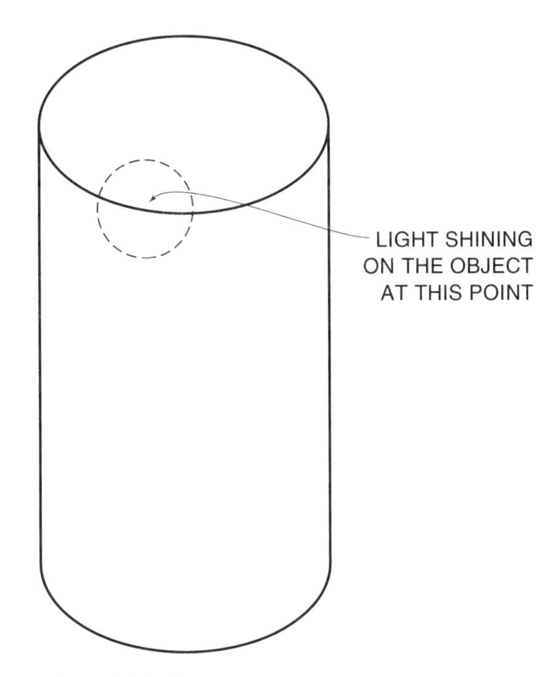

FIGURE 21.9 ■ Exercise 1

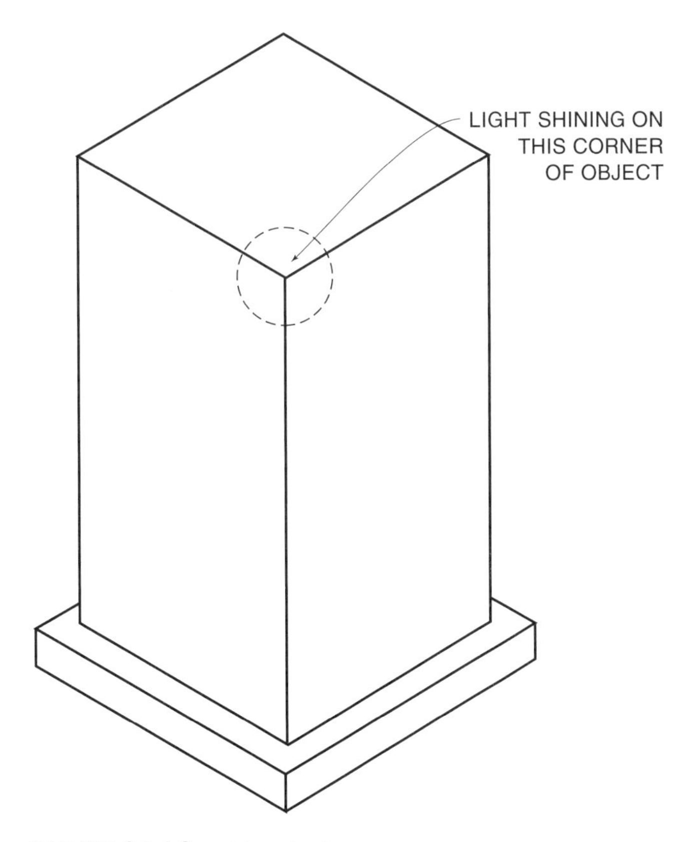

FIGURE 21.10 ■ Exercise 2

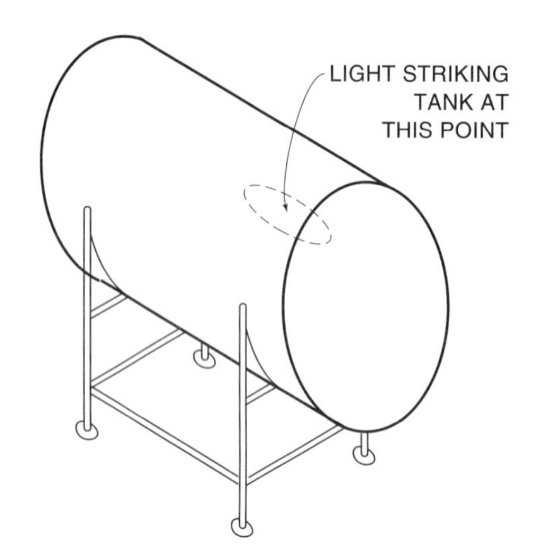

FIGURE 21.11 ■ Exercise 3

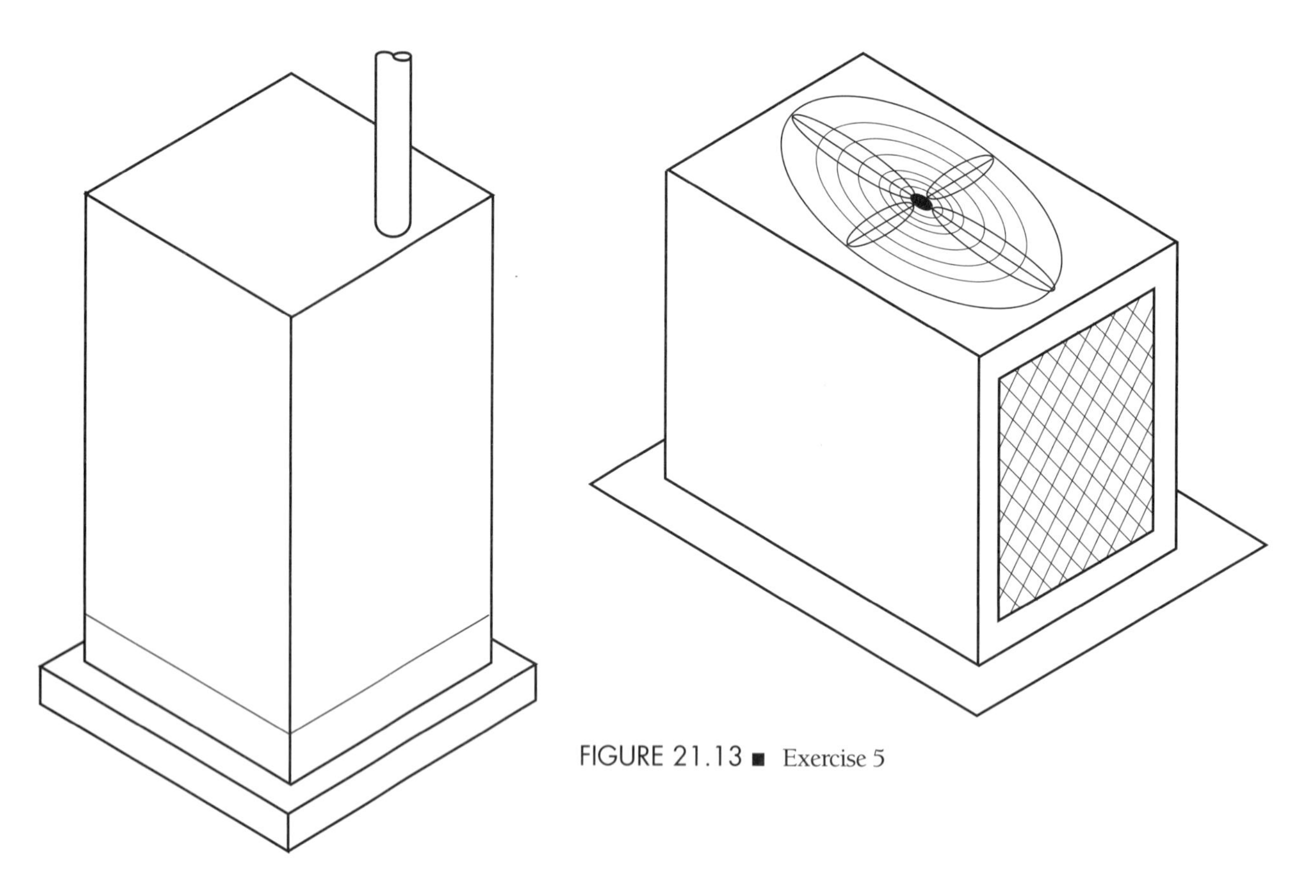

FIGURE 21.13 ■ Exercise 5

FIGURE 21.12 ■ Exercise 4

CHAPTER 22

Drawing Construction Elevations

OBJECTIVES

After studying this chapter, you should be able to:

- describe typical information found on exterior elevations
- locate and discuss cutting plane lines and symbols used to denote cutting planes
- identify interior wall elevations
- describe typical information found on interior wall elevations
- describe information found on mechanical, electrical, and plumbing elevations

22.1 INTRODUCTION

Many types of elevations are required in building construction. Outside elevations of a building include front, rear, and side elevations. Interior wall elevations include kitchen cabinet elevations and built-in specialties. The mechanical, plumbing, and electrical sections in a set of building plans include elevations of equipment, equipment rooms, and other features. The larger the job, the more elevations are required to show how the finished product will look, the locations of the equipment, and the specialty items required with the equipment.

22.2 EXTERIOR ELEVATIONS

Exterior elevations are orthographic views of the various sides of a building. Often, the architect identifies the view by the direction it faces. Thus, if the front elevation faces north, it may be called the north elevation, while the rear view would be called the south elevation. Other architects simply identify the front, rear, and side views of the drawings without including the direction each one faces as part of the title.

The architect or drafter shows the elevations of the various sides of the building in the architectural section of the plans. These drawings are developed from the floor plans of the building (Figure 22.1). The doors, windows, and so forth are located at the same place on the elevation as on the floor plan. Types of windows and doors are drawn in detail so that the owner and the builder will know what they will look like. The windows and doors are also described in detail in the schedule section of the plans. Dimensions for windows and doors are usually given in these schedules.

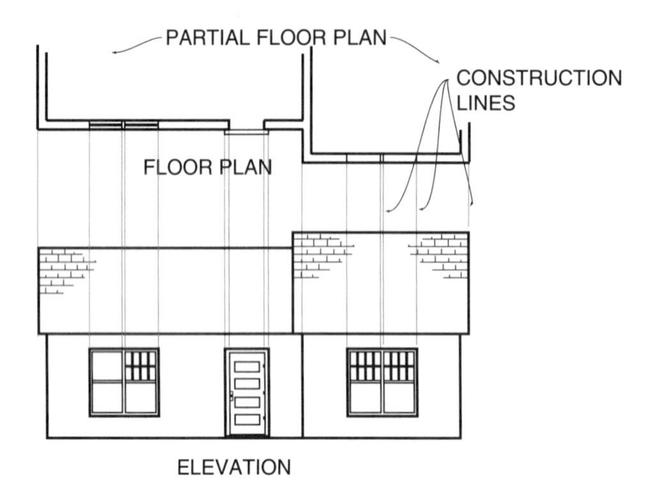

FIGURE 22.1 ■ Developing an elevation from a floor plan

On Sheet A-8 of the plans at the end of this textbook is an example of an elevation developed from a partial floor plan (see "Plan of 'B' Wall"). The elevation of the wall, developed from the partial floor plan, is shown in the lower left corner of the sheet.

Sheet A-2 of this set shows the exterior elevations of a small bank and trust company. The front elevation is not shown on Sheet A-2 as such, but the note under the rear elevation states that the front elevation is the same as the rear elevation except that it is reversed.

The cutting planes for sections drawn at this point are represented by circles with two numbers located inside, with a long line projecting through the height of the building. The top number in the circle represents the section number, and the bottom number gives the page number where the section is drawn.

Sheet A-5 of the drawings shows transverse and longitudinal sections of a bank building. These sectional drawings supplement the information on the exterior and interior building elevations, including detailed information on the doors, glass, and walls. The windows are fixed glass, requiring no schedule, and the doors are shown on the floor plan (Sheet A-1). The doors are numbered and are scheduled on Sheet A-8. If windows other than fixed glass had been used, the windows would have been numbered on the floor plan (and sometimes on the elevation plan). The windows would have been scheduled on the same sheet as the door schedule.

Other items that are usually shown on the outside elevation plans include hose bibbs, downspouts, openings in the outside wall, and all pieces of equipment mounted on the wall. Any features that can be seen on the finished outside wall should be shown on the outside elevation drawing.

22.3 INTERIOR ELEVATIONS

Interior elevations are very important to the owner and the builder. Interior elevations are usually drawn for cabinets and special built-in fixtures. They normally show everything that will be located on the finished wall. The set of construction drawings at the back of the book includes interior elevations of the teller's counter. These elevations, shown on Sheet A-7, are drawn in conjunction with the floor plan so that all dimensions for constructing the cabinets are shown on one of the two drawings. Another wall elevation is shown on Sheet A-8.

Interior elevations of this type are usually drawn for each wall in the building that requires special construction or special features on that wall. Interior wall elevations are usually drawn from one adjoining wall to the other adjoining wall. Items shown include electrical outlets, chair rails, wall finishes, shelving, crown molding, baseboards, telephone outlets, and special wall-mounted devices, such as safes, fire extinguishers, and special wall features. An interior elevation of a wall of a bathroom might show built-in cabinets, a mirror, lighting, toilet fixtures, electrical receptacles, and shelving. Specific details are shown on interior elevations to help the builder and tradespeople know what the finished product should look like. On larger jobs, several sheets of drawings may be required to accommodate all the required wall sections.

22.4 MECHANICAL ELEVATIONS

Mechanical elevations, like architectural elevations, show how equipment should look when the installation is completed. The engineer or drafter draws the equipment and shows all connections to the equipment. Elevations in the mechanical section include elevations of cooling towers, air-handling equipment, compressors, chillers, special ductwork, ductwork connections, piping, condensing units, and any other mechanical equipment requiring detailed instruction for installation.

Interior wall elevations are sometimes required to show supply and return air outlet installations and how they fit into the room decor. An equipment elevation of a condensing unit is shown on Sheet M-1 of the construction plans at the end of this textbook. Drawing 5 on Sheet M-1 shows the condensing unit and the mounting arrangement. The larger the job, the more mechanical elevations are required to show the required information.

22.5 ELECTRICAL ELEVATIONS

The electrical engineer and drafter use elevation drawings to show the locations and to describe the mounting of electrical equipment. It may be advantageous for the engineer or drafter to draw interior wall elevations of an electrical room to show where each piece of equipment is to be located (Figure 22.2).

A wall elevation inside the building can show the locations of electrical outlets, telephone connection jacks, and wall-mounted lighting fixtures.

An exterior elevation drawing may be required to show the electrical service wiring, the weatherhead, and the electrical meter location (see Figure 22.3). This drawing can also show how the service is to be grounded.

22.6 PLUMBING ELEVATIONS

Plumbing elevations are as important as any other elevations; however, not as many elevations are usually required in the plumbing trade. On larger jobs, the engineer or drafter sometimes draws elevations of toilet spaces and equipment room walls to show the locations of plumbing equipment.

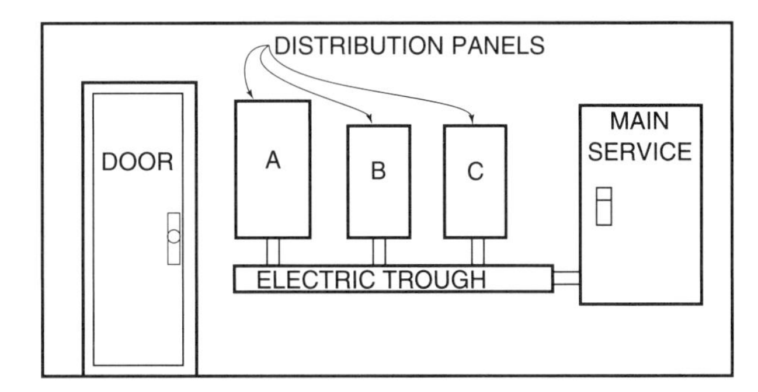

FIGURE 22.2 ■ Elevation of the north wall of an electrical room

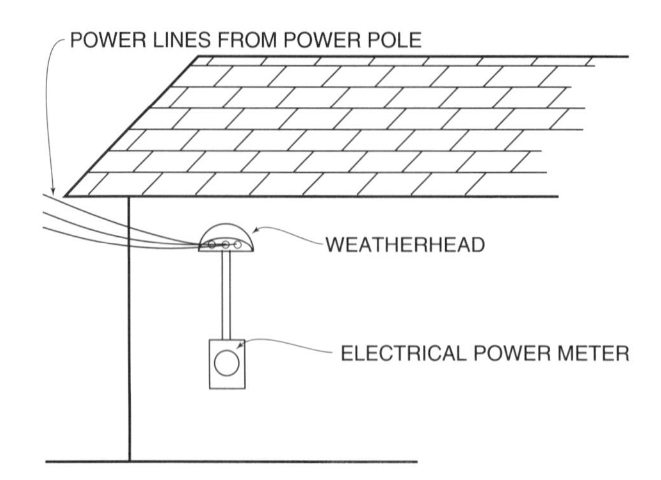

FIGURE 22.3 ■ The electrical weather head provides a waterproof service entrance for the electrical wires to the structure

SUMMARY

- The larger the job, the more elevations are required to show what the finished product should look like.
- Exterior elevations can be identified by the view (front, rear, right side, or left side) or by the direction faced (north, south, east, or west).
- Elevation views are developed from the floor plans of the building.
- Cutting planes are used to cut away sections of the building and show details of construction.
- The exterior elevation should show everything that can be seen on the finished outside wall.
- Interior elevations are usually drawn from one wall to the other to show specific details of construction on that wall.
- Mechanical and electrical elevations show how equipment should appear when installation is completed.
- Plumbing elevations are drawn to show the locations and details of plumbing equipment.

REVIEW QUESTIONS

1. An exterior drawing of the front of a building facing south can be titled ______________________ __________ or ______________________________.
2. A detailed description of windows and doors is given in the ________________ section of the plans.
3. Circles denoting the locations of cutting planes contain two numbers: The top number represents the ______________________________ and the bottom number represents the ______________________________.
4. Hose bibbs and downspouts are shown on the (exterior, interior) elevation drawings.
5. Cabinets and built-in features are shown on the (exterior, interior) elevations.
6. Interior elevations are usually drawn from (wall to wall, floor to ceiling, outside to inside).
7. Electrical outlets, safes, and chair rails are shown on (exterior, interior) elevation drawings.
8. Special ductwork is drawn on the (mechanical, electrical, plumbing) elevations.
9. Wall-mounted lighting fixtures are shown on the (mechanical, electrical, plumbing) elevations.
10. Toilet equipment is shown on the (mechanical, electrical, plumbing) elevation drawings.

CHAPTER 23

Drawing Construction Details

OBJECTIVES

After studying this chapter, you should be able to:

- explain why details are needed for coordination on the job
- identify commonly used architectural details and discuss information on them
- identify commonly used mechanical details and discuss information on them
- tell how to locate details on drawings
- describe the process by which details are usually developed

23.1 INTRODUCTION

The construction and equipment for modern buildings have become so complex that no single person can plan the entire building. The architect, engineer, and designers/drafters are dependent on each other for answers to questions as to strength, space, construction, mechanical apparatus, design, and appearance of the finished product. The architect is usually associated with architectural engineers (structural, mechanical, plumbing, and electrical) who work closely to coordinate the design with the various trades through joint meetings and open discussions. It is not uncommon to use freehand details to give the information needed for coordination. After being approved, these freehand details are often drawn with drafting instruments as details that are placed on the working drawings. Construction technicians and workers use the details to see how the various systems were designed and to choose equipment that will satisfy the requirements of the building.

This chapter will discuss details that are commonly used in the building construction industry. In earlier chapters, the design responsibilities were broken down into architectural (designing the building) and engineering (mechanical, plumbing, electrical and structural) areas. Site work plans and structural plans are included under the major heading of architectural. We will discuss the site plan section and the structural plan section under separate headings rather than grouping them under the architectural (major) heading.

Each of these areas has a need for details. Most of these details are drawn to scale; however, isometric and oblique drawings, which are not drawn to scale, are often used. This chapter will discuss information to be shown in a detail and will describe how details should be arranged on the working drawings.

23.2 WHAT DETAILS SHOULD BE DRAWN

The architects and the engineers, along with their designers/drafters, must decide what information is needed and what procedures must be followed to install the equipment. Decisions must also include coordination with other construction areas to verify that no equipment will conflict with equipment from another trade and that work done by one trade will not interfere with that done by another trade. Decisions as to what details are required are usually made during the design process.

If the drawing is too small for the required information to be shown on the floor plan, a detail of the area in question is added, enlarging a small portion of the drawing and relaying the information to the builder.

23.3 SITE DETAILS

The site plan section sometimes consists of demolition plans plus the new grading plan and the new work to be done on the site. Some details that are usually needed are as follows:

1. Storm drainage pipe installation detail
2. Catch basin details
3. Manhole details
4. Roadway paving details
5. Curb and gutter details
6. Details for planting plants and trees
7. Details of specialty items, such as fountains, pools, and planters
8. Walls (brick, stone, wood, etc.)

Any or all of these details are required to complete the site work.

23.4 STRUCTURAL DETAILS

Structural details are necessary to show how the structural members are joined in order to establish safe and acceptable joints. Some of the structural details are as follows:

1. Joining of major structural steel members
2. How smaller structural members join the major structural members
3. How structural members are connected to building members
4. How structural members join footings

Structural details should include notes and dimensions so that the installer knows how to make the installation.

23.5 ARCHITECTURAL DETAILS

Some details that are commonly used in the architect's office are the following:

1. Window and window trim
2. Cornices and moldings
3. Fireplace and mantel trim
4. Doors and door trim
5. Bookcases and shelving
6. Built-in cabinets
7. Stairs and trim

The architectural detail should include such information as what types of materials are used, how different members are joined together, necessary dimensions for different members, and finishes for each member.

23.6 MECHANICAL DETAILS

Some details that are commonly used in the mechanical engineer's office are as follows:

1. Mechanical unit details
2. Fan details
3. Equipment mounting bases
4. Unit hangers details
5. Roof-mounted equipment
6. Supply and return air terminals
7. Detail of special valves
8. Duct connections
9. Piping details
10. Pump details
11. Damper details
12. Insulation details
13. Coil mounting details
14. Louver details
15. Equipment connection details

Mechanical details should include such information as the type of materials used, how different members are joined together, necessary dimensions for different members, connection points, and hangers or mounting devices.

23.7 PLUMBING DETAILS

Plumbing details, like mechanical details, are often repeated from job to job. The plumbing engineer and the plumbing technician must decide what details are needed to relay the necessary information. Some of the commonly used details in the plumbing section are as follows:

1. Interior and exterior clean-out details
2. Water heater piping detail
3. Roof drain details
4. Floor drain details
5. Piping diagrams (details)
6. Riser diagrams (details)
7. Yard hydrant details
8. Manhole details
9. Equipment connection details

Like the site, structural, architectural, and mechanical details, the plumbing details should contain all the information needed to install the detailed equipment properly. Dimensions may be required, and if so, they should be accurate and complete.

23.8 ELECTRICAL DETAILS

The electrical section must have details. The engineer and design technician must decide what details are required. Some of the commonly used details are as follows:

1. Details for mounting and connecting special equipment
2. Electrical panel details
3. Special wiring diagrams (details)
4. Power service diagrams (weatherheads, etc.)
5. Mounting pads for electrical equipment
6. Installation details for generators
7. Mounting details for parking lot lighting
8. Details of various lighting fixtures

The electrical plans are more diagrammatic than the other plans. For this reason, the electrical section may require more details to relay information.

Like other details, the electrical details may require accurate and complete dimensions.

23.9 LOCATING DETAILS ON THE SHEET

For smaller buildings, details are usually located on floor plan sheets or other sheets where there is space available. The details are usually identified within the titles. They should be indexed in some manner so that they can be found easily. On larger jobs, details are titled and numbered for identification and are shown on a separate sheet.

23.10 DRAWING DETAILS

After a decision has been made as to the type of detail, the drafter proceeds with the drawing. Images are drawn with fine lines that are easily erased. When the drawing has been developed with light lines, a soft pencil with a fairly dull point (No. 2, for example) is used to draw over the construction lines, showing the object with crisp construction lines. Extension and dimension lines are often shown, giving the exact dimensions that are required.

SUMMARY

- Details help architects, engineers, and designers/drafters coordinate designs with the various tradespersons.
- Details sometimes start as freehand sketches that are placed on the working drawings after coordination conferences.
- Site details include demolition plans and grading and site work details.
- Structural details show how structural members are to be joined and usually include notes and dimensions.
- Architectural details list types of materials, how members are joined, and dimensions and finishes for members.
- Mechanical details list types of materials, how members are joined, dimensions, connections, and mounting devices.
- Plumbing details include information needed to install and give sizes of equipment.
- Electrical details contain diagrams to relay information and include dimensions.
- Details for small buildings are often placed on the floor plan sheets or wherever space is available; for large buildings, details are usually titled, numbered, and placed on a separate sheet.

REVIEW QUESTIONS

1. How do details help the various persons working on a job to coordinate their designs?
2. When is it decided what details should be drawn?
3. List three types of details often found in the site plan section.
4. List three types of details often found in the structural section.
5. List three types of details often found in the architectural section.
6. List three types of details often found in the mechanical section.
7. List three types of details often found in the plumbing section.
8. List three types of details often found in the electrical section.
9. Where are details usually located if the building is small?
10. Where are details usually located if the building is large? How are details identified on the drawings?
11. Discuss the steps a drafter follows in proceeding with a drawing of a detail.

CHAPTER 24

Freehand Sketching

OBJECTIVES

After studying this chapter, you should be able to:

- describe three types of drawings that often begin as freehand sketches
- demonstrate the proper way to use guidelines for lettering notes on a plan
- demonstrate at least three factors to consider when locating information on a drawing
- list at least three factors to consider when locating notes on a plan
- demonstrate the procedure for shading a drawing to show contrast between the different parts or planes
- demonstrate how to shade an isometric or an oblique drawing to help create a three-dimensional effect

24.1 INTRODUCTION

The purpose of this chapter is to make the reader aware of the various aspects of freehand sketching techniques. No two people are alike. Everyone has natural skills that can be developed. Some people have the ability to do beautiful freehand sketching, while others must struggle to become proficient. In this chapter, you will learn acceptable methods for developing the needed skills for freehand sketching.

After learning the various techniques, you need to practice, practice, practice. Freehand sketching skills are learned by doing, and the information in this chapter is only an introduction. The student becomes proficient by learning new techniques and improving his or her skills each day.

24.2 TYPES OF DRAWINGS CREATED BY FREEHAND SKETCHING

Orthographic projection is probably the most common type of drawing done with freehand sketching. Orthographic projection is discussed in detail in Chapter 16. This type of drawing consists of three basic views (plan, front, and side views). The front and side views are sometimes called elevation views.

In the HVAC field, a technician has many opportunities to use freehand sketching to produce plans of existing conditions. These plans are then taken back to the office, where a drafter creates a scale drawing of the space. Elevation drawings can be drawn freehand and used to produce scaled drawings at a later date.

Freehand sketching is also used for making oblique drawings. Oblique drawings created with drafting instruments are discussed in detail in Chapter 17. The same procedures used for instrument drawings can be applied to produce freehand oblique drawings.

Isometric drawings are a third type of freehand sketches. Isometric drawings are drawn using the same procedures described in Chapter 18.

24.3 FREEHAND SKETCHES ARE NOT DRAWN TO SCALE

In freehand sketching, the drawings are not drawn to scale; however, these drawings must be kept in proportion to reflect the relationship of one part of each drawing to another and to accurately show the difference in sizes.

The architect's scale is often used to help the freehand drafter keep the different components of the freehand drawing in proportion. Once an approximate scale has been selected, the architect's scale can be used to establish dimensions for the drawing.

24.4 LINES USED FOR FREEHAND SKETCHING

Many types and weights of lines are drawn on freehand sketches. Refer to Chapter 15 for a discussion of the various types of lines used on working drawings. Figure 15.1 describes commonly used sizes of drafting pens and the line weights they produce. Figure 15.2 is a schedule of lines containing names, descriptions, and examples of the various types of lines found on drawings. Freehand sketches use the same types and weights of lines as working drawings.

24.5 LETTERING FOR FREEHAND SKETCHING

The beginning drafter soon learns that the lettering of notes on a drawing is very important in freehand sketching. The same lettering and notation procedures that were discussed in Chapter 19 are used in freehand sketching and drafting.

The more proficient the drafter is at lettering, the better the drawings look. See Chapter 19 for information on forming letters properly for freehand sketching and for manual drafting.

Guidelines should always be used for hand-lettering a plan (see Chapter 19). These lines help the drafter to make the letters the same size and properly aligned.

Technicians often develop special flairs in their systems of lettering. Even with basic uppercase letters, drafters can achieve a unique appearance by developing variations of the standard block or the standard sloped-block letters. A few examples of these modifications are shown in Figure 24.1.

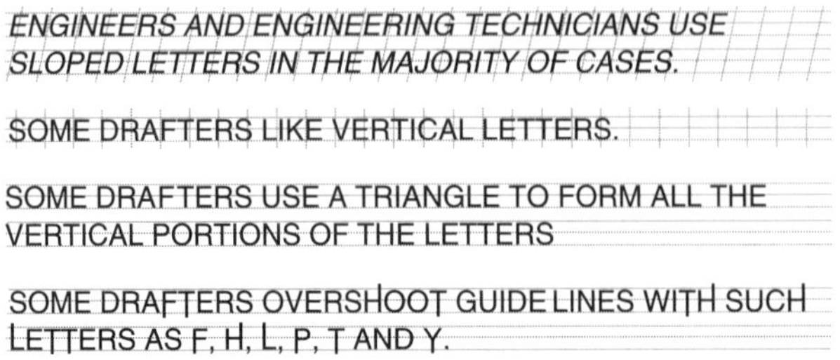

ENGINEERS AND ENGINEERING TECHNICIANS USE
SLOPED LETTERS IN THE MAJORITY OF CASES.

SOME DRAFTERS LIKE VERTICAL LETTERS.

SOME DRAFTERS USE A TRIANGLE TO FORM ALL THE
VERTICAL PORTIONS OF THE LETTERS

SOME DRAFTERS OVERSHOOT GUIDE LINES WITH SUCH
LETTERS AS F, H, L, P, T AND Y.

FIGURE 24.1 ■ Lettering

24.6 LOCATING NOTES ON FREEHAND SKETCHES

Notes on drawings are very important. Freehand sketches and drawings are noted in the same manner as other drawings (see Chapter 19).

In most cases, the drafter or technician completes the drawing before lettering the required notes. The organization and location of notes should be planned in advance of putting them on the plan. How to add notes to a freehand sketch is discussed in Chapter 19.

24.7 LOCATING DRAWINGS ON A PAGE

Planning prior to starting the actual drawing process is very important. The sheet must be neat and well organized so that it is easily read. The drawings should not be crowded or placed in spaces that are not suitable. To be sure that an object or building will fit on the sheet, it should be drawn to an approximate scale and then located on the sheet.

To locate the object being drawn in the middle of the sheet, follow these steps: Locate the center of the object (or building) based on its dimensions, find the center of the sheet, and then place the center of the object on the center of the sheet.

24.8 DIMENSION AND EXTENSION LINES

As discussed in Chapter 11, construction drawings cannot be scaled for accurate dimensions because of the stretching or shrinking that takes place as copies are made of the working drawings. It is the responsibility of the drafter to write the dimensions on working drawings so that the dimensions can be read directly from the plan. The freehand drafter should show all important dimensions on the freehand sketch so that the correct dimensions are available to the drafter. Chapters 15 and 25 describe ways to show dimensions on the plan.

24.9 SHADING AND CROSSHATCHING FREEHAND DRAWINGS

Shading and crosshatching are discussed in detail in Chapter 21. Shading and crosshatching pertaining to freehand sketching will be briefly discussed here.

The term *shading* has several meanings. One meaning is "the act of 'coloring,' crosshatching, or adding special markings to various surfaces of an object to make these surfaces stand out from the remainder of the drawing." Some of this special treatment of surfaces is shown in Figure 24.2. Figure 24.3 is an example of crosshatching, and Figure 24.4 shows special treatment to represent how the actual material looks in an air-handling unit. Poché (pronounced "poh-shey") is commonly used by architects to describe special treatment (see Chapter 21). In Figure 24.5, an example of poché is shown.

A special transparent sheet with a series of dots or dashes is available for shading surfaces. This transparent material has a sticky backing that adheres to the working drawing. An architect or engineer uses this manufactured material because it is easy to use and is a good, consistent way to shade or highlight surfaces on working drawings.

The second variation of shading is by highlighting surfaces in isometric or oblique drawings to make them look more realistic and to achieve a three-dimensional appearance. Even though the drawing is shown on a single plane, it has a three-dimensional effect. Examples of shading by highlighting are shown in Figure 24.6.

To shade an isometric or oblique drawing, the drafter should visualize how the object would look with a spotlight trained on the object from an angle. The irregular shapes would cast shadows on the object where the rays of light did not strike the surfaces. In shading, the coloring or shading is placed on the object on the same surfaces where shadows would be cast if the object were seen in three dimensions. If the shading is done properly, the effect is the same as if the object were seen naturally with the spotlight being directed from an angle onto the object. Several examples of shading isometric and orthographic drawings are shown in Figures 17.4, 17.5, 17.6, 21.7, and 21.8.

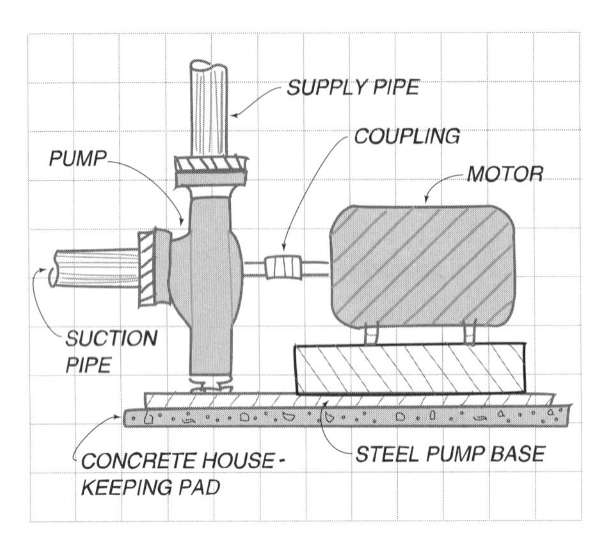

FIGURE 24.2 ■ Special treatment of surfaces (freehand)

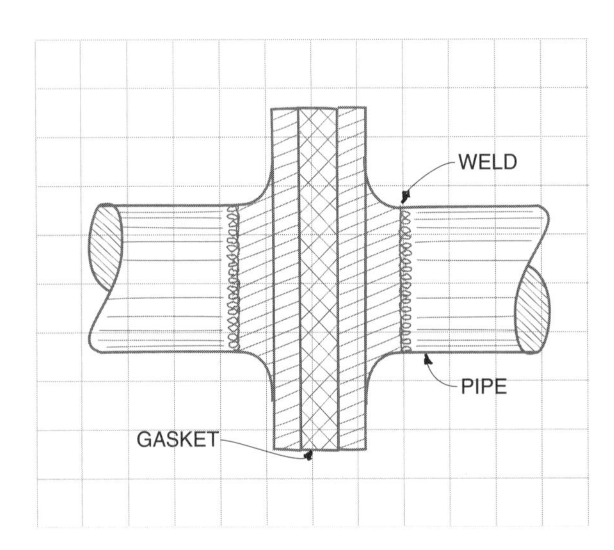

FIGURE 24.3 ■ Freehand drawing showing different surfaces crosshatched

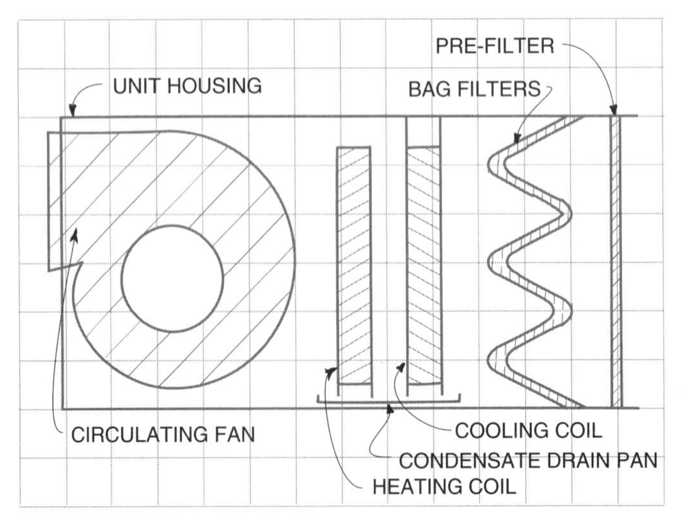

FIGURE 24.4 ■ Crosshatching various components of an air-handling unit

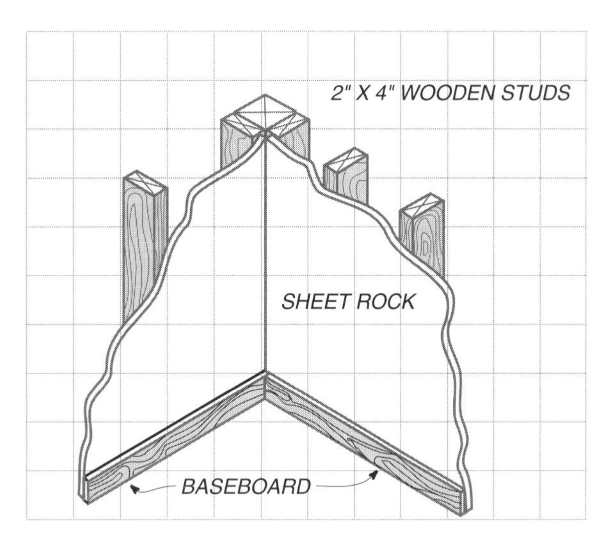

FIGURE 24.5 ■ Architectural detail with poché on wood studs

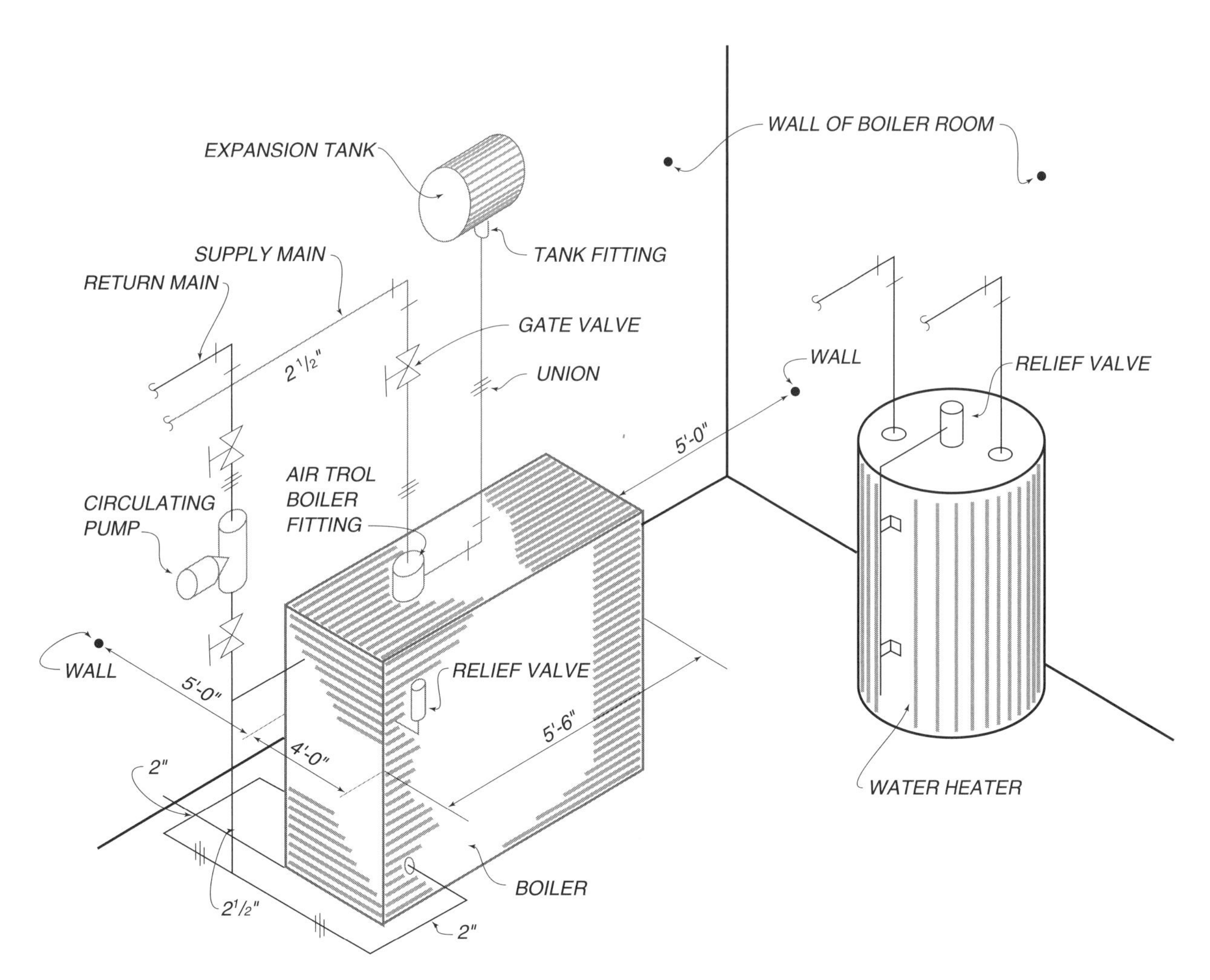

FIGURE 24.6 ■ Highlighting shading to improve drawing

SUMMARY

- Freehand sketching skills are learned by practicing learned techniques.
- Orthographic projection, oblique drawings, and isometric drawings can be drawn with freehand sketching.
- Freehand sketches are not drawn to scale, but they must be kept in proportion to show accurate relationships of the various parts of the drawing.
- Guidelines are used in hand-lettering a plan to keep letters in a straight line and about the same size.
- Planning prior to starting the actual drawing process is very important.
- Dimensions should be written on the drawings and sketches, since construction drawings cannot be scaled for accurate dimensions.
- Poché, shading, and crosshatching make various surfaces stand out on the object being drawn.

REVIEW QUESTIONS

1. What are three types of drawings that often begin as freehand sketches by the technician in the field?
2. Why must freehand sketches be in proportion even though they are not drawn to scale?
3. What is the best way to learn freehand sketching skills?
4. Why is it important to write dimensions on the plans instead of relying on reading the scale?
5. Which is usually done first: the drawing or the notes?
6. What are some of the steps to take when locating an object or building on a sheet?
7. Why does the drafter often use a commercially prepared transparent paper for shading?
8. To shade an isometric or oblique drawing, how does the drafter visualize the object?

ADDITIONAL STUDENT EXERCISES: FREEHAND SKETCHING

1. Within the confines of the enclosed space (Figure 24.7), freehand-sketch an equipment room containing a boiler, air-handling unit, and a round water heater. Use rectangles to represent the boiler and air-handling unit. The air-handling unit is to be approximately one-third larger than the boiler.
2. In the space provided in Figure 24.8, freehand-sketch an isometric drawing of an outside condensing unit, a cooling coil, and a round water heater. (Turn the sheet to 90 degrees to have the graph paper lines accommodate isometric drawings.) Shade the units to improve the appearance of the drawings.

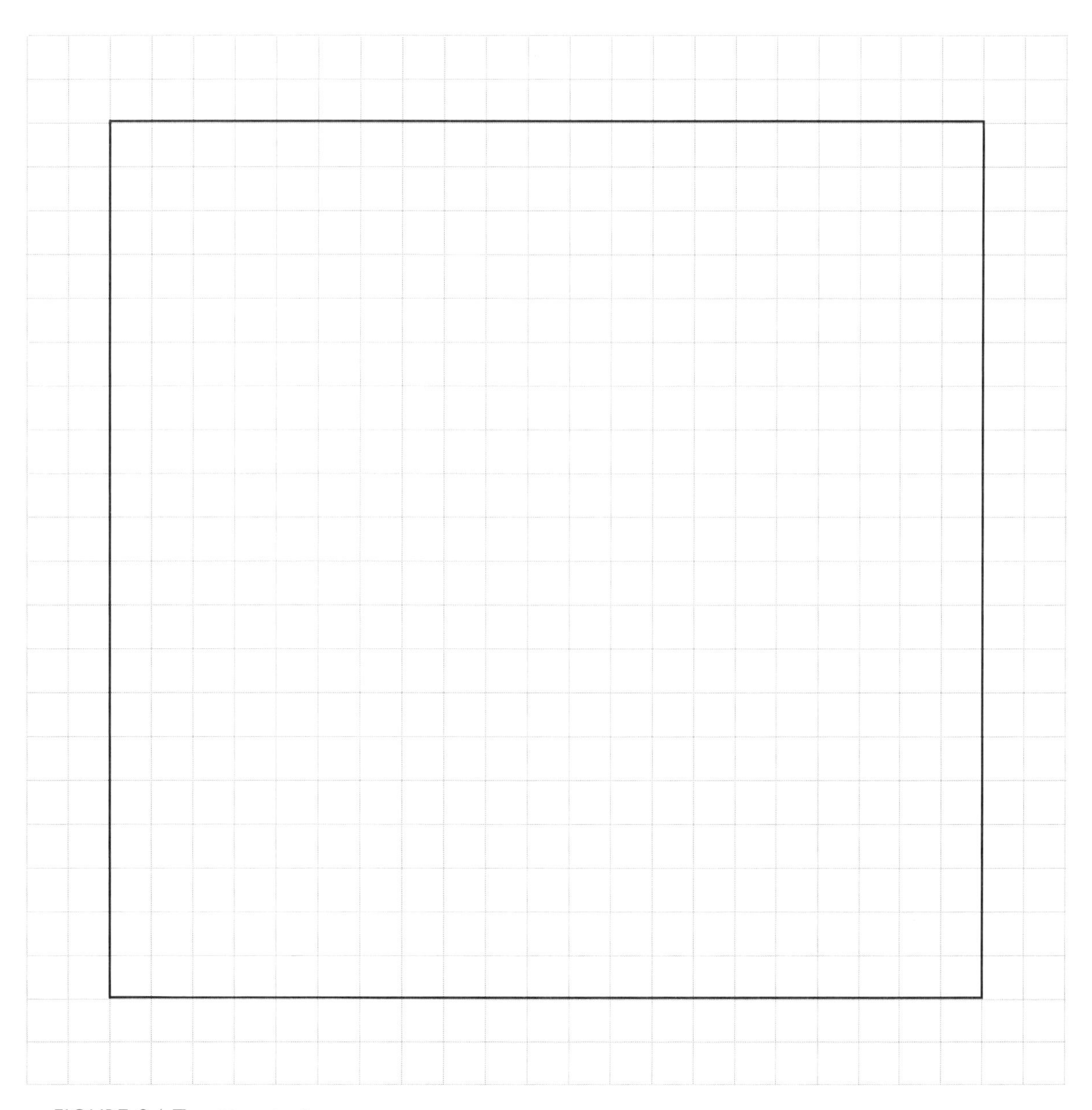

FIGURE 24.7 ■ Exercise 1

FIGURE 24.8 ■ Exercise 2

CHAPTER 25

Drafting with Instruments

OBJECTIVES

After studying this chapter, you should be able to:

- discuss reasons why design, accuracy, and speed affect the success of a project
- identify commonly used manual drafting instruments
- describe and identify time-saving procedures for drafting
- discuss and demonstrate ways to draw air-conditioning components
- read and explain a simplified plan for an equipment room
- demonstrate the ability to draw ductwork and fittings with manual drafting instruments according to directions

25.1 INTRODUCTION

The success or failure of a building construction project can often be determined by the quality of the construction plans. If the plans are correctly drawn with no dimensional mistakes and if all the elevations, sections, details, and schedules are correct, the construction personnel will have little trouble with construction. If the plans have drafting errors, dimensional mistakes, or if the drawings cannot be read, the builder will have to spend precious construction time trying to find correct answers or tearing out work that is not correct. Even at best, mistakes will be made; however, these mistakes can be held to a minimum if drawings are checked and rechecked for accuracy and clarity.

For the successful completion of any project, architects, engineers, designers, and drafters are constantly concerned with three factors: design, accuracy, and speed. This chapter will concentrate on the accuracy, good design, neatness, drafting speed, and common sense needed on the plans to relay the necessary information to the builder. Manual drafting instruments will be described, and drafting procedures will be illustrated.

25.2 COMMONLY USED MANUAL DRAFTING INSTRUMENTS

Years ago, all working drawings were produced using drawing boards and T-squares. Refer to Figure 8.1 of this textbook for an illustration of a drawing board and T-square, along with an illustration of a drawing board with a drafting machine and parallel bar. These devices are widely used in drafting rooms today to establish horizontal lines. With triangles placed on the T-square and parallel bar, vertical lines and angular lines can be drawn. The drafting machine shown in Figure 8.1 is used to draw horizontal lines and vertical lines (90 degrees off horizontal), along with any angular lines in the 360-degree field. The angular lines can be adjusted with the drafting machine head. Once the angular line has been selected, the head is locked into place so that the angular line can be drawn without the use of triangles.

Additional manual drawing instruments shown in Figure 8.2 are as follows:

1. Dividers (often called "points")—used to transfer dimensions or distances on drawings.
2. Compass—has one sharp metal point and one graphite lead point. The compass is used to draw circles and arcs.
3. Triangles—usually come in pairs (a 30/60-degree triangle and a 45-degree triangle). These triangles, used with the T-square or the parallel bar, are used to draw vertical lines and angular lines.
4. Protractor—used to measure angles (see Figure 4.9).
5. Plastic template—sheet of plastic with cutouts of circles, squares, triangles, and other geometric figures, which allows the drafter to insert a pencil or pen in the cutout and draw the figure on the plans. Using the template saves drafting time.
6. Architectural scale and engineering scale—used to reduce the size of a drawing (see Chapters 5 and 6).
7. Drafting pencils (wood and mechanical) and drafting pen—used to render a drawing.

Other manual drafting equipment includes erasing shields and lettering guides. Erasing shields are thin sheets of metal with cutouts (Figure 25.1). For erasing, the shield is placed over the portion of the drawing that is not to be erased. Through the cutout, the unprotected part of the drawing can be erased without disturbing the remainder of the drawing. Lettering guides are used to draw guidelines so that letters will have the same height. (See Chapter 19 for a discussion of lettering guides.)

25.3 DRAWING LINES ON WORKING DRAWINGS

In drafting, the use of different kinds of lines helps the blueprint reader read and interpret the drawing. These lines should be drawn straight and should end abruptly. When two lines intersect at a corner, they should end at the corner and not overshoot or leave a gap at the intersection (Figure 25.2). The quality of the lines throughout the drawing should be uniform. All object lines should be the same, and all hidden lines, center lines, extension and dimension lines, leaders, and so forth should be the same at all locations on the drawing.

Chapter 15 is devoted to drafting lines. Figure 15.1 shows common line weights (width and darkness), and Figure 15.2 is a line schedule, also known as an alphabet of lines. Other figures in Chapter 15 illustrate the use of different types of lines.

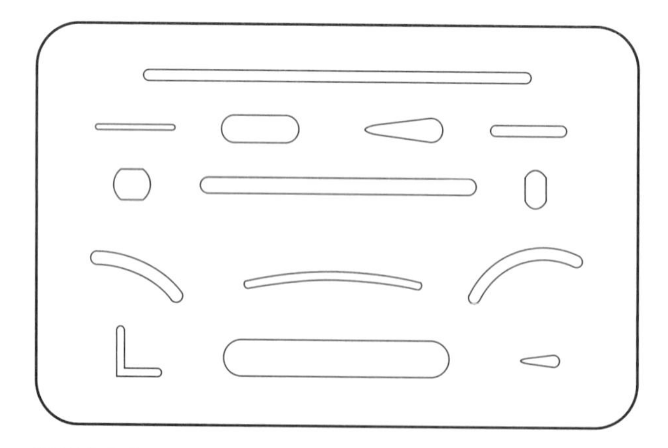

FIGURE 25.1 ■ Erasing shield

All circles and arcs should be drawn with uniform lines, and the circle should be complete, with no starting and ending points (Figure 25.3). Arcs should have definite starting and ending points.

25.4 TIME-SAVING DRAFTING PROCEDURES

With experience, the drafter learns shortcuts that save both time and effort. Figures 25.4 through 25.7 demonstrate some of these time-saving procedures.

Figure 25.4 shows how to bisect an angle using a compass. Figure 25.5 illustrates how to locate the center of a circle if the center is unknown. Figure 25.6 shows how to divide a line segment into equal parts. The example shown divides the line segment into twelve equal parts, but any number of parts can be found by changing the angle of the scale or by changing the scale.

A circle can be divided into equal segments as shown in Figure 25.7. First, the estimated distance is set on the divider or compass. The divider or compass is "walked" around a quarter of the circle. If the distance is incorrect, the divider or compass is adjusted up or down (by estimation), and the instrument is again "walked" around the quarter of the circle. This procedure is repeated until the distance is determined to be

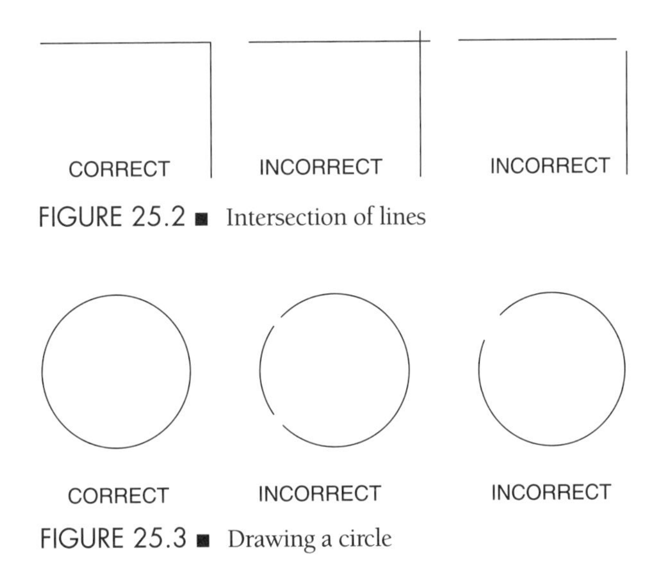

FIGURE 25.2 ■ Intersection of lines

FIGURE 25.3 ■ Drawing a circle

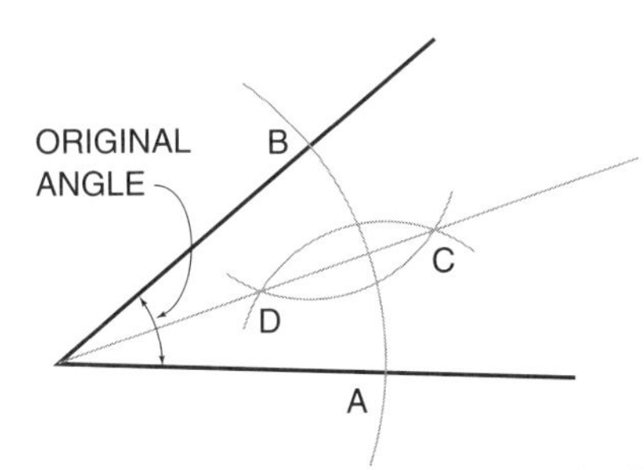

NOTE: (1) WITH A COMPASS, SWING AN ARC WITH POINT OF COMPASS ON POINT OF ANGLE. (2) SWING AN ARC WITH COMPASS POINT AT POINT "A." SWING THE SAME ARC WITH COMPASS POINT AT "B." DRAW A CONSTRUCTION LINE BETWEEN POINTS "C," "D" AND THE POINT OF THE ANGLE. THIS CONSTRUCTION LINE DIVIDES THE ORIGINAL INTO TWO EQUAL PARTS.

FIGURE 25.4 ■ Bisecting an angle

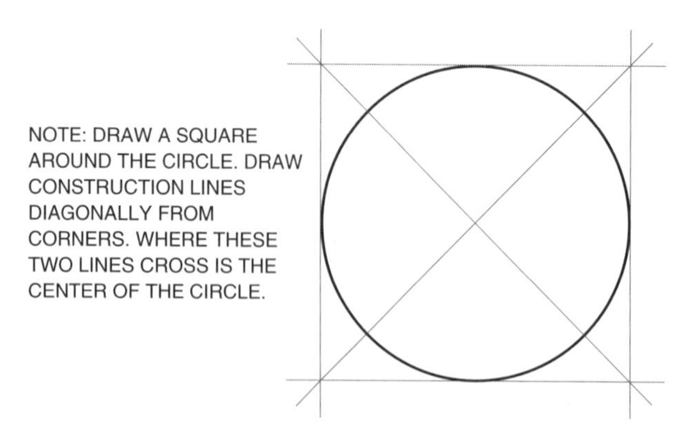

FIGURE 25.5 ■ Finding the center of a circle

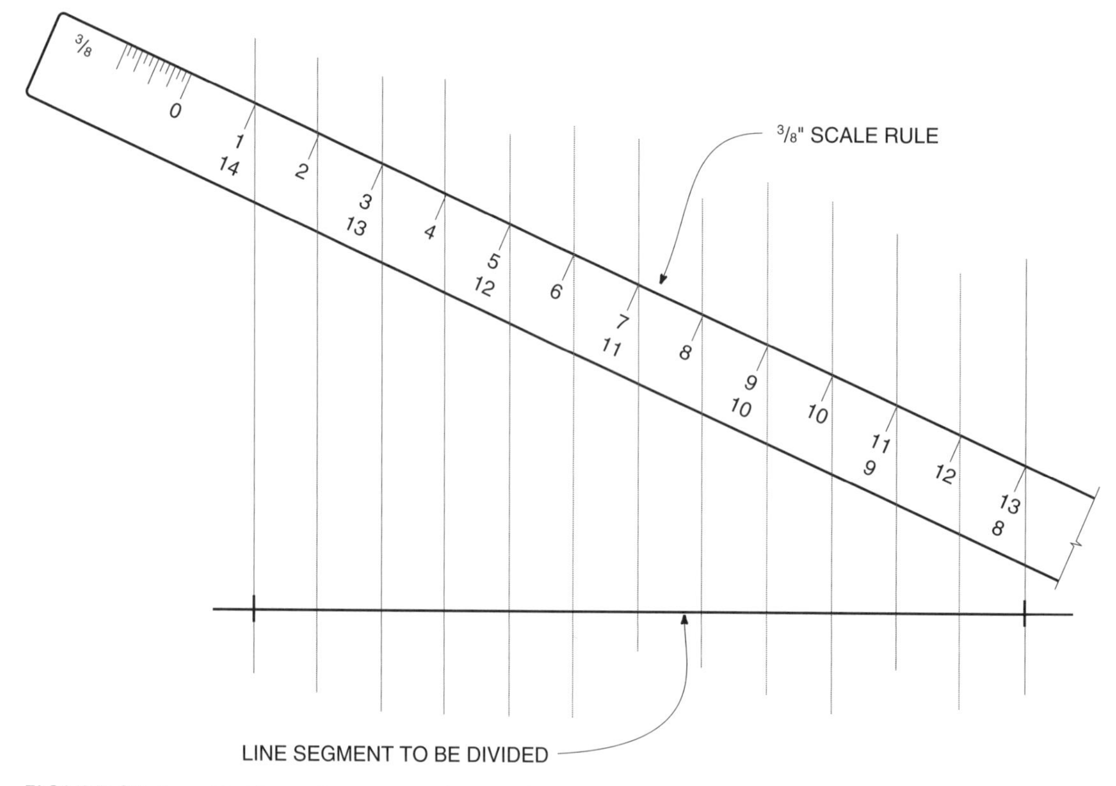

FIGURE 25.6 ■ Dividing a line into twelve equal segments

correct. To check the accuracy of the determined distance, the divider or compass is then "walked" around the entire circle. This procedure can also be used to divide line segments into equal segments.

25.5 DRAWING HEATING AND AIR-CONDITIONING COMPONENTS

The ductwork elbow is one of the most common fittings used in a duct system. Figure 25.8 shows a pictorial view of a 90-degree elbow with the various parts of the elbow identified. A 45-degree and a 30-degree elbow have the same name as a 90-degree elbow. Each side of the elbow is called a "cheek." The short radius on the inside of the ell is called the "throat," and the outside radius is called the "heel" or "wrapper."

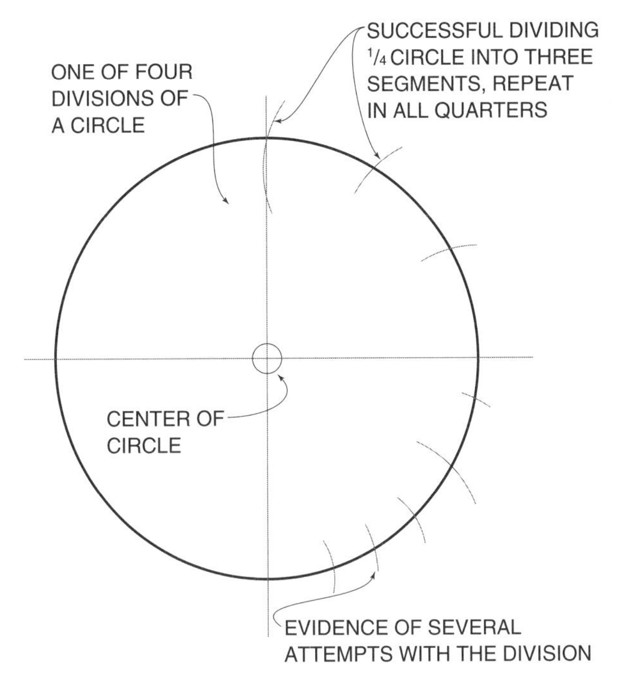

FIGURE 25.7 ■ Dividing a circle into twelve equal segments

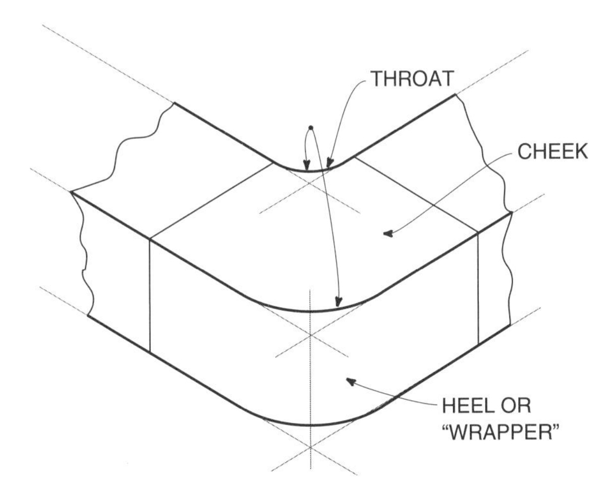

FIGURE 25.8 ■ Parts of a 90° square or rectangular elbow

The procedure for drawing the elbow is shown in Figure 25.9. The throat radius (R-1) is usually one-half the width of the ductwork, with a minimum of six inches. The R-1, or "throat," radius is drawn to connect the two sections of ductwork. The larger radius (R-2 or "heel") is drawn to connect the outside of the elbow. In some cases, a square throat is required (Figure 25.10).

A splitter fitting for a rectangular duct system is shown in Figure 25.11. Drawing a splitter fitting is very similar to drawing an elbow. The throat radius (R-1) is established, and the main duct is divided into segments representing the quantity of air to be supplied in each duct. The large radius (R-2) is drawn from this division point to complete the "heel" portion of the takeoff.

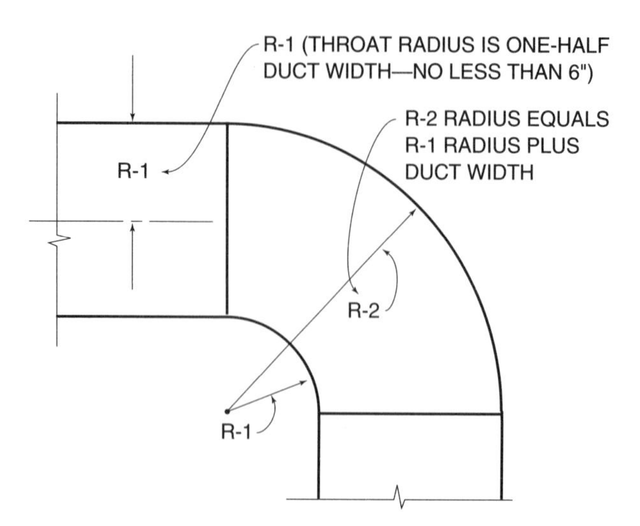

FIGURE 25.9 ■ Drawing a 90° elbow

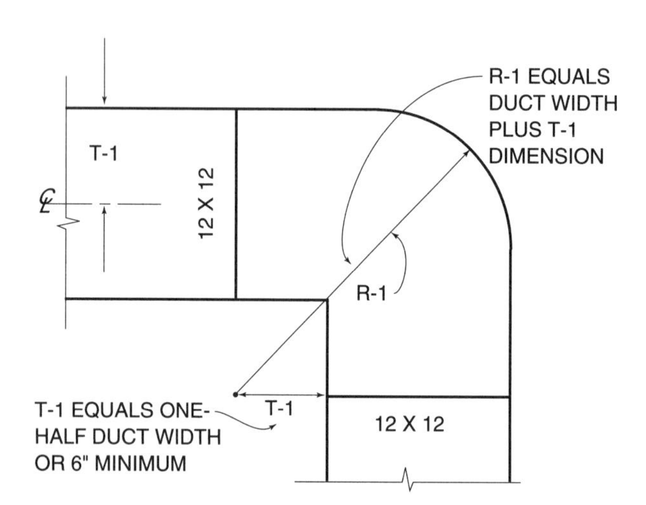

FIGURE 25.10 ■ Round elbow with square throat

The procedure for drawing the single splitter fitting is shown in Figure 25.11. A three-way splitter damper is illustrated in Figure 25.12. In both cases, the main duct is divided to represent the percentage of air to be supplied in each branch. The inside radius (R-1) is drawn, and the larger radius (R-2) is drawn to connect the divided segment of the main duct into the various segments downstream to the main duct.

The drawing procedure for a ductwork offset is shown in Figure 25.13. Here again, the small radius (R-1) is usually set at six inches, and the large radius (R-2) is set as required to accommodate the duct size without restricting the air flow. A straight line is used to connect the arcs and to complete the fitting.

The ductwork transition can reduce the size on one side, both sides (central transition), bottom, or top. The side shown on the plan (plan view or elevation view) shows how the transition is made on that side, but the other transitions must be noted (FOT = flat on top, FOB = flat on bottom). See Figure 25.14 for an example of a ductwork transition that is flat on top.

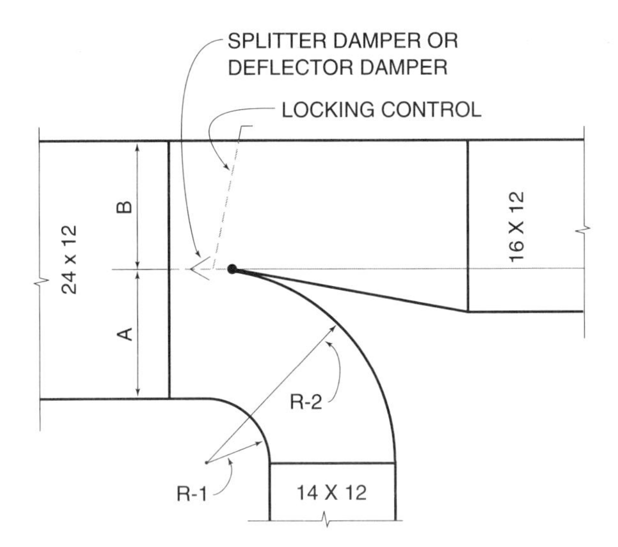

FIGURE 25.11 ■ Single splitter fitting

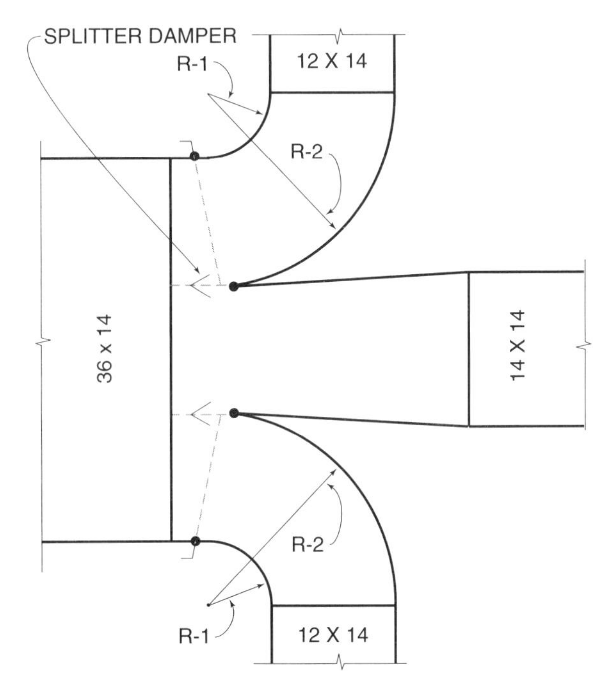

FIGURE 25.12 ■ Three-way splitter damper

25.6 EQUIPMENT ROOM PLAN

Figure 25.15 is an example of a simple mechanical equipment room with a boiler, air-handling unit, hot water circulating pump, expansion tank, and hot water piping. Although the plan does not show all details that are required to complete the system, enough information is given to locate the major equipment. The general location of the hot water piping within the equipment room is given. The ductwork has been shown, along with major fittings and duct sizes.

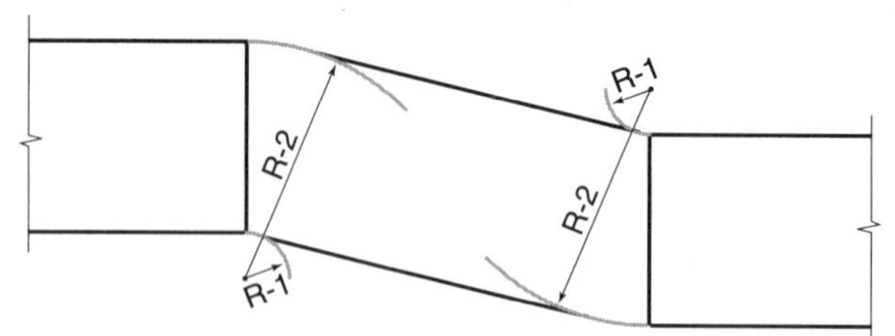

FIGURE 25.13 ■ Drawing an offset fitting

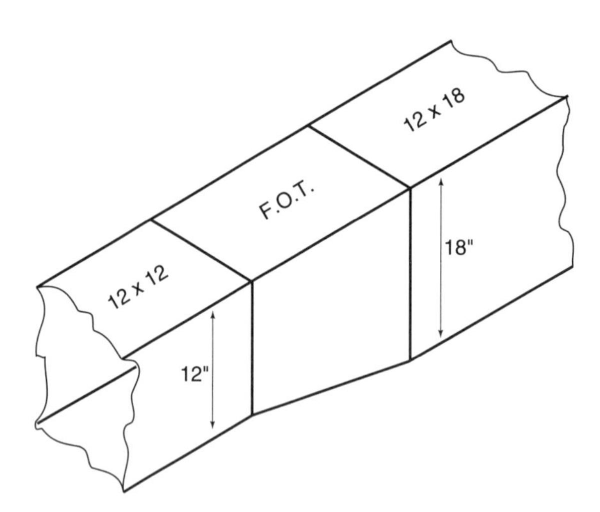

FIGURE 25.14 ■ Ductwork transition that is flat on top

Major dimensions are given so that the equipment can be located in the space. Notice that service space has been provided around each unit. Housekeeping pads have been shown under each piece of equipment, and floor drains (installed by the plumbing contractor) have been shown so that piped drain piping can be extended from the equipment by the mechanical contractor.

The drafter will be required to draw sections and details of these pieces of equipment to show specialties, such as the safety valve and control devices on the boiler, the cold water makeup for the boiler, the hanging of the expansion tank, the boiler fitting, and piping connections to the boiler.

The piping connection to the pump should be detailed so that the valving method for mounting and other details can be shown.

The air-handling unit should have details to show how the hot water piping is connected to the heating coil. A separate detail of the heating coil may be necessary to show the coil connections that would include the controls, bypass, valves, and so forth.

The air-handling unit would normally have a cooling coil and chilled water or refrigerant piping. For clarity in the description of the drawing, this cooling system has been omitted.

All lines on this plan were drawn with the drawing instruments described in Section 25.2 and illustrated in Chapter 8. All the straight lines were drawn with a parallel bar and triangles. Circles representing piping elbows turned up and down were drawn with a circle template. The arcs necessary to draw the ductwork elbows were drawn with a circle template and with a compass.

SUMMARY

- Working drawings can be drawn manually with a T-square and drawing board or with a drafting machine and parallel bar.
- Other manual drawing instruments include dividers, compasses, triangles, protractors, templates, erasing shields, and lettering guides.
- Lines on drawings should conform to a standard Table of Line Weights and Alphabet of Lines.
- Experienced drafters learn time-saving procedures by using compasses, dividers, and other manual drafting instruments.
- The drafter can gain valuable experience by manually drawing the various components of a heating and air-conditioning system.

REVIEW QUESTIONS

1. Define and describe the following drafting instruments:
   A. T-square
   B. Drafting machine
   C. Parallel bar
   D. Dividers
   E. Compass
   F. Triangles
   G. Template
   H. Scale rules
   I. Drafting pencil
   J. Erasing shield
   K. Lettering guide
2. Formulate directions for the following procedures, and demonstrate each with an example:
   A. Drawing straight lines that intersect at a corner
   B. Drawing a circle
   C. Bisecting an angle using a compass
   D. Dividing a line segment into several parts
   E. Dividing a circle into equal segments by using a compass
3. Draw a ductwork elbow and then identify the parts.
4. Describe the procedure for drawing the following:
   A. Single splitter fitting
   B. Three-way splitter fitting
5. List the major components of a simple mechanical equipment room.

ADDITIONAL STUDENT EXERCISES: DRAFTING WITH INSTRUMENTS

1. Refer to Figure 25.15 to answer the following questions:
   A. Is there a chimney on this drawing?
   B. Is there an expansion tank?
   C. Is there a housekeeping pad under the boiler?
   D. Are flexible duct connections shown on ductwork?
   E. Is there a splitter damper in the supply duct?
   F. Does the boiler supply hot water to the second floor?
   G. How many gate valves are shown on this drawing?
   H. Are any unions shown?
   I. Is the ductwork located underfloor?

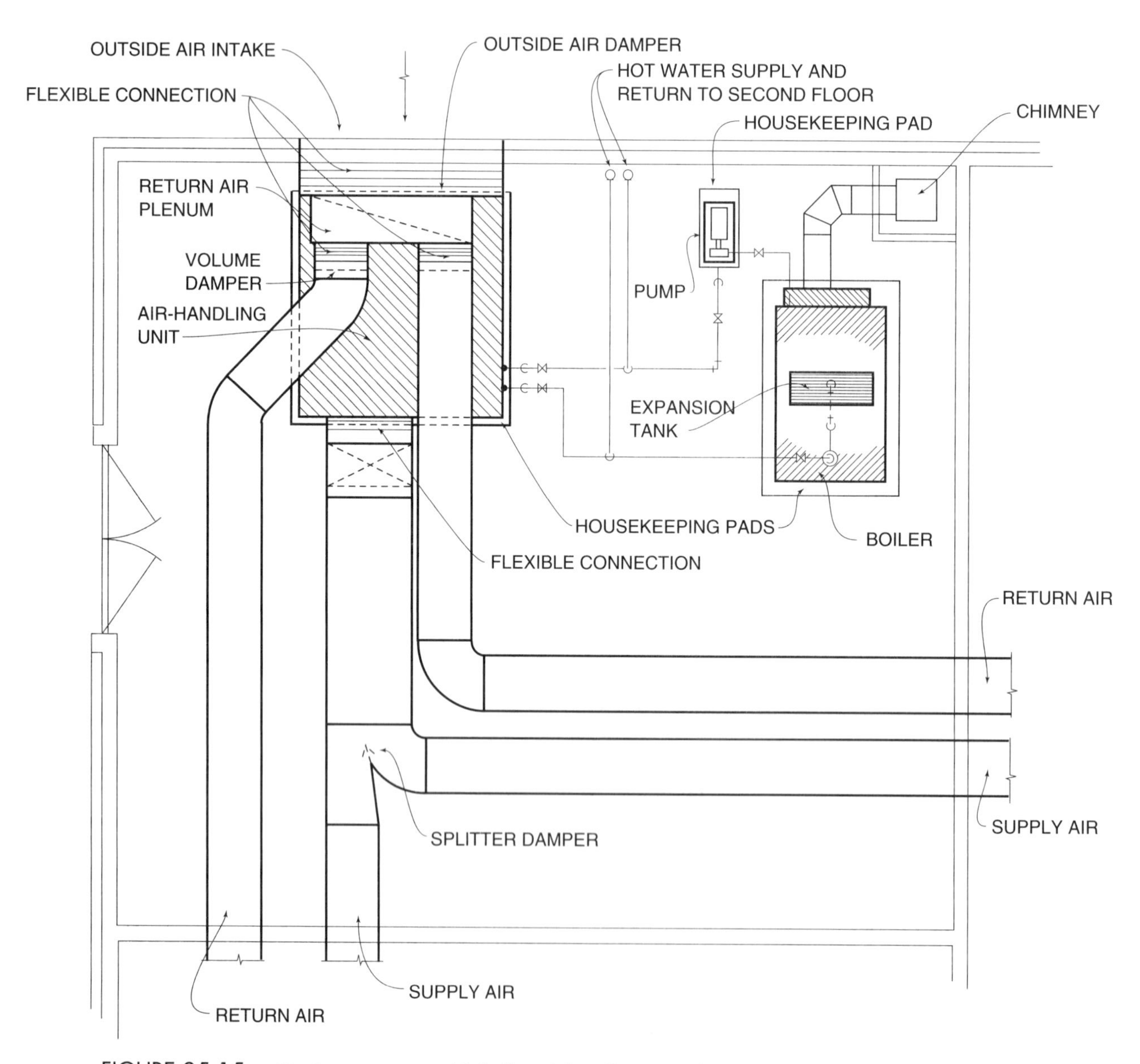

FIGURE 25.15 ■ Equipment room with boiler, air-handling unit, hot water pump, expansion tank, ductwork, and hot water piping

2. (A) Draw a 90-degree elbow on the end of the indicated duct in Figure 25.16A. The elbow is to have a round "throat" and a round "heel."

   (B) Figure 25.16B is an offset section of ductwork. Swing the minimum and the maximum arcs, and connect these arcs with a straight line.

   (C) Using Figure 25.16C, complete the transition that is to be flat on top.

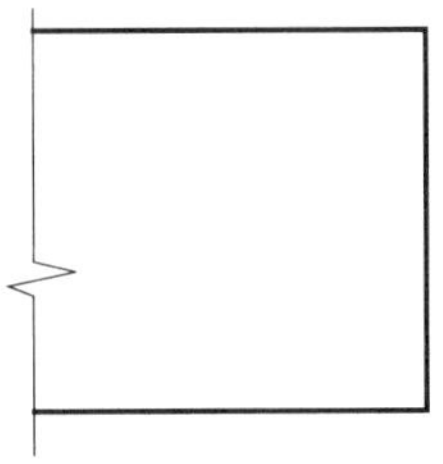

(A) 90° DEGREE, ROUND THROAT ELBOW. THROAT RADIUS IS EQUAL TO 1/2 THE WIDTH OF DUCT.

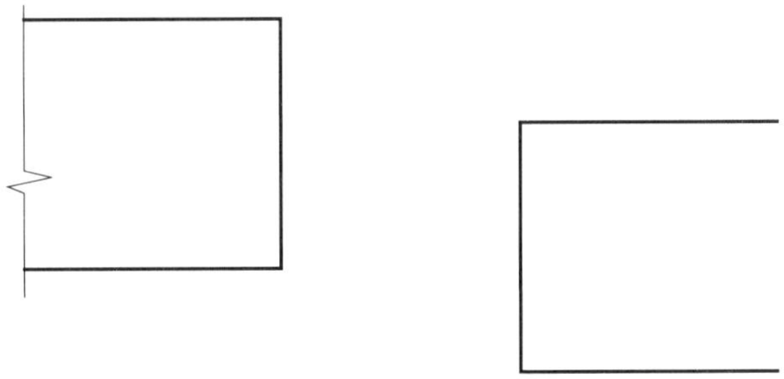

(B) OFFSET WITH THE SMALL RADIUS EQUAL TO 1/2 THE WIDTH OF DUCT.

(C) DOUBLE TRANSITION WITH THE DUCT FLAT ON TOP.

FIGURE 25.16 A–C ■ Exercise 2

3. Refer to Figure 25.17. Draw the indicated 90-degree elbows, the splitter fitting, the offset, and the transition.

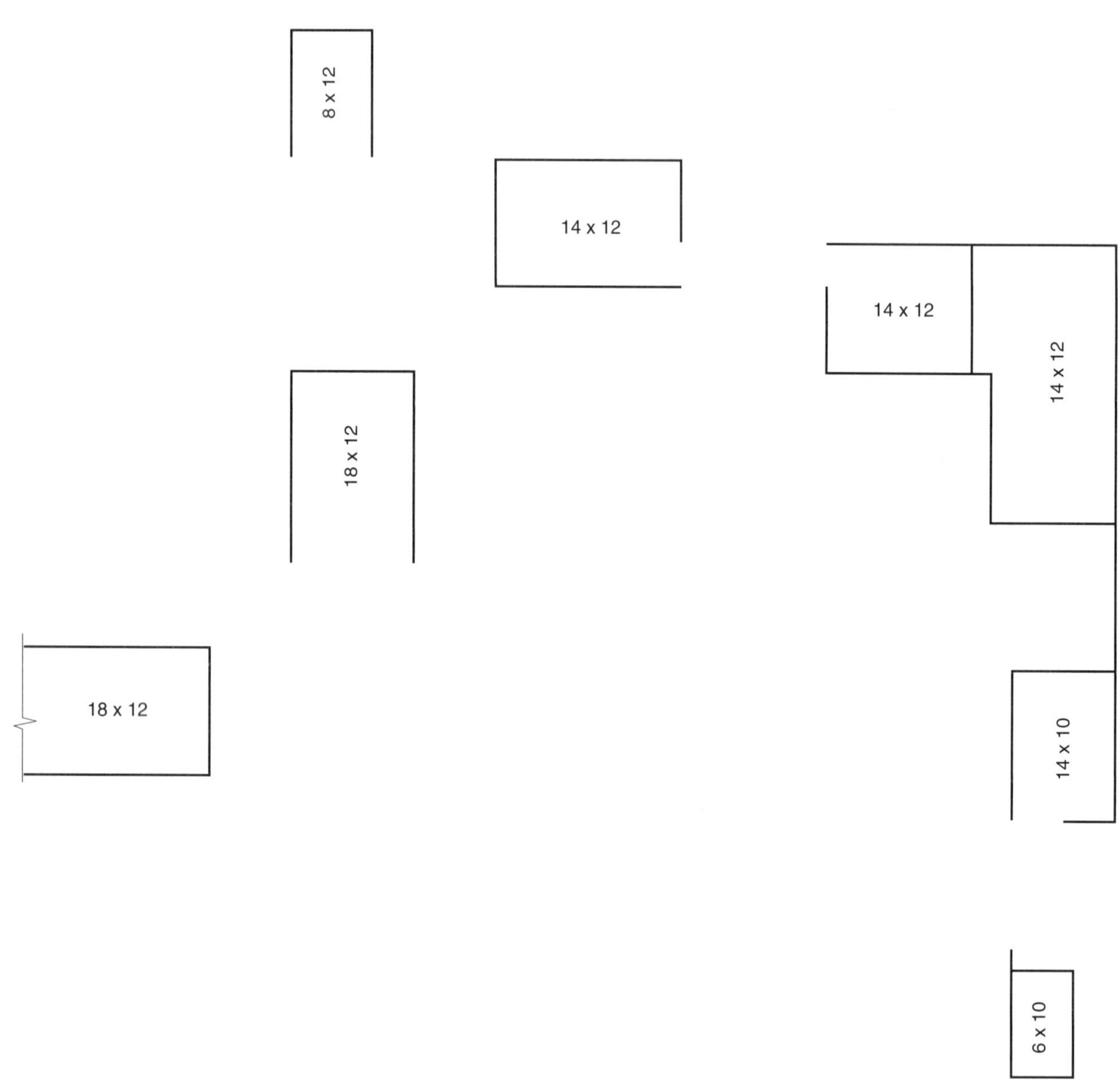

FIGURE 25.17 ■ Exercise 3

4. Refer to Figure 25.18. Use the symbols shown in Appendix D to draw the piping connections to the heating coil. Connect the heating coil with a three-way valve in the H.W. "OUT" line. Show a union in the hot water connection "IN" and "OUT." The bypass line off the three-way valve shall have a union and a ball valve for balancing flow. Show a gate valve in the "OUT" line, upstream of the three-way valve. Install a globe valve in the H.W. line upstream of the bypass line of the three-way valve.

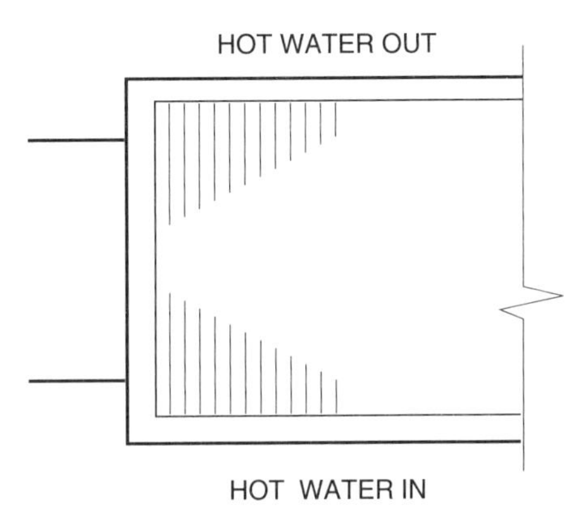

FIGURE 25.18 ■ Exercise 4

PART III

READING AND INTERPRETING ARCHITECTURAL PLANS

CHAPTER 26

Reading Architectural Plans

OBJECTIVES

After studying this chapter, you should be able to:

- identify parts of the architectural floor plan
- describe the information found on the roof framing plan
- locate and discuss the importance of the transverse and longitudinal sections
- locate and discuss the importance of the wall sections and details

---

NOTE: Refer to the plans for "A Small Bank and Trust Company" (at the end of this textbook) while studying this chapter.

26.1 INTRODUCTION

An HVAC technician needs to be able to read and interpret complete architectural plans. In reviewing these plans, it is suggested that each sheet be read similar to the way one reads a book. The technician should visualize the sheet in squares of approximately six inches and then read each section from left to right for a general idea of the information in that part. When all parts of the sheet have been reviewed in this manner, the technician will have a general idea of what information is given on the sheet, what the building looks like, and what systems have been designed for the building. In this chapter, students are given an opportunity to look at actual architectural plans to find various features, details, building materials, and construction procedures for a building.

26.2 SITE PLAN AND LOCATION PLAN

The site plan and location plan are shown on Sheet 1 of the plans at the end of the book. The larger drawing is the location plan. This plan gives the location of the project by listing the U.S. highway and state route intersection. (The highway and route numbers have been omitted.) This plan also shows the proposed shopping center to be built on the property as well as the present and future drives.

The site plan shows the building located on the site along with the drives, parking, planters, and so forth. Contour lines are shown along with the benchmark. The benchmark is a known elevation point, usually a concrete pad or a metal stake driven in the ground. This benchmark determines all elevations needed in the grading of the site and all elevations to be used for the building.

26.3 FOUNDATION PLAN

The foundation plan shows the foundation for the building. Dimensions are given for the outside walls of the building, and sections are indicated so that the builder will know how to install the building foundation. This plan is critical because if the foundation is not properly installed, the building walls may fail. If the dimensions are not correct, the building may not fit the requirements of the plan.

26.4 ARCHITECTURAL FLOOR PLAN

The architectural floor plan is shown on Sheet A-1 at the end of this textbook. This sheet contains information about the various features of the building, such as walls, windows, doors, floors, overhangs, types of walls, plumbing fixture locations, and so on. The HVAC design technician must be familiar with all these items in order to design a suitable system. The installing technician must know how the building is put together and where the various items are to be located for maximum efficiency.

On Sheet A-1, sectional drawings show where sections have been cut through the walls. The sections are indicated by circles with arrows pointing in the direction that the section is cut. For example, there are three sections through the right-hand wall and several sections through the front wall. These are shown on Sheets A-4 and A-5 and will be discussed later in this chapter.

26.5 WALL SECTIONS

On Sheet A-2, wall sections are cut to allow elevations to be shown on Sheets A-4 and A-5. On the elevations sheets, the HVAC technician can measure the square footage of wall spaces for windows, doors, overhangs, and other features of the building that affect the heating and air-conditioning load. On the front and rear elevations of this set of plans, a note under the title indicates that the front and rear elevations are the same except they are reversed. This is true of many small buildings, such as small bank buildings. The front and rear are the same, but the two side elevations are different, as can be seen by examining the plans.

26.6 ROOF FRAMING PLAN

Sheet A-3 shows the roof framing plan. This is the structural plan. The sizes of steel members are given, along with the locations of steel joists. This information is necessary for the person designing the system, who needs to know whether to hang the ductwork under the steel joists or between them. The joists are the steel structural members that support the roof. In some cases, air-conditioning units are located on the roof, and steel must be sized large enough to carry the extra weight of the air-conditioning units. The sections shown through the building are located on Sheet A-5.

26.7 TRANSVERSE AND LONGITUDINAL SECTIONS

The transverse sections and longitudinal sections give a cross-section of the building as it will be constructed. These sections show what the steel looks like and what spaces are available for ductwork and piping. The transverse and longitudinal sections also give ceiling heights, the construction and locations of beams, and the location of the roof floor. They show exactly what the building looks like as it is cut along the cutting plane. These transverse and longitudinal sections are a very important part of the architectural plan.

26.8 WALL SECTIONS AND DETAILS

Sheet A-6 is titled Wall Sections and Details. These sections, which have been referred to on other sheets, show how the various components are put together to form the building. Sheet A-6 gives elevations and dimensions on a larger scale so that the builder knows exactly how the building is to go together. The HVAC technician, whether designing the system or installing it, uses these parts of the plan to decide where to put ductwork and how to install it properly.

26.9 ENLARGED FLOOR PLANS AND INTERIOR ELEVATIONS

An enlarged floor plan and the elevation of the tellers' area are shown on Sheet A-7. This sheet shows the design of the tellers' stations and includes details and dimensions for the cabinetmakers. This drawing helps the HVAC contractor design the heating/cooling system that will best serve the tellers.

26.10 SCHEDULES, SECTIONS, AND ELEVATIONS

The Room Finish Schedule, the Door Schedule, and the Color Product Schedule are shown on Sheet A-8. The partial plan for the public lobby glass wall and doors is shown, along with the elevation of the wall. Other details for other parts of the lobby wall are also shown.

SUMMARY

- The location plan gives details as to the intersection where the building is to be located.
- The site plan shows how the building is to be placed on the site, along with drives, parking areas, planters, and other features.
- The benchmark is an elevation point to which all elevations are referenced.
- The foundation plan shows dimensions of the building, with sections that show how the foundation is to be built.
- The architectural floor plan shows various features of the building and includes sectional drawings.
- Wall sections give dimensions of windows, doors, and other features of the building.
- The roof framing plan gives information about the structural requirements of the building.
- Transverse and longitudinal sections give cross-sections of the building at various locations.
- Schedules give specific information on doors, colors, and other details of the building.

REVIEW QUESTIONS

1. Describe the method suggested in Section 26.1 for getting an overview of the plans.
2. What kind of information is given on the location plan?
3. What kind of information is given on the site plan?
4. What types of information are given on the wall sections?
5. Why is the information on the roof framing plan especially important to the HVAC technician?
6. Describe the kinds of information given on the transverse and longitudinal sections.
7. What are some other parts of the plans that furnish important information to the builders and contractors?

ADDITIONAL STUDENT EXERCISES: READING ARCHITECTURAL PLANS

A. The Site Plan—Sheet 1

Using the set of building plans at the end of this textbook, answer the following questions or perform the described tasks for the site plan (Sheet 1).

1. Which is closer to the bank building: the U.S. highway or the state route?
2. On which side of the building is the drive-up window located: north, south, east, or west?
3. Have parking spaces been provided for customers of the bank?
4. What kind of walkway leads to the bank entrances?
5. What is the width of the power line easement?
6. What is the finished floor elevation of the building?
7. Is there any existing pavement on the site other than the road and highway?
8. Does the shopping center exist or is it proposed?
9. Is there an exit/entrance on the north side of the building?
10. What is the size of the existing water main?
11. Is the size of the existing sanitary main given?
12. Are standard symbols used for the domestic water main and the sanitary water main?
13. Draw an example of the standard symbols for the domestic water main.
14. Draw an example of the standard symbol for the sanitary sewer main.
15. How many sanitary manholes are shown on the site plan?
16. Is the sanitary sewage pumped up to the sanitary sewer main?
17. How deep should the concrete curb be below the finished pavement?
18. Where is the sewage lift pump located?
19. How many brick panels are shown for each walkway to the building?
20. What is the elevation of the benchmark?
21. How big is the plot of land that the bank is to be built on?
22. Who furnished information about the site to the architect?
23. What is the elevation of the drive-up window pavement?
24. How much higher is the bank floor than the drive-up window pavement?
25. Can any part of the building be located on the power line easement?

B. The Foundation Plan—Sheet 2

Using the set of building plans at the end of this textbook, answer the following questions or perform the described tasks for the foundation plan (Sheet 2).

1. What section number shows the column located near the center of the building? (Find the section drawing and then list the numbers—in the top and bottom half of the circle—that indicate that section.)

2. What is the size of the concrete footing for the column near the center of the building?
3. What do the broken lines shown on each side of the foundation wall represent?
4. Do the columns supporting the canopy over the drive-up window have the same type of footing as the column inside the building?
5. What type of construction is used in the vault floor?
6. How thick is the vault floor?
7. What is the size of reinforcement bars used in the construction of the vault?
8. What is the door that is shown inside the vault used for?
9. Is carpeting shown in the vault?
10. What is the size of the concrete footings under the vault?
11. How many reinforcement bars are required in the vault footings, and what size are they?
12. What is the wall construction around the drive-up window?
13. What is the wall construction around the depository?
14. How thick is the wall around the depository?
15. Why is the wall under the depository reduced to 8" thick?
16. What material is used for foundation walls?

C. The Floor Plan—Sheet A-1

1. How many teller stations are shown?
2. What are the inside dimensions of the conference room?
3. How much is the floor slab depressed to accommodate the brick pavers at the tellers' counter?
4. How many check desks are shown?
5. In what room is the roof access ladder located?
6. What are the dimensions of the lounge (Room 102)?
7. Does the kitchen unit have a sink?
8. Does the building have two entrances?
9. What scale was used to draw the original working drawing? (Remember that the plans in your textbook have been reduced, so the scale is no longer valid.)
10. What scale was used to draw the elevation in the employee lounge?
11. What scale was used to draw the detail showing the depression for brick pavers in the lobby?
12. What kind of columns is used in the building?
13. What is the overall width of the entrance?
14. How is the roof overhang shown?

D. The Elevation Plan—Sheet A-2

1. What material is used for the fascia?
2. From what material is the gravel stop fabricated?
3. What kind of glass is used?

4. How thick are the glass panels?
5. What is the special brick that is installed between the brick panels on the outside wall?
6. Looking at the brick details, describe a sailor course.
7. Is the drive-up window unit a manufactured product?
8. Who installs the drive-up window unit?

E. The Structural Plan—Sheet A-3

1. What are the sizes of the two structural beams located near the center of the building?
2. What is the size of the pipe columns?
3. What is the size of the steel joists used in the center of the building?
4. How many of these joists are required?
5. What is the size of the steel joists used in the outer portion of the building?

F. The Reflected Ceiling Plan—Sheet A-4

1. What type of ceiling is used in the main part of the building?
2. What material is used around the perimeter of the ceiling?
3. The round circle with double crosshatching at 90 degrees is a symbol representing speakers. How many speakers are shown on this job?
4. What is the ceiling height in the main lobby?
5. What is the ceiling height over the teller area?
6. What is the ceiling height in the storage area?
7. What is the ceiling height of the outside soffit?
8. On this sheet, what materials are shown in the outside wall section?

G. Transverse and Longitudinal Sections—Sheet A-5

1. Does the cashier's cabinet exist on these sections?
2. Are hangers for the acoustical ceiling shown?
3. What scale was used to draw these sections? (Remember that the scale is no longer accurate because plans have been reduced to fit in this textbook.)
4. Is the drive-up window shown?
5. Is compacted fill shown on the plan?

H. Architectural Sections—Sheet A-6

1. What is the finish material on the fascia?
2. In Section 5/A-1, A-2, and A-3, what is the size of the steel beam structural member that supports the bar joists?
3. What is the size of the bar joists shown in Sections 4 and 5?

4. Are bar joists used in the soffit construction?
5. What type of insulation is shown on the roof?
6. What kind of roof deck is shown?
7. What is meant by sailor coursing of brick?

I. Elevations—Sheet A-7

1. Has knee space been provided for the tellers?
2. Are foot rests provided?
3. How many tellers' stations are shown?

J. Schedules—Sheet A-8

1. What information is shown on the Room Finish Schedule?
2. What information is given on the Product Schedule?
3. What information is given on the Door Schedule?
4. Why is the elevation of a wall, along with a plan for the wall, in the public lobby shown on this sheet?

CHAPTER 27

Ductwork Plans

OBJECTIVES

After studying this chapter, you should be able to:

- read and use a sheet metal chart to determine what gauge metal to use for various sizes of ductwork
- explain how linear measurements on ductwork and fittings are obtained
- use a ductwork takeoff form correctly
- read and use a chart for measuring ductwork insulation
- discuss the various factors that determine the cost of ductwork and insulation, including the waste factor and labor costs

27.1 INTRODUCTION

For the majority of air-conditioning and heating jobs, ductwork is included as part of the installation. Several kinds of ductwork are commonly used in today's market, including conventional sheet metal ducts, fiberglass duct-board ducts, and flexible ducts. Special duct systems can be fabricated from fiberglass and other modern materials.

This chapter will discuss a method for estimating the amount of sheet metal required to fabricate ductwork in the shop for a job. The discussion will include how to take off (measure) the ductwork from a set of plans and how to estimate the number of pounds of metal required to fabricate the ductwork. The method recommended in this chapter has been proven to be satisfactory and simple to use.

The estimating of prefabricated sheet metal ducts, fiberglass duct-board ducts, and flexible duct materials will not be included in this chapter. Manufacturers of these products suggest methods for estimating such duct systems.

An approved method for estimating the amount of insulation required for a ductwork system will also be discussed in this chapter. There are several methods for estimating ductwork insulation. The method described in this chapter is a proven system that is simple to use.

27.2 WHAT KIND OF DUCTWORK IS SHOWN ON THE PLANS

The kind of ductwork to be used on the job is usually noted and described on the plans or described in the specifications. In some cases, the kind of ductwork is described in both places. One set of construction plans may show several different kinds of ductwork.

Sheet Metal Duct Estimating Factors Chart

1	2	1	2	1	2	1	2	1	2	1	2	1	2	1	2
0–12" 26 GA.															
12	2.00	30	6.5	56	12.13	55	13.75	81	20.25	107	26.75	77		103	
13	2.17	31	6.72	57	12.35	56	14.0	82	20.5	108	27.0	78	22.8	104	30.4
14	2.34	32	6.94	58	12.56	57	14.25	83	20.75	109	27.25	79		105	
15	2.50	33	7.15	59	12.80	58	14.5	84	21.00	110	27.5	80	23.4	106	30.9
16	2.67	34	7.37	60	13.0	59	14.75	85	21.25	111	27.75	81		107	
17	2.84	35	7.59	**31–60" 22 GA.**		60	15.0	86	21.5	112	28.0	82	23.9	108	31.5
18	3.0	36	7.8	36	9.0	61	15.25	87	21.75	113	28.25	83		109	
19	3.17	37	8.02	37	9.25	62	15.5	88	22.0	114	28.5	84	24.5	110	32.1
20	3.34	38	8.24	38	9.5	63	15.75	89	22.25	115	28.75	85		111	
21	3.5	39	8.45	39	9.75	64	16.0	90	22.5	116	29.0	86	25.1	112	32.7
22	3.67	40	8.67	40	10.0	65	16.25	91	22.75	117	29.25	87		113	
23	3.84	41	8.89	41	10.25	66	16.5	92	23.0	118	29.5	88	25.7	114	33.3
24	4.0	42	9.1	42	10.5	67	16.75	93	23.25	119	29.75	89		115	
13–30" 24 GA.		43	9.32	43	10.75	68	17.0	94	23.5	120	30.0	90	26.3	116	33.9
		44	9.54	44	11.0	69	17.25	95	23.75	**61–90" 20 GA.**		91		117	
19	4.11	45	9.76	45	11.25	70	17.5	96	24.0			92	26.7	118	34.4
20	4.34	46	9.97	46	11.5	71	17.75	97	24.25	67	19.66	93		119	
21	4.55	47	10.2	47	11.75	72	18.0	98	24.5	68	19.9	94	27.4	120	35.0
22	4.77	48	10.42	48	12.0	73	18.25	99	24.75	69		95		121	
23	1.97	49	10.64	49	12.25	74	18.5	100	25.0	70	20.4	96	28.0	122	35.6
24	5.2	50	10.85	50	12.5	75	18.75	101	25.25	71		97		123	
25	5.42	51	11.85	51	12.75	76	19.0	102	25.5	72	21.0	98	28.6	124	36.2
26	5.64	52	11.29	52	13.0	77	19.25	103	25.75	73		99		125	
27	5.85	53	11.5	53	13.25	78	19.5	104	26.0	74	21.6	100	29.2	126	36.8
28	6.07	54	11.72	54	13.5	79	19.75	105	26.25	75		101		127	
29	6.29	55	11.91			80	20.0	106	26.5	76	22.1	102	29.8	128	37.3

Key

Columns numbered 1 are the total of duct dimensions: width plus height or W + H = column no. 1 (see text).
Columns numbered 2 are the weight of metal required to make one linear foot of the indicated duct size (see text).
Additional weights must be added to the totals to account for waste from fabrication (see text).

FIGURE 27.1 ■ Duct takeoff factors

27.3 SHOP-FABRICATED SHEET METAL DUCTWORK

Large air-conditioning and heating contractors have their own sheet metal shops that produce all the ductwork used by the company.

To estimate the cost of fabricating and installing ductwork, a dependable method must be developed. Figure 27.1 is the Sheet Metal Duct Estimating Factors Chart. This chart, properly used, helps a drafter or takeoff person to estimate the cost of a duct system.

27.4 USING THE DUCT FACTOR CHART

The Duct Factor Chart (Figure 27.1) is used to aid the estimator in determining the correct metal gauge and weight per linear foot for duct sizes. Figure 27.1 has eight columns labeled "1" and eight columns labeled "2." The numbers listed in Column 1 are obtained by adding together the two dimensions of a square or rectangular duct. The numbers in Column 2 represent the weight of metal that is required to make one foot of ductwork of the size given and the gauge metal selected.

The gauge metal to use is determined by the larger dimension of the rectangular duct size. (For example, the "larger dimension" of a duct 24" × 36" is 36 inches.) The gauges are given in the lined-in rectangular blocks on the ductwork chart. Four different gauges are shown on the chart: 26 gauge (for larger duct size dimension of 0"–12"), 24 gauge (for larger duct dimension sizes of 13"–30"), 22 gauge (for larger duct dimension sizes of 31"–60"), and 20 gauge (for larger duct dimension sizes of 61"–90"). Metal gauges refer to the thickness of the metal. Figure 27.2 is a table of metal gauges, thicknesses, and weights per square foot.

The following example demonstrates the way to use the duct factor sheet. A duct 12 inches by 14 inches is described as 12" × 14". (The 12 inches refers to the side of the duct that faces the viewer.) The total of these dimensions is 12" + 14" = 26". This 26" is found in Column 1 on the chart (see Figure 27.1).

Since the larger dimension of the duct size is 14", the duct should be fabricated from 24-gauge metal. Move from the figure in Column 1 (at the number 26) to the right, to Column 2. The number 5.64 is found in Column 2 beside the 26 in Column 1. This number (5.64) represents the pounds of metal required to make one linear foot of 12" × 14" ductwork using 24-gauge metal.

This same procedure is followed for each size of ductwork shown on the plans.

Galvanized and Hot-Rolled Mild Steel Sheetmetal Thickness and Weights				
U.S. Gauge	Thickness Inches		Weight Pounds per Square Foot	
	Hot-Rolled	**Galvanized**	**Hot-Rolled**	**Galvanized**
24	.0239	.0276	1.00	1.5625
22	.0299	.0336	1.25	1.40625
20	.0359	.0396	1.50	1.65625
18	.0478	.0516	2.00	2.15625
16	.0598	.0635	2.50	2.65625
14	.0747	.0785	3.125	3.28125
12	.1046	.1084	4.375	4.53125
10	.1345	.1382	5.625	5.78125

FIGURE 27.2 ■ Sheet metal gauges and thicknesses for galvanized and hot-rolled mild steel sheets

27.5 MEASURING THE LINEAR FEET OF DUCTWORK REQUIRED

On the ductwork chart (see Figure 27.1), information under Column 2 lists the pounds of metal required to fabricate one linear foot of ductwork of the indicated size. To find the number of linear feet of ductwork during the sheet metal takeoff, special care should be taken in measuring linear feet of straight ductwork. Each partial foot less than six inches should be omitted. Each partial foot six inches or more should be counted as a full foot.

27.6 MEASURING LINEAR FEET FOR DUCT FITTINGS

When ductwork fittings are fabricated, considerably more sheet metal is wasted in the process than is normally wasted in fabricating straight ductwork. To offset the additional waste of sheet metal during fabrication of ductwork fittings, each fitting in the duct system should be measured from each direction. When this is done, the number of linear feet for the fitting is essentially doubled (see Figures 27.3 through 27.8).

When the duct dimension is changed in a fitting (such as a transition fitting), each duct size is used to determine the linear feet of ductwork for the fitting (Figure 27.3).

When the fitting is measured from each direction, the extra metal (waste) required for the fitting has been allowed for in the takeoff. Examples of ways to measure elbow fittings are shown in Figures 27.3 through 27.8. The measuring procedure is the same for a square throat elbow and a round radius elbow. If the duct size is changed with the elbow, both duct sizes are used when showing the length of duct for the elbow (Figures 27.5 and 27.6). An offset is measured like a transition (Figures 27.4 and 27.7).

NOTE: One rule is important enough to repeat: If the duct size changes in a fitting, one linear measurement across the fitting is made using one duct dimension and the second measurement across the fitting is made using the other duct dimension.

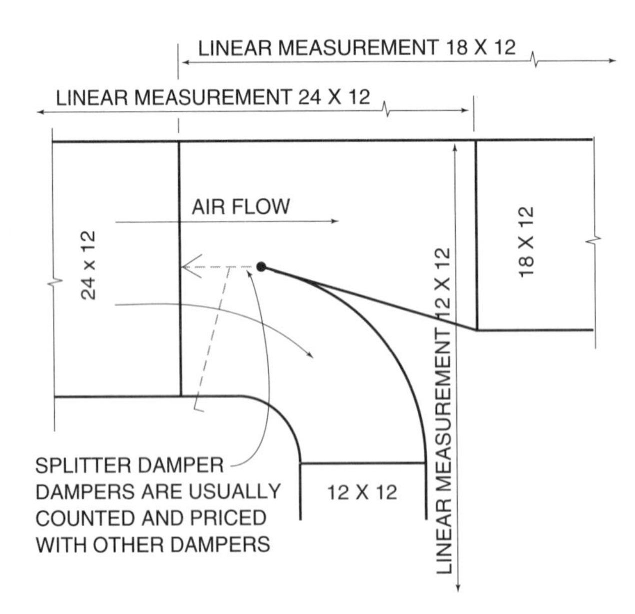

FIGURE 27.3 ■ Measuring a splitter fitting for estimating ductwork costs

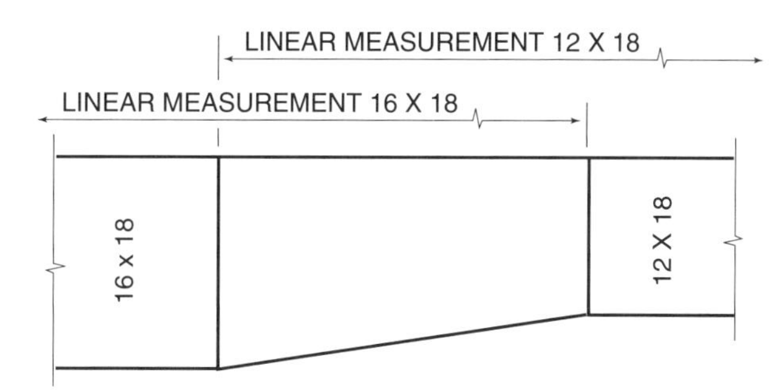

FIGURE 27.4 ■ Measuring a transition fitting for estimating ductwork costs

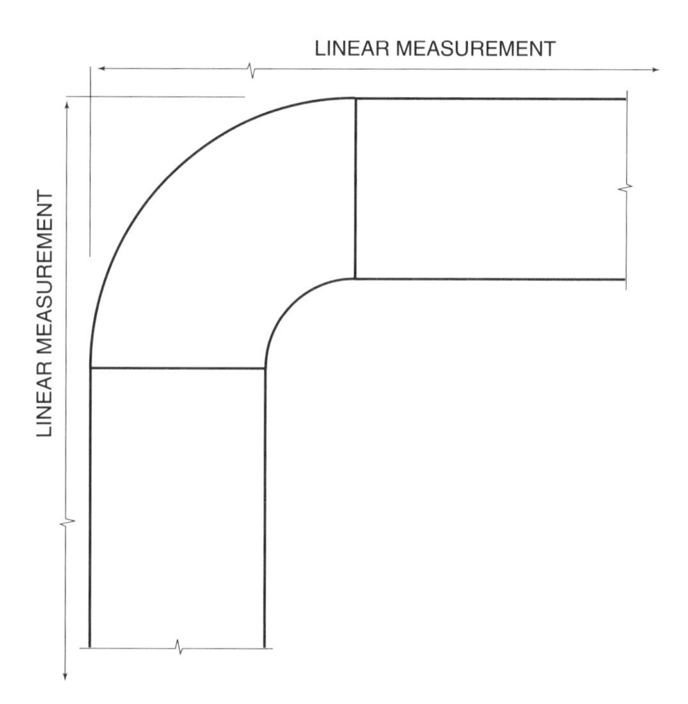

FIGURE 27.5 ■ Measuring a round radius elbow for estimating ductwork costs

The splitter fitting with two branches is measured somewhat differently from the other fittings described. Each duct size is measured as shown in Figure 27.8. The linear measurement of the straight ductwork is measured, and this measurement is extended in a straight line to where a new duct size is given. For example, the 12" × 12" straight duct is measured, and the measurement is extended through the fitting to the 16" × 12" duct joint as shown (see Figure 27.8). This is repeated for each duct size. Measuring the fitting four times adds enough metal for the fabrication and installation of this more complicated duct fitting.

27.7 DUCTWORK TAKEOFF FORM

The takeoff person must have a form that keeps up with the duct takeoff and facilitates tabulation of the total weight of metal needed to fabricate the ductwork. The takeoff sheet can be an original design or a variation of the takeoff sheet shown in Figure 27.9. This sheet is used in the examples in this chapter.

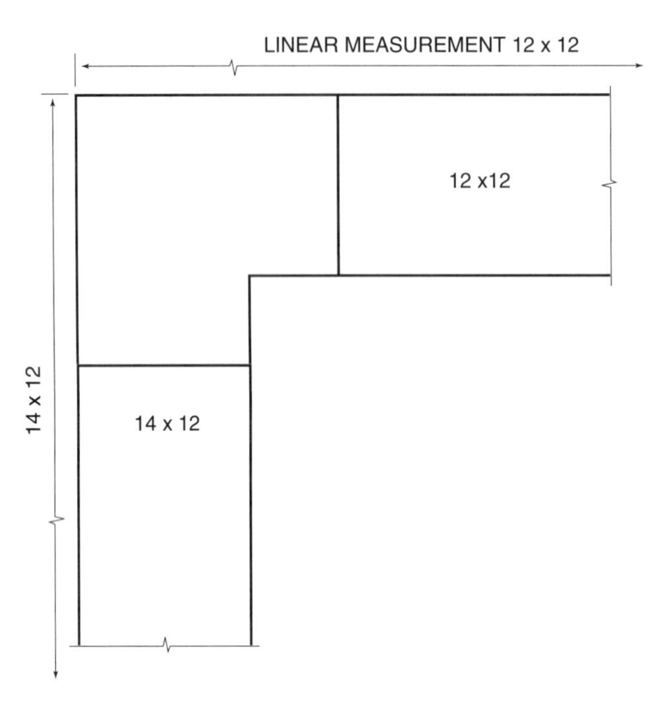

FIGURE 27.6 ■ Measuring a square elbow for estimating ductwork costs

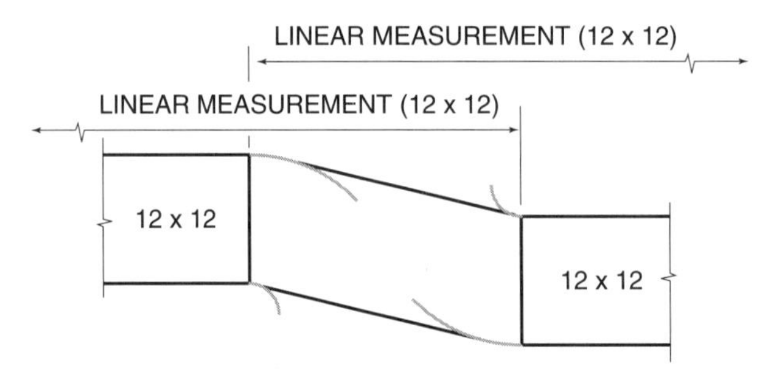

FIGURE 27.7 ■ Measuring an offset fitting for estimating ductwork costs

Column A on the sheet indicates the duct dimensions. Column B provides space to total the two duct dimensions. Column C provides space for the measured length of the duct size indicated in Column A. This total measured length of duct uses the double measuring of the fittings described in Section 27.6. The gauge of metal required for the duct size is recorded in Column D. Column E provides space to enter the weight of metal required to fabricate one foot of the duct size indicated. This weight is listed in Column 2 of Figure 27.1. Column F provides space for the total weight of the duct size indicated. Column F contains the poundage resulting from multiplying the total measured length (shown in Column C) times the weight of metal required per linear foot of duct (shown in Column E). Column G is used to indicate the total pounds of metal required for a specific length and size of ductwork. The formula $C \times E = G$ can be used to determine the total pounds of metal required to make the measured length of ductwork of the measured size.

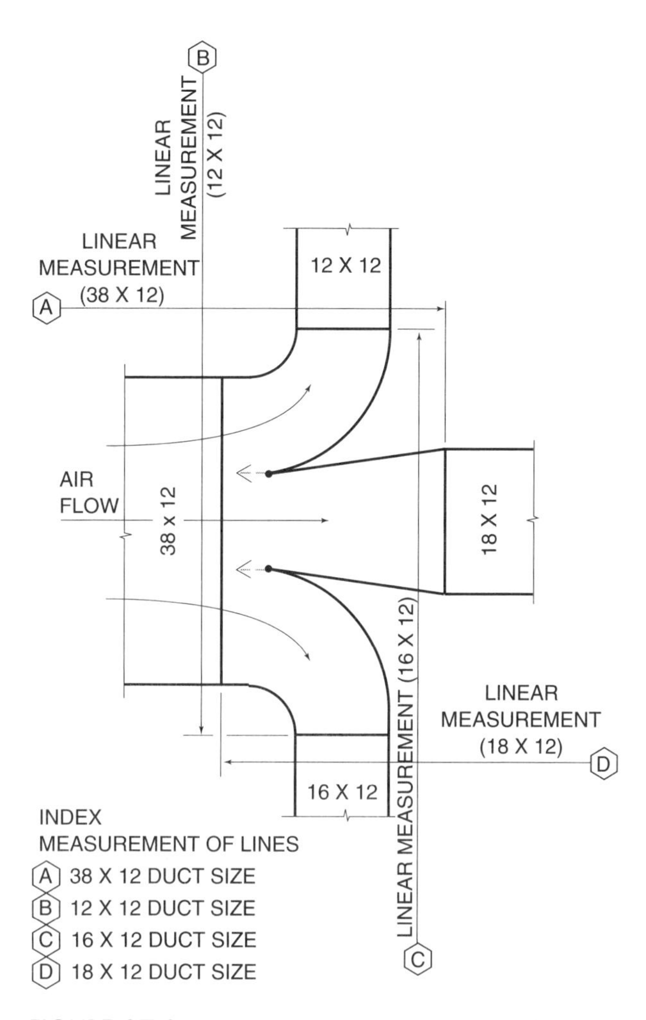

FIGURE 27.8 ■ Measuring a three-way splitter for estimating ductwork costs

27.8 DETERMINING THE TOTAL WEIGHT OF METAL REQUIRED FOR DUCTWORK

The procedure explained in Section 27.7 is repeated for each size of ductwork shown on the design plans. When all ductwork has been taken off and the multiplication of linear feet times weight per linear foot has been completed, the total weight of metal can be calculated. To obtain the total of metal weight required for the job, add together all the numerals in Column G on all takeoff sheets.

When the total weight for the metal has been determined, a percentage must be added to that total weight to allow for the overall waste in fabricating the ductwork and for the miscellaneous metal required for hangers and trim needed for hanging the ductwork on the job. (Remember that some extra metal has been figured in the fitting takeoff to account for the additional metal needed for making the fittings versus straight ductwork.)

The extra percentage added for waste and miscellaneous metal must be added to the total weight of metal tabulated on the takeoff sheet. The waste percentage depends on the degree of efficiency of the sheet metal

A DUCT SIZE	B DUCT LENGTH PLUS WIDTH	C DUCT MEASURED LENGTH	D METAL GAUGE	E WEIGHT PER LINEAR FOOT	TOTAL WEIGHT
				TOTAL LBS. METAL	

FIGURE 27.9 ■ Duct takeoff form

shop, and the company makes this percentage figure available to the ductwork technician. Some companies figure a waste factor of approximately 30 percent. If special hangers or sheet metal specialties are needed to install the ductwork, these items must be taken off (measured) and priced separately from the ductwork.

An example of the metal takeoff and the adding of the waste factor is as follows:

NOTE: The metal takeoff shows a total of 2000 pounds of metal. With a 30% waste factor, the waste poundage is 2000 × 30% (or 600 pounds of metal). When this 600 pounds is added to the original 2000 pounds of metal, the grand total of metal needed for the job is 2600 pounds.

27.9 COST FOR MATERIALS FOR SHOP-FABRICATED DUCTWORK

By multiplying the grand total poundage of metal times the cost of the metal per pound, the total cost of the metal required for the ductwork can be determined. For example, 2600 pounds × $1.00 per pound = $2600.00 for the cost of the metal.

27.10 LABOR COSTS FOR SHOP-FABRICATED DUCTWORK

The cost of fabrication and installation of the ductwork can be determined by multiplying the grand total poundage of metal times the labor cost factor for fabrication and installation. This labor cost factor should be available from the company. For example: 2600 pounds of metal × $1.10 per pound labor cost = $2860.00.

27.11 DUCTWORK INSULATION

Several different materials are used to insulate ductwork. The more popular materials are fiberglass, mineral wool, and cellular insulation. There are two basic types of insulation: exterior insulation (sometimes called wrapped insulation) and interior insulation that is glued inside the ductwork (called duct liner). Although this chapter is not concerned with the *materials* or *types* of insulation, a method for taking off and estimating the cost of insulation will be explained.

Insulation takeoff procedures are the same regardless of the material or type of insulation. Determining the total square feet of insulation is the job of the takeoff person. It is important to keep the figures for the different types and different materials separate because the costs vary for these different insulations. To calculate accurately, prices for each kind of insulation must be figured separately.

27.12 METHOD FOR TAKING OFF DUCTWORK INSULATION

The process for the linear takeoff of insulation is identical to the ductwork takeoff of sheet metal. The duct size, the lengths of straight duct, and the two-way takeoff for fittings are the same as described for the sheet metal takeoff (Figures 27.3 through 27.8).

27.13 READING THE INSULATION CHART

The duct insulation factors chart (Figure 27.10) shows Column 1 and Column 2 and their continuations. Column 1 represents the total duct dimensions (length plus width). A duct 18" by 12" totals 30 inches. Find the number 30 in Column 1 of the chart. To the right of this number, in Column 2, the number 5.0 is listed. This number indicates that 5 square feet of insulation are needed to cover 1 linear foot of ductwork 18 inches by 12 inches in size. This procedure is repeated for all ductwork on the plan.

Duct Insulation Factors Chart

1	2	1	2	1	2	1	2	1	2
12	2.0	37	6.17	62	10.32	87	14.52	124	20.7
13	2.17	38	6.33	63	10.5	88	14.7	126	21.0
14	2.33	39	6.5	64	10.65	89	14.85	129	37.9
15	2.5	40	6.67	65	10.85	90	15.0	130	
16	2.66	41	6.85	66	11.0	91	15.2	131	38.5
17	2.83	42	7.0	67	11.15	92	15.35	132	
18	3.0	43	7.16	68	11.3	93	15.52	133	39.0
19	3.27	44	7.35	69	11.5	94	15.7	134	
20	3.34	45	7.5	70	11.65	95	15.85	135	39.6
21	3.5	46	7.68	71	11.84	96	16.0	136	
22	3.67	47	7.84	72	12.0	97	16.2	137	40.2
23	3.83	48	8.0	73	12.18	98	16.35	138	
24	4.0	49	8.16	74	12.33	99	16.5	139	40.8
25	4.17	50	8.32	75	12.5	100	16.7	140	
26	4.34	51	8.5	76	12.7	102	17.0	141	41.4
27	4.5	52	8.66	77	12.84	104	17.4	142	
28	4.67	53	8.84	78	13.0	106	17.7	143	42.0
29	4.83	54	9.0	79	13.2	108	18.0	144	
30	5.0	55	9.18	80	13.4	110	18.4	145	42.6
31	5.17	56	9.35	81	13.5	112	18.7	146	
32	5.35	57	9.5	82	13.7	114	19.0	147	43.2
33	5.5	58	9.65	83	13.85	116	19.4	148	
34	5.67	59	9.85	84	14.0	118	19.7	149	
35	5.84	60	10.0	85	14.2	120	20.0	150	
36	6.0	61	10.16	86	14.35	122	20.4	151	

KEY

Columns numbered 1 are the total of duct dimensions: width plus height or W + H = column no. 1 (see text).

Columns numbered 2 are the square footage of insulation required to insulate one linear foot of duct of the indicated size.

Additional square footage of insulation must be added to the grand totals to account for waste in fabrication (see text).

FIGURE 27.10 ■ Duct insulation factors

27.14 INSULATION TAKEOFF FORM

An insulation takeoff form is needed to keep the takeoff orderly and correct (Figure 27.11). This form is similar to the one described for sheet metal takeoff. Column A is for the duct dimension (duct size). Column B provides space for the total of the duct dimensions (length plus width). Column C provides space for the measured length of duct. This measured length of duct contains the measured length of the fittings (measured in both directions). Refer to Figures 27.3 through 27.8 for examples. The duct length should be measured as described for sheet metal takeoff. Column D provides space to enter the square feet of insulation per linear foot of duct for the duct size listed (taken from the insulation chart in Figure 27.11). Column E provides space to enter the total square feet of insulation needed for the indicated duct size.

When the measurements in Column C and Column D have been recorded on the takeoff sheet, the takeoff person multiplies linear feet (Column C) times square feet of insulation (Column D) to obtain the square footage of insulation required. This figure is entered in Column E. In other words, Column C × Column D = Column E.

After the multiplication step, the amounts in Column C are totaled to determine total square footage of different types of insulation required for the job. *Special care must be taken to separate different kinds of insulation in the takeoff because the different insulations have different prices per square foot.*

A percentage is usually added to the total square footage of each type of insulation to account for waste that occurs during the installation process. This percentage varies from shop to shop; however, 25 percent to 30 percent is sometimes used as a reasonable amount to be figured for waste.

27.15 COST FOR INSULATION MATERIALS

To determine the total cost of insulation material, multiply the total square feet of insulation (including the waste factor) by the cost of the various types of insulation required. Sometimes, it is necessary to add an additional cost for special requirements, such as plaster, special fasteners, and so forth. This cost is part of the insulation material cost.

The labor cost for installing the insulation is determined by the air-conditioning and heating shop. Each shop establishes a cost figure per square foot of insulation installed.

27.16 COSTS OF LABOR FOR INSTALLING INSULATION

When the cost per square foot of insulation is multiplied by the total square feet of insulation, the labor cost for installing the insulation is obtained. Because labor costs vary for various types of insulation, different types of insulation should be calculated separately.

Add any special requirements to the labor cost. This cost must be calculated to suit job requirements and conditions.

SUMMARY

- The kind of ductwork is noted and described on the plans, in the specifications, or in both places.
- The sheet metal gauge is determined by its thickness.
- The estimation of size and cost of ductwork, fittings, and insulation is easier if the takeoff person understands and uses appropriate charts.
- The amount of sheet metal needed must include an extra amount to cover the waste factor during fabrication, especially of fittings and transitions.
- Different types of ductwork and insulation materials must be figured separately because of price variations.
- The percentage of waste factor and labor cost for fabrication varies from shop to shop.

DUCT SIZE	A DUCT LENGTH PLUS WIDTH	B DUCT MEASURED LENGTH	D SQ. FT. PER LINEAR FOOT	E TOTAL SQUARE FEET

FIGURE 27.11 ■ Insulation takeoff form

REVIEW QUESTIONS

1. List three materials used for fabricating ductwork.
2. Where is the kind of material for ductwork specified?
3. What does the term "takeoff" mean?
4. What is meant by the "gauge" of sheet metal?
5. Describe (or draw an explanatory sketch) how fittings and transitions are measured.
6. As used in this chapter, what does the formula $C \times E = G$ mean?
7. Why is extra material required for fittings and transitions?
8. Why should costs of different materials used for ductwork or insulation be figured separately?
9. What are the two basic types of insulation?
10. What are the five column headings of the ductwork takeoff form used in this chapter?
11. What are the five column headings of the insulation takeoff form used in this chapter?
12. Where does the takeoff person get the labor cost for fabricating and installing ductwork and insulation?

ADDITIONAL STUDENT EXERCISES: TAKEOFF PRACTICE

1. Refer to Figure 27.12 to complete this exercise. Use the architect's scale and the ductwork takeoff sheet to determine the total weight of metal required to fabricate the linear duct system shown. The duct system is drawn to 1/4" = 1'-0".
2. Refer to Figure 27.12 to complete this exercise. Use the insulation takeoff chart to estimate the square feet of insulation required for the linear duct system.

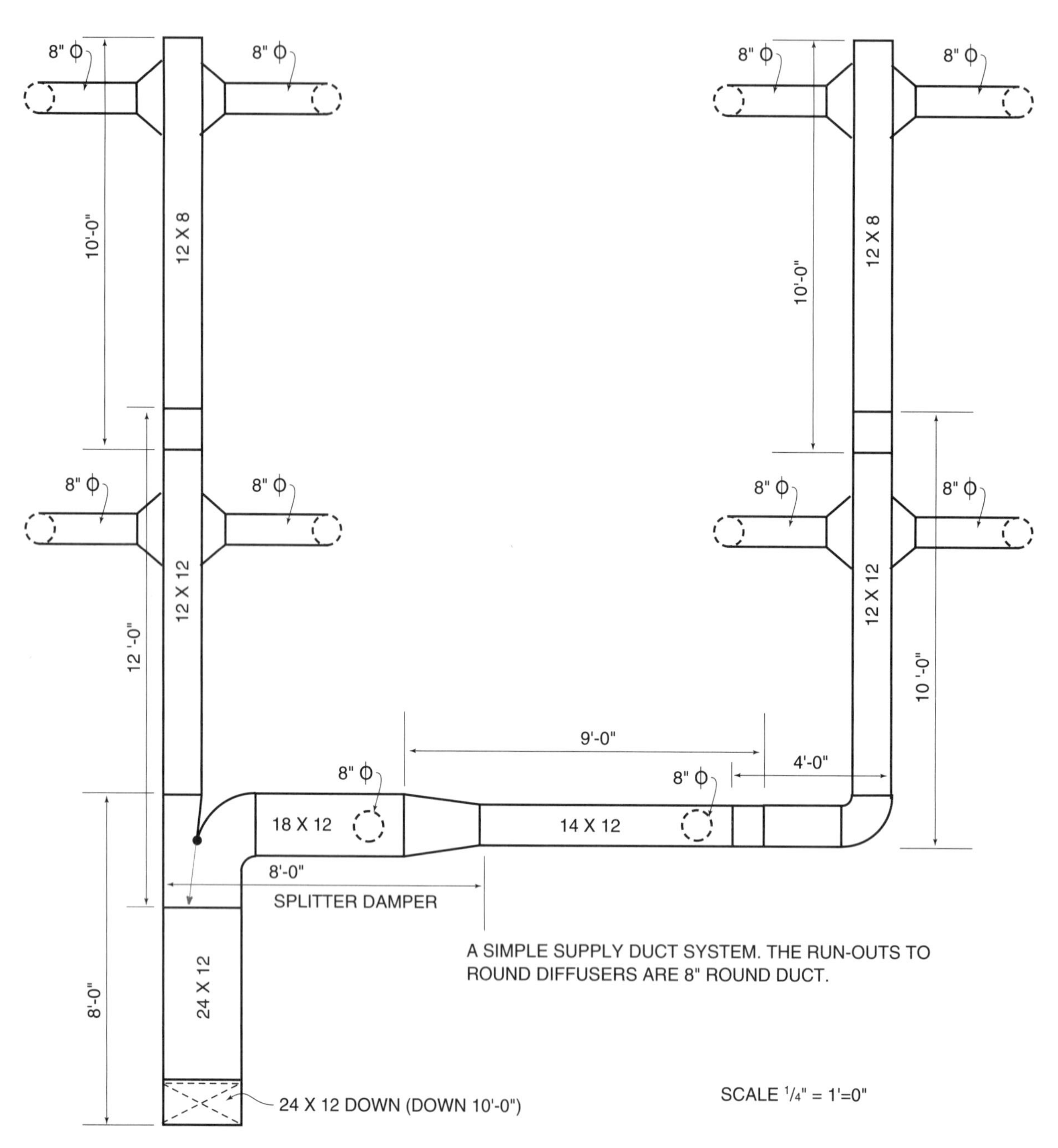

FIGURE 27.12 ■ Exercise 1

CHAPTER 28

Reading Mechanical Plans

OBJECTIVES

After studying this chapter, you should be able to:

- locate the title and scale used on the mechanical plan sheet
- scan a set of plans for an overview of the mechanical system
- locate heating and air-conditioning units, ductwork, and supply and return outlets
- identify and explain the Air Distribution Schedule
- locate the mechanical specifications
- identify switches, thermostats, piping, and other components of the mechanical system
- answer specific questions about details of the mechanical system

NOTE: Refer to the plans for "A Small Bank and Trust Company" (at the end of this textbook) while studying this chapter.

28.1 INTRODUCTION

The first thing that a technician should do upon receiving a set of plans for a building is to review all the plans, scanning each one to see what drawings are available, to locate major information, and to get a general overview of the building. A review of the specifications (see Chapter 13) will help the technician learn more about the building and what is expected of the builder.

The set of building plans located at the end of this textbook is typical for a small building. This chapter will refer to these plans to demonstrate the arrangements and locations of various portions of information. Architectural plans for the same building were discussed in Chapter 26. In Chapter 28, the discussion will be directed to the reading of the mechanical plan. Sheet M-1 shows the mechanical floor plan for the building.

Considerable attention will be devoted to ductwork because it is such a vital part of a building project. The student is encouraged to review Chapter 27 for a method for "taking off" ductwork and ductwork insulation for estimating purposes.

Student exercises located at the end of this chapter are based on information shown on Mechanical Sheets M-1 and M-2.

28.2 READING THE MECHANICAL PLANS

The first thing to find on the sheet is the title of the drawing. On Sheet M-1, the title is "Floor Plan." If the building had multiple floors, the title would indicate which floor is represented on the drawing. The scale used for the drawings is given as part of the title. In this case, the plan has been reduced, making the 1/4" = 1'-0" scale incorrect. Because of this discrepancy, the only dimensions that can be relied on are the ones specified on the plans. In actual practice, this drawing would not be issued as a construction drawing for builders and technicians to work from.

After determining the title of the plan and the scale used for the drawing, the technician should review the entire plan to study all the details and construction information that is given on the sheet. If there are multifloor and multisheet sets of plans, it is important to know, in general, where various details, schedules, sections, and so forth are located. It is also important to obtain an understanding of the entire mechanical system.

In reviewing this plan, it is suggested that the technician read each sheet like reading a book. This procedure is described in Chapter 26.

28.3 LOOKING FOR SPECIFIC SYSTEMS

After the general overview of the plan and the mechanical design, each specific part of the air-conditioning and heating system should be reviewed. In most cases, the air-conditioning and heating units are reviewed first. On this building, the units are a part of a split system, air-to-air heat pump. The outside unit is located on the roof, and the inside unit is located in a mechanical space above the ceiling to the right of the front entrance. A floor plan of this mechanical space is shown in the upper left-hand corner of the mechanical plan. (See Sheet A-5, Drawing Section 10/2.A-1, A-2, and A-3 for the location of the mechanical space above the ceiling.)

28.4 FOLLOWING THE DUCTWORK

The inside section of the heat pump has a 30" × 18" supply duct extended toward the top of the plan. (Note: Where the north arrow is not indicated, the top of the sheet is normally considered north.) This supply duct splits into two 15" × 18" ducts with a splitter damper, and these two ducts continue to serve the air supply to the entire building. The supply air outlets are shown with the air quantities (cfm) indicated.

The return air duct connects to the inside section of the heat pump, extends to the front entrance, and then turns down and connects to an 18" × 18" return air grille mounted in the wall six inches below the ceiling.

The duct sizes are shown on the plans. In Chapter 27, it was pointed out that the first dimension shown on ductwork is the side of the ductwork that is being seen and the other dimension is the second number. In other words, the 30" × 18" duct shown for the supply ductwork is 30 inches wide and 18 inches long.

28.5 THE SUPPLY AND RETURN AIR OUTLETS

In the table at the top center of the mechanical plan is the Air Distribution Schedule. This schedule gives the description of the supply and return outlets, including a typical manufacturer and catalog number. In many jobs, the design engineer or technician is not allowed to mention specific manufacturers for the specified equipment. In such cases, a detailed description of the desired air distribution outlet must be included in the specification.

28.6 WRITTEN NOTES AND INSTRUCTIONS

It is sometimes advantageous to have written notes and instructions placed on the plans. On this job, these notes have been titled "General Notes."

28.7 OTHER INFORMATION SHOWN ON THE MECHANICAL PLAN

In addition to the information described above, the mechanical plan describes and shows electric heaters, timed override switches, night set-back thermostats, and refrigeration piping. Also included on the mechanical plan are such items as drawings of equipment space and sections. Plan 2/M-1, on Sheet M-1, is a plan of the equipment space located over the Women's Toilet. Section 4/M-1 is a longitudinal section of the inside heat pump unit.

28.8 MECHANICAL SPECIFICATIONS

As noted in Chapter 13, some small jobs have the specifications printed on a separate sheet and bound with the regular plans. For example, the specifications for the plans at the end of this textbook are bound in this way. On larger jobs, all specifications are bound in a separate book that accompanies the plans when they are turned over to the contractor.

SUMMARY

- Information regarding the scale of the drawing and specific information on the sheet is found on the title of the sheet.
- It is important for the technician to gain an understanding of the entire mechanical system before concentrating on any of its parts.
- The types of air-conditioning and heating units and their locations are shown on the mechanical plans.
- If there is no directional arrow on the plans, the top of the plan is considered north.
- The technician should study the supply and return air ductwork throughout the system.
- When ductwork sizes are stated, the first number refers to the visible side of the ductwork and the second number refers to the other dimension.
- The Air Distribution Schedule describes supply and return outlets for typical units or includes a detailed description in the specification.
- Written notes and instructions may be placed on the plans.

REVIEW QUESTIONS

1. Upon receiving a set of plans, what information should the technician look for first?
2. What information is found on the title of the drawing?
3. Why is the scale listed on the drawings at the end of the textbook not accurate?
4. What kind of heating and air-conditioning units are used in the small bank and trust company?
5. Where are the units located?
6. Where is the mechanical space shown on this mechanical plan?
7. If there is no directional arrow on the plan, which direction is considered north?
8. What are the dimensions of the two supply ducts that serve the air supply to the building?
9. What are the dimensions of the main return air ducts?
10. On a duct with dimensions of 30" × 18", what does the 30" refer to?
11. What information is given in the Air Distribution Schedule?
12. If the designer is not allowed to mention specific manufacturers, how are descriptions of outlets given?
13. What other types of mechanical equipment are shown on the mechanical plan?
14. On small jobs, where may mechanical specifications be placed?
15. Where are the mechanical specifications usually placed on large jobs?

ADDITIONAL STUDENT EXERCISES: READING MECHANICAL PLANS

Refer to Sheet M-1 in the back of this textbook to answer the following questions:

1. How many exhaust fans are located on the roof?
2. What is the cooling capacity of the heat pump?
3. What rooms have electric heaters installed?
4. Where are the refrigeration lines shown to be installed?
5. What are the sizes for the refrigeration lines?
6. How many thermostats are needed for the job? Name the types of thermostats needed.
7. What is the symbol for the supply outlet in the conference room? What is the symbol for the supply outlet for the east entrance?
8. What is the total air (cfm) that the system is designed to distribute throughout the building?
9. Are dampers installed in the duct takeoffs for balancing the air quantities?
10. What is the minimum clearance around the outside heat pump unit?
11. A time clock is shown as part of the control system. Where is this clock located?
12. Does the time clock have an override switch? If so, where is this switch mounted?
13. What is the size of the condensate drain line?
14. Are roof curbs fabricated on the job?
15. Is the outside heat pump unit located on the roof?
16. Does the exhaust fan have a damper?
17. Where are the starter and disconnect switches for the inside heat pump unit located?
18. Are any square or rectangular duct offsets required?
19. How high (above the floor) should the thermostat be mounted?

Refer to Sheets M-1 and M-2 of the plans at the end of this textbook to answer the following questions:

1. How many fresh air intakes are shown?
2. What size flexible ductwork is used?
3. What is the size of the largest duct shown?
4. What air distribution and return air grilles are specified?
5. Is there a disclaimer stated regarding the plans?
6. List the codes, agencies, and so forth that the contract work must comply with.
7. What are special requirements for the exhaust fan?
8. Is the inside heat pump unit floor mounted?
9. Describe the unit casing.
10. Is the air-handling unit insulated?
11. Describe the supply air fan.
12. What special features should the compressor have?
13. What special fittings are to be installed in the equipment lines off the compressor?
14. What type piping is used for the condensate drain line?

15. What type and thickness of insulation is used on the refrigerator lines?
16. What agency is specified for standards for constructing the ductwork?
17. What is the maximum spacing for the horizontal duct hangers (supports)?
18. Where are flexible duct connectors used?
19. What is the time clock used for?
20. Does the time clock also control the exhaust fan "on-off" schedule?
21. Who does the testing of the equipment upon completion of the job?
22. Is a warranty required? If so, describe it.

CHAPTER 29

Reading Electrical Plans

OBJECTIVES

After studying this chapter, you should be able to:

- explain how the various types of lighting fixtures are identified on the electrical plan
- discuss the information about electrical wiring as shown on the electrical plan
- identify and explain symbols, notes, and schedules on the electrical plans
- define electrical specifications
- discuss reasons why coordination by the various tradespersons is necessary

NOTE: Refer to the plans for "A Small Bank and Trust Company" (at the end of this textbook) while studying this chapter.

29.1 INTRODUCTION

As noted in Chapter 28, the first thing that a technician should do upon receiving a set of plans for a building is to review them, scanning each one to get an overview of the building. The electrical system, the air-conditioning and heating system, and the plumbing system for a building are very closely related. Design technicians and installation technicians must coordinate the systems of all trade areas to have a well-designed and smooth installation on the job. Chapter 26 discussed the architectural plans that are at the end of this textbook. Chapter 28 concerned the reading of the mechanical plans. This chapter will deal with the reading of the electrical plans.

On larger buildings, the electrical plans usually contain plans for lighting, power distribution (showing receptacles and special connections), panel schedules, and other schedules and details pertaining to the electrical system. Since the construction plans at the end of this book are for a small building, all the necessary information is shown on one sheet.

29.2 READING THE ELECTRICAL PLANS

The larger drawing on Sheet E-1 is the floor plan. The lighting plan and the power distribution plan are consolidated on the same floor plan. Lighting fixtures are shown with rectangles for fluorescent fixtures and circles for other types of fixtures. Each lighting fixture is identified by a letter of the alphabet. These letters are shown on the lighting fixture schedule on the plan.

The power distribution is also shown on the plan. The electrical receptacles use the standard symbol (see Chapter 7 and Appendix D). The hash marks on the circuit lines indicate the number of conductors (wires) required. All the circuits are indexed (numbered), and these numbers indicate the number of the circuit breaker in the distribution panel. A schedule for Panel A is shown on the plan, with each circuit numbered. The service to the panel is shown with a separate detail. The main circuit breaker is located inside Panel A.

29.3 THE ELECTRICAL WIRING

The electrical wiring is shown with an arched line to electrical devices and arched lines between lighting fixtures or between receptacles. The exact locations of the wiring and conduit are left to the installing technician. On each wire are hash marks indicating how many wires are required with the circuit. The arrowhead at the end of the wiring symbol indicates that the conduit is to be extended to the panel. The small number located at the arrowhead gives the number of the circuit breaker to which the wiring is to be connected. The panel detail located on the lower left-hand side indicates the circuit breaker numbers. The wiring shown with broken lines and the receptacles connected with these broken lines are alternate (marked ALT). The ALT marking means that the contractor should give a separate price for this work, and the owner can decide whether to include this work in the contract.

29.4 ADDITIONAL INFORMATION SHOWN ON AN ELECTRICAL PLAN

The electrical plan on Sheet E-1 contains additional information that should be noted at this point. A lighting fixture schedule on the plan describes the lighting fixtures to be used on the job. Each fixture shown on the floor plan has a letter that corresponds to the fixture schedule.

A symbol schedule identifies the symbols used on the plan. The power riser diagram is shown in the lower right-hand corner. A general note is also shown on the plan, which gives instructions to the contractor. A note at the top right corner of the plan restricts the use of the plans.

29.5 ELECTRICAL SPECIFICATIONS

The electrical specifications, like the mechanical specifications, contain written instructions and descriptions about construction equipment and procedures. The specifications for this job were bound with the General Contract specifications and are not available with this text.

29.6 UNDERSTANDING THE PLANS

As previously noted, it is very important for persons reading the plans to understand the plans thoroughly before attempting to estimate the cost of the job or to construct the job. After construction has started, the various trades often have to do their installations in the same limited space. For this reason, coordination on the job between the different contractors is necessary to complete the job satisfactorily.

SUMMARY

- Coordination among the various tradespersons is necessary for a successful construction project.
- Lighting fixtures are identified by symbols and letters on the lighting schedule.
- Electrical receptacles, circuits, and circuit breakers are shown on the power distribution plan.
- Specific information regarding the electrical wiring is given on the plan, but the exact location of wiring and conduit is usually left up to the installing technician.
- The exact description of each lighting fixture is given on the fixture schedule.
- Electrical specifications give instructions about equipment and procedures and are sometimes listed under the General Contract specifications.
- It is important to understand plans thoroughly before attempting to estimate a job or construct a project.

REVIEW QUESTIONS

1. Why should the technician scan the entire set of plans before beginning any work?
2. What kind of lighting does a rectangle indicate? A circle?
3. Draw the standard symbol for an electrical receptacle.
4. Where is the main circuit breaker located?
5. What do the hash marks on the circuit lines represent?
6. Who decides the exact locations of wiring and conduit?
7. What does ALT mean? When would ALT be found on the plans?
8. What information is given on the note at the top right corner of the plans on Sheet E-1?
9. Why are no electrical specifications shown on the plans on Sheet E-1?

ADDITIONAL STUDENT EXERCISES: READING ELECTRICAL PLANS

Refer to Sheet E-1 in the back of this textbook to answer the following questions:

1. Is any electric baseboard heating to be installed on this job?
2. Is there an exhaust fan on the job?
3. Is the electrical wiring shown on the floor plan?
4. Does the job have a telephone space board?
5. Does the outside section of the heat pump have a disconnect switch?
6. How is the outside lighting controlled?
7. How many telephone outlets are shown on the electrical plans?
8. How many above-the-counter receptacles are shown?
9. Is there a separate space for the amplifier and tuner for the drive-up window?
10. Are any electrical receptacles located in the floor?
11. Why are some of the electrical receptacles (marked ALT) and the "home runs" shown with broken lines?
12. Why is the soffit lighting (outside) put on a time switch?
13. Where is the telephone space (wiring board) located?
14. Does the mechanical contractor do the control wiring for the HVAC equipment?
15. How many lighting tubes are to be installed in each Type A fixture?
16. Is there a duct-mounted electric heater on the job?

Refer to Sheet E-1 of the plans at the end of this textbook to answer the following questions:

1. How many Type A lighting fixtures are shown?
2. How many circuits are being used for lighting?
3. What is the connected load for the baseboard heater?
4. What electrical service is connected to Electrical Panel "A"?
5. What electrical load is connected to the duct heater?
6. What circuits are connected to the outside section of the heat pump?
7. What size wires are used to connect the electrical service to the circuit breaker for the outside section of the heat pump?

8. Is the electrical wiring that serves the outside section of the heat pump installed in conduit?
9. Who makes final electrical connections to the mechanical equipment?
10. What is the maximum number of circuit breakers that can be put on the electrical panel?
11. How many Type F fixtures are shown?
12. Where are electrical receptacles mounted flat in the corners?
13. Where is the volume control for the speaker system in the lobby located?
14. The second volume control is located in the office near the storage room. How many speakers does it serve?

CHAPTER 30

Reading Plumbing Plans

OBJECTIVES

After studying this chapter, you should be able to:

- explain why coordination by the various trades is necessary
- list the various plumbing systems found on the plumbing plans
- identify the various types of piping on the plans
- explain what information is found on the Plumbing Fixture Schedule
- discuss the various line configurations used to show specific types of piping
- identify the parts of the plumbing riser diagram

NOTE: Refer to the plans for "A Small Bank and Trust Company" (at the end of this textbook) while studying this chapter.

30.1 INTRODUCTION

As noted in Chapters 28 and 29, the first thing that a technician should do upon receiving a set of plans for a building is to review them, scanning each plan to get an overview of the building. Understanding the mechanical, electrical, and plumbing plans is very important because the three trades will be working in the same area to install their equipment, and coordination of their work will be required. Studying each of the plumbing systems one at a time will help the technician understand the entire system more thoroughly.

The major systems found in a complete plumbing plan are as follows: sanitary drainage system, plumbing vent system, domestic hot and cold water system, roof drainage system, fire protection sprinkler system, and compressed air system. In a hospital or medical building, plumbing also includes the medical gas systems (oxygen, medical air, nitrogen, and vacuum). All these systems will occupy the same space that is allotted for the mechanical and electrical systems. It is essential, therefore, that the systems be understood and coordinated. The set of plans at the end of this textbook is typical for a small building, with only one sheet required for the plumbing system. This chapter will focus on the plumbing plan for this building.

30.2 LOOKING FOR SPECIFIC SYSTEMS

On the plan, the size and location of each piping system is shown. The domestic cold water piping includes the service piping to the building and the distribution piping inside the building. The water heater and the domestic hot and cold water piping are shown.

The heavier lines represent the drain lines. The lines connected to the plumbing fixtures are the sanitary waste lines. The sanitary piping is located underground except for the part that turns up to connect to the plumbing fixtures.

The roof drainage piping is connected to the roof drains. This piping is extended, usually above the ceiling, to a point where it can be turned down and installed below the floor. These lines are drawn the same as the sanitary drain lines, and they can be identified as storm drain lines because they are connected to the roof drains.

Another piping system is the plumbing vent piping. This piping is connected above the trap on each fixture to let air enter and leave the piping system when water is introduced into the drainage system. (Remember that water and air cannot occupy the same space.)

30.3 READING THE PLUMBING PLANS

The floor plan shows the location of the plumbing fixtures, and the fixtures are numbered to correspond to the numbers in the Plumbing Fixture Schedule. The schedule has the name of the fixtures, the manufacturer and model number of each fixture, and the connection size for all the piping connected to the fixture. There is a space for notes that pertain to the fixtures.

The domestic cold water piping is shown with a light solid line having long dashes and a dot. The domestic hot water is shown with the same weight line as the domestic cold water lines except that the line is a series of long dashes with two dots. These piping systems have all the fittings and valves needed for a complete system.

The sanitary drainage piping is shown as a heavy solid line. All the various fittings and related equipment associated with the drainage piping are shown. The plumbing vent piping is shown with a series of dashes. The weight of these dashes is the same as the weight for the sanitary drain line.

The roof drainage system consists of a series of roof drains with related piping and fittings. This piping is shown with the same weight lines as the sanitary drain lines.

Notes are used to direct the technician during the installation. Because this is a small job, the specifications are shown on the drawing. These specifications are titled "Plumbing Notes and Specifications."

A plumbing riser diagram on the plan shows the method for installing the various piping systems serving the plumbing fixtures. The pipe sizes are shown on this diagram along with the method for connecting them. This diagram has the sanitary piping, vent piping, and domestic hot and cold water. Each vent riser is identified with an "R" number or riser number.

The connection to the water heater is included. A larger connection detail for the water heater is shown in the upper left-hand corner of the drawing. This detail shows the piping and fittings necessary for connecting the piping to the heater.

Two other details are shown on the plan. The installation of the exterior clean-out is shown with notes about the installation. The other detail shows the method for installing the emergency roof drain. The purpose of this drain is to drain water off the roof if the other drains fail or if, for some reason, water builds up on the roof to a depth of three inches.

SUMMARY

- Understanding mechanical, electrical, and plumbing plans is necessary for each tradesperson.
- The plumbing plan includes sanitary drainage, plumbing vents, domestic hot and cold water, roof drainage, fire protection sprinkler, and compressed air systems.
- In a hospital or medical facility, the medical gases (oxygen, medical air, nitrogen, and vacuum) are part of the plumbing.
- The plumbing plan shows whether piping is underground or aboveground and includes the connections and locations of piping.
- The Plumbing Fixture Schedule identifies the manufacturer, model number, and other pertinent information about each fixture.

- Certain line weights and configurations denote various types of piping.
- The riser diagram shows how various piping systems are to be installed.
- Details show specific information on connections, installation, and fittings.

REVIEW QUESTIONS

1. Why does the plumbing technician need to be familiar with the mechanical and electrical plans?
2. List six parts of the complete plumbing system.
3. List four types of medical gases found in hospitals and medical facilities.
4. List a significant fact about each of the following:
   A. Domestic cold water piping
   B. Sanitary waste lines
   C. Roof drainage piping
   D. Plumbing vent piping
5. Draw the distinctive lines that indicate each of the following:
   A. Domestic cold water piping
   B. Domestic hot water piping
   C. Sanitary drain piping
   D. Plumbing vent piping
6. On the "A Small Bank and Trust Company" plans, where are the specifications located?
7. What does an "R" number indicate?
8. What information is found on the Plumbing Fixture Schedule?
9. What is the purpose of the emergency roof drain?

ADDITIONAL STUDENT EXERCISES: READING PLUMBING PLANS

Refer to Sheet P-1 in the back of this textbook to answer the following questions:

1. How many water closets are required for the job?
2. How many lavatories are required for the job?
3. How many roof drains are shown?
4. How many exterior clean-outs are shown on the plan?
5. Are the domestic hot and cold water lines that are located inside the building insulated? If so, how thick should fiberglass insulation be?
6. Are all the roof drains the same?
7. Are the roof leaders insulated?
8. What is the size of the plumbing vent through the roof?
9. How many yard hydrants are required for the job?
10. Does the water heater have a safety relief valve?
11. What is the size of the cold water service?
12. What is the size of the sanitary sewers?

13. Does the cold water service have a shutoff valve?
14. Are roof leaders to be installed without regard to ductwork?
15. How many floor drains are shown?
16. Who wires the water heater?
17. What kind of piping is used for the roof drain piping?
18. Are domestic water lines sterilized?
19. Do hot and cold water service lines to the plumbing fixtures have air chambers?
20. Describe the emergency roof drain.
21. Where is the emergency roof drain located?
22. Do outside clean-outs have concrete pads?

CHAPTER 31

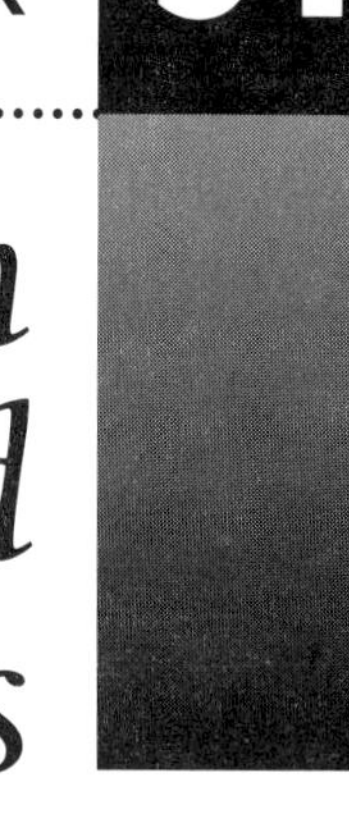

Introduction to Load Calculations

OBJECTIVES

After studying this chapter, you should be able to

- use blueprints to size air-conditioning equipment
- use blueprints to size heating equipment
- determine what envelope information is needed for cooling/heating load calculations
- identify internal heat loads
- determine ductwork loads
- determine when it is necessary to zone equipment

31.1 INTRODUCTION

Doing load calculations can be a challenge. This chapter is designed to give the reader an overview of what to expect when tackling the load-calculating process. The chapter will not do a load calculation. Using the information from the blueprints will make the task much easier and more understandable. The purpose of this chapter is to introduce you to interpreting this important information before applying it to sizing air-conditioning and heating equipment.

Blueprints can be used to size air-conditioning and heating equipment. The term used for sizing this equipment is *load calculation*. The outcome of a cooling load calculation is measured in British thermal units (BTU) and is converted into tons of air conditioning. One BTU is the amount of heat removed from or added to a pound of water to change its temperature by one degree Fahrenheit. One BTU will change dry air temperature about four degrees. When cooling, the BTU measurement is the amount of heat removed from a conditioning space in one hour. Sometimes, it is expressed as BTUH, with the H indicating hour. BTUs can be expressed in other forms, such as BTUs per day, month, or year. The BTU per hour is the most common form of expression and it is understood to be an hourly term unless stated differently. One ton of air conditioning is equal to 12,000 BTUs of heat being removed per hour. The final expression of a cooling load calculation is usually in tons.

The term for sizing heating equipment is also measured in BTUs. In this case, it refers to the amount of heat being added to a space. The heat can be added through a variety of heating options, such as the following:

- Gas heating
- Electric heating
- Oil heating

- Heat pumps
- Water or steam heating (hydronic)
- Coal or wood heating

The source of heat is not important, provided that the correct number of BTUs are supplied to the space. Except for electric heat and heat pumps, the aforementioned heating systems are generally measured in BTU input of the heating equipment. The input heat rating of gas, oil, hydronic, and coal heating is greater than its heat output. This is due to losses in the heat exchanger and losses associated with the venting of combustion products. Electric heating is measured in kilowatts or KW. One watt of electrical energy is equal to approximately 3.4 BTUs of heat; therefore, a kilowatt (1000 watts) is equal to about 3400 BTUH of heat output. An individual heat strip for a residential electric heater is rated at 5 KW @ 240 volts. This is equivalent to 17,200 BTUH of heat.

(5000 watts × 3.4 BTU = 17,200 BTUH)

This chapter will discuss the various pieces of information that you will need to glean from the blueprints in order to complete a cooling and heating load calculation. A load calculation can be completed using a *Manual J* or a *Manual N* load calculation form. These load calculation manuals are available from the Air Conditioning Contractors of America (ACCA). Similar load calculation forms are available from other sources. *Manual J* is for residential load calculations, and *Manual N* is for commercial load calculations. The American Society of Heating, Refrigeration, and Air-Conditioning Engineers (ASHRAE) also has a load calculation process. The load calculations can be done with pencil and paper or on computer. It is best to use the basic pencil and paper format to learn the process and then advance to a quicker computer-based calculation.

31.2 HEAT TRANSFER BASICS

In order to understand this chapter, it is important to review heat transfer theory. Heat will travel from hot to cold. Hot to cold is relative. Heat will transfer from an 80° F outdoor condition to a 75° F indoor space (Figure 31.1). Heat removal is required to keep the space comfortable. In winter, heat will travel from a 75° F indoor space to a colder outdoor condition (Figure 31.2).

Heat is transferred by conduction, convection, and radiation. Let us review how heat is transferred.

Adding heat is required to keep a space comfortable. One form of heat transfer is known as conduction. Heat transfers through the building material molecules like a chain reaction. Some materials transfer heat at a slower rate, and these are known as insulators. Some materials, such as metal or glass, transfer heat quickly. Heat is also transferred by convection and radiation.

Convection is the transfer of heat when the air in the space creates a slow movement pattern—with warm air rising and cold air falling. This is known as natural convection. A blower moving air across a cooling or heating heat exchanger is known as forced convection. Convection also occurs in fluids such as water. The warmer water will rise to the top and colder, denser water will fall to the bottom of a vessel. This creates a circular convection pattern.

Radiation is the heat transfer by heat or solar rays through the air. The best example of this is the solar radiation of the sun heating a space directly through windows or indirectly by heating an attic space. Radiation also occurs when the human body is near an object whose temperature is different from the surrounding air temperature. For example, a building may have an indoor temperature of 70° F and the inside window temperature may be 35° F. A person sitting near the window will feel cold even though the room's ambient temperature is a comfortable 70° F. Radiation heat transfer occurs when the inside surface temperature is greater or less than 1° of the surrounding air temperature (Figure 31.3).

Radiation effects are difficult to evaluate, but they do have an impact on human comfort. In this case, the occupant would probably increase the room temperature in order to overcome the effects of the radiation losses from the body. Window covers or double-pane windows will reduce the radiation loss effect and make

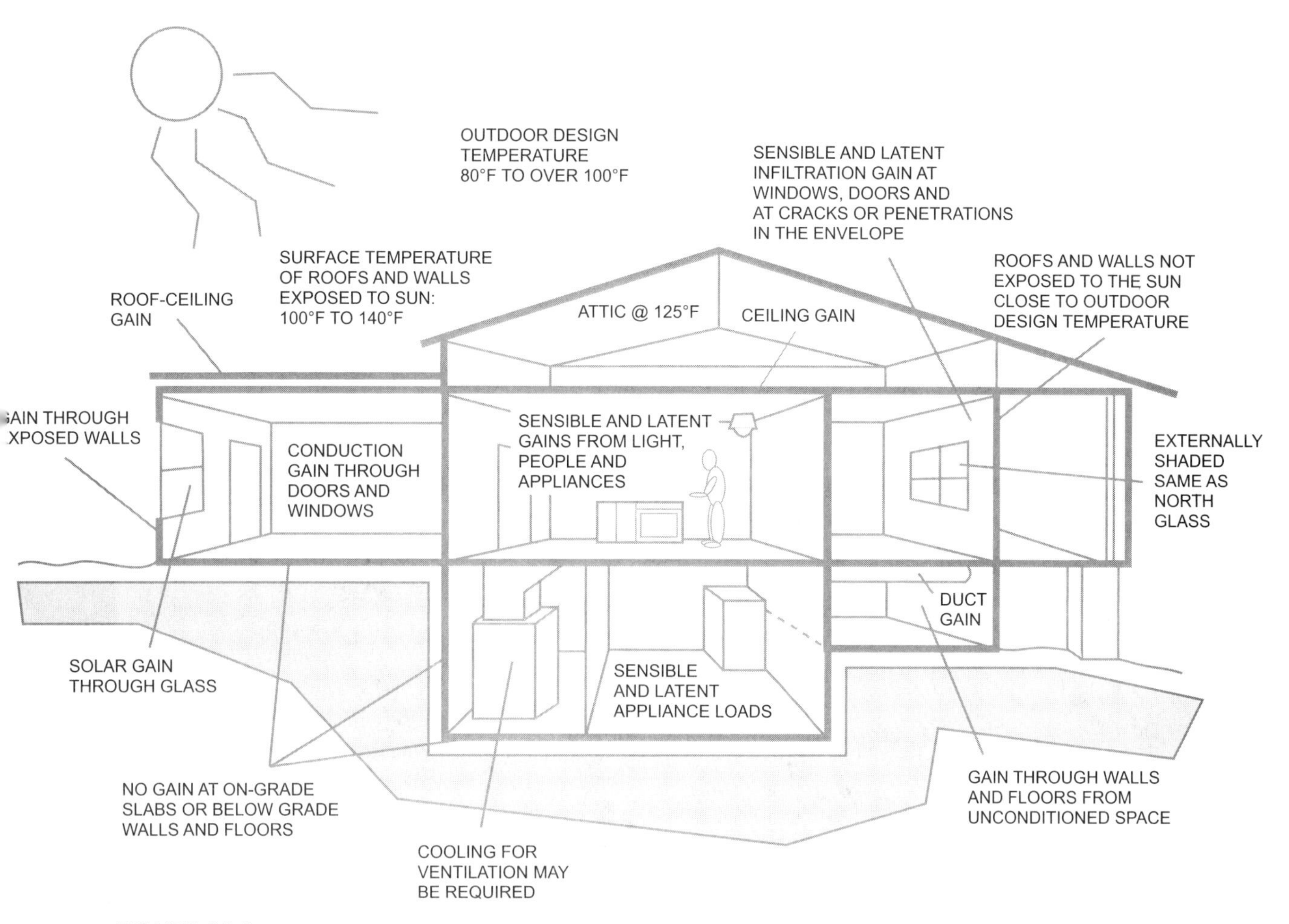

FIGURE 31.1 ■ Heat transfers from outside to inside. Notice all the areas where heat is being transferred into the house *(Courtesy: Air Conditioning Contractors of America, Manual J)*

the occupant feel more comfortable. The opposite condition occurs in the summer months when radiation heat transfers to the body.

31.3 WHAT YOU WILL NEED TO DO A LOAD CALCULATION

For a basic load calculation, the following pieces of information can be obtained from the blueprints or occupancy classification.

- Design conditions
- Building orientation
- Floor area
- Foundation type
- Floor insulation
- Ceiling area

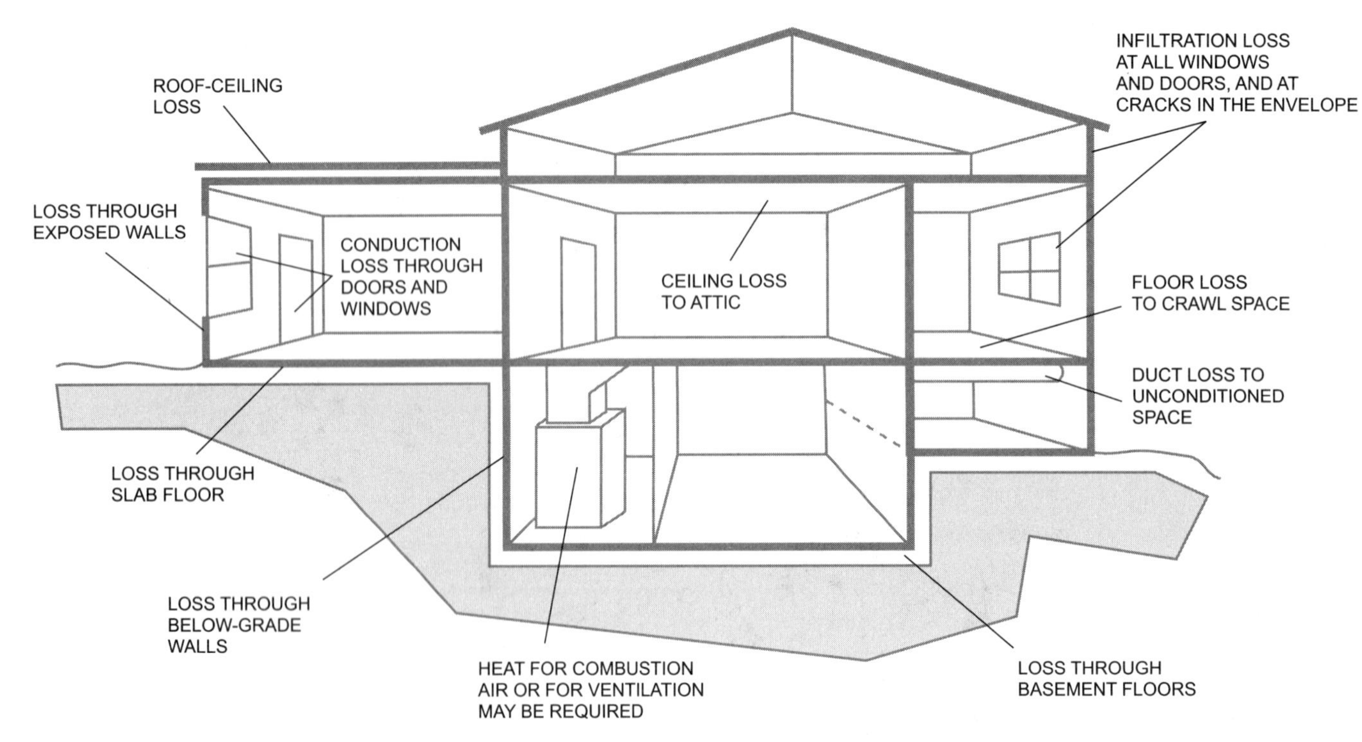

FIGURE 31.2 ■ Heat transfers from inside to outside. Notice all the areas where heat is being transferred from the house *(Courtesy: Air Conditioning Contractors of America, Manual J)*

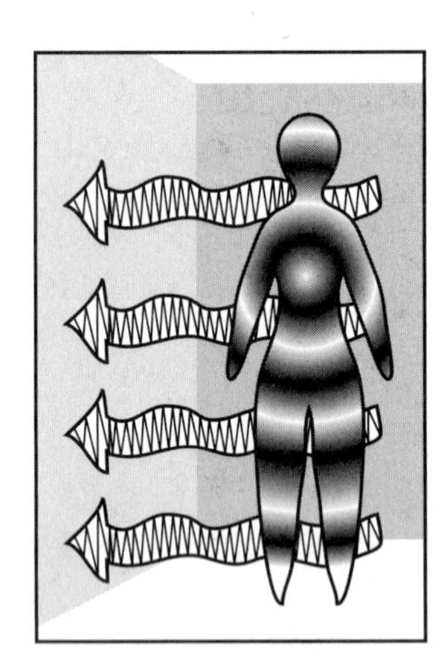

FIGURE 31.3 ■ Radiation effects on the human body transferred from the house *(Courtesy: Air Conditioning Contractors of America, Manual J)*

- Ceiling insulation
- Net wall area
- Wall insulation
- Window area
- Window construction
- Shading factor
- Door area
- Door construction
- Duct design
- Number of occupants
- Infiltration rate
- Ventilation rate
- Miscellaneous load

31.4 DESIGN CONDITIONS

Design conditions prescribe the outdoor and indoor temperatures and relative humidity that the cooling and heating system is designed to handle. The outdoor design conditions for summer and winter will vary with the climate. The indoor conditions are standard and are not dependent on outdoor conditions. For example, the outdoor design conditions for Houston, Texas, are:

Summer: 95° F
Winter: 33° F

In contrast, the outdoor design conditions for Denver, Colorado, are:

Summer: 91° F
Winter: 1° F

The inside design conditions for most climates are the same:

Summer: 75° F, 55% RH
Winter: 70° F

These design conditions can be obtained from ASHRAE or ACCA *Manual J* or *Manual N* calculation manuals.

As expected, warmer climates will require larger cooling equipment (BTUH of heat removed) and colder climates will require larger heating equipment (BTUH of heat added).

31.5 BUILDING ORIENTATION

Building orientation means the direction a building is facing related to north, south, east, and west. Review the textbook set of plans and locate *Sheet 1, Site Plan*. This plan shows the arrow for the north direction at the bottom of the page. The site orientation is important because of the heat gain from the sun during the cooling season. This is called solar heat gain. The walls facing east and west will receive more solar heat gain for the building than the south and north walls. The north wall receives the least amount of heat gain of all the walls. It is interesting to note that the heat gain of an east wall and a west wall are about the same. The heat gain on an east wall may not seem as intense, however, due to cooler temperatures in the morning. Solar heat gain is measured in BTUs per hour per square foot of wall area.

From an air-conditioning energy conservation perspective, it is best to minimize east and west windows. Windows facing north will have the least heat gain. Southern windows will need a properly designed building overhang or awning to reduce heat gain in the summer and allow solar penetration in the winter. A two-foot to three-foot overhang or eave over the south-facing glass will shade most locations or latitudes. Specific window and wall shading can be designed for each latitude to maximize energy-saving features of this construction.

31.6 FLOOR AREA

Calculating the floor area will be important in determining heat gain or loss through the floor. The floor area is found by multiplying the outside length and width of the structure to obtain the overall area in square feet. Formulas for these calculations are found in Appendix A of this book. Square feet is sometimes written as ft-sq or ft^2.

If the ceiling is flat, the ceiling area will be the same as the floor area. Many structures are not square or rectangular; therefore, they must be broken into subsets in order to find the total area of the structure (Figure 31.4).

Multiply	30' × 32'	=	960 square feet
	28' × 60'	=	1680 square feet
	Total		2640 square feet

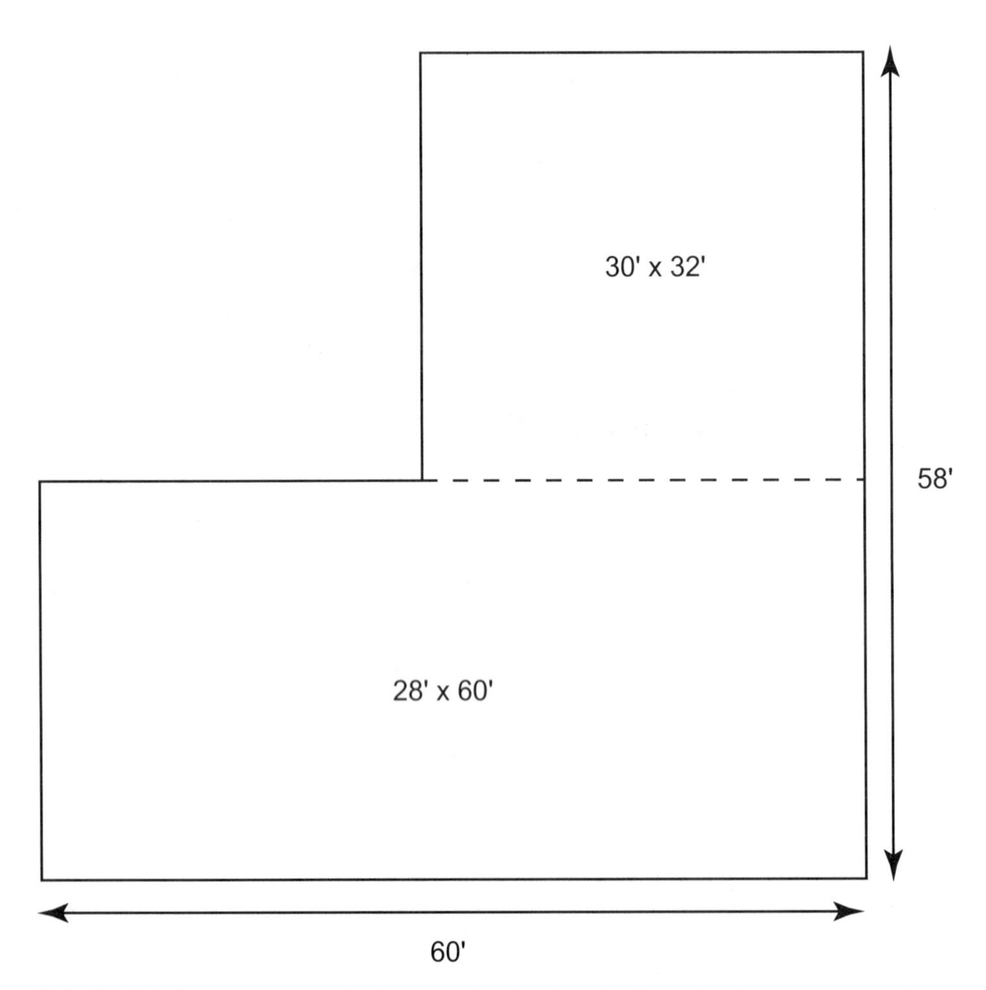

FIGURE 31.4 ■ Determine the floor area by dividing the floor plan into two rectangles

31.7 FOUNDATION TYPE

Determine the foundation type in order to calculate the load on the floor. A structure built on a concrete slab will have no heat gain in the summer. Slab construction will have most of its heat loss around the perimeter of the structure. Soil offers some resistance to heat transfer; therefore, there is insignificant heat transfer from all but the sides of the concrete slab.

A structure built off the ground with an exposed air space under the floor can have heat loss in the wintertime. Buildings that are skirted will reduce air infiltration and—to some degree—heat loss. A structure built over a conditioned basement will have no heat transfer, since the temperature of the basement is close to the temperature of the first floor (Figure 31.5).

31.8 FLOOR INSULATION

A building constructed over an unconditioned air space may be insulated. Common insulation material is fiberglass batt material secured between the floor joist spaces (Figure 31.6). Mineral or rock wool batts are also used. Spray-on cellulose may be used if the building is well above the ground. Certain types of cellulose spray applications can add to the fire resistance of the structure. Fiberglass and mineral wool batts do not offer fire-resistance protection to the floor structure.

Review the blueprints for the thickness and R-value of the insulation material. This will be important to the load calculation. The R-value of the insulation material can be determined from a chart similar to the one shown in Figure 31.7. The R-value is the thermal resistance of the material to heat transfer. Higher R-value materials reduce heat transfer.

Review textbook blueprint *A-6, Wall Sections 4 and 5.* No slab insulation is indicated on the drawing. In cold climates, the perimeter of the slab might be insulated against heat loss. In extremely cold climates, the bottom and perimeter of the slab may also be insulated.

31.9 CEILING AREA

If the ceiling is flat, it will have the same area as the floor area. The area is measured in square feet. The ceiling area will need to be measured if it is a cathedral ceiling or splayed type or an unusual design. Appendix A is useful for nonstandard shapes, such as those found in a cathedral ceiling. If this is a multiple-story structure, only calculate the ceiling area that is on the top story of the building. There should be no heat transfer between stories that have the same temperature; therefore, no heat transfers from the ceiling to the floor above (Figure 31.8).

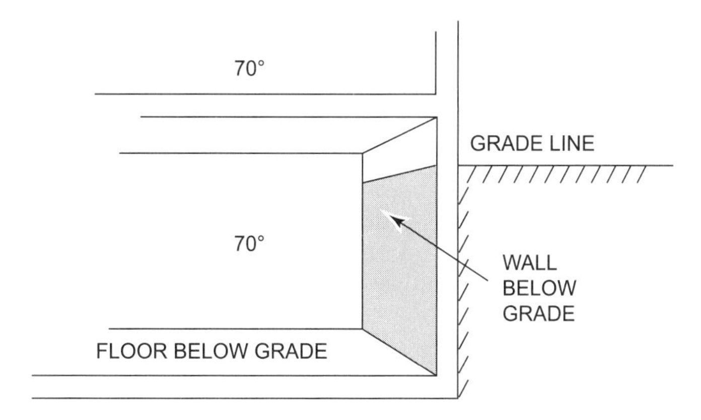

FIGURE 31.5 ■ No heat transfer will occur between floors because the basement and first story are at or near the same temperature *(Courtesy: Air Conditioning Contractors of America, Manual J)*

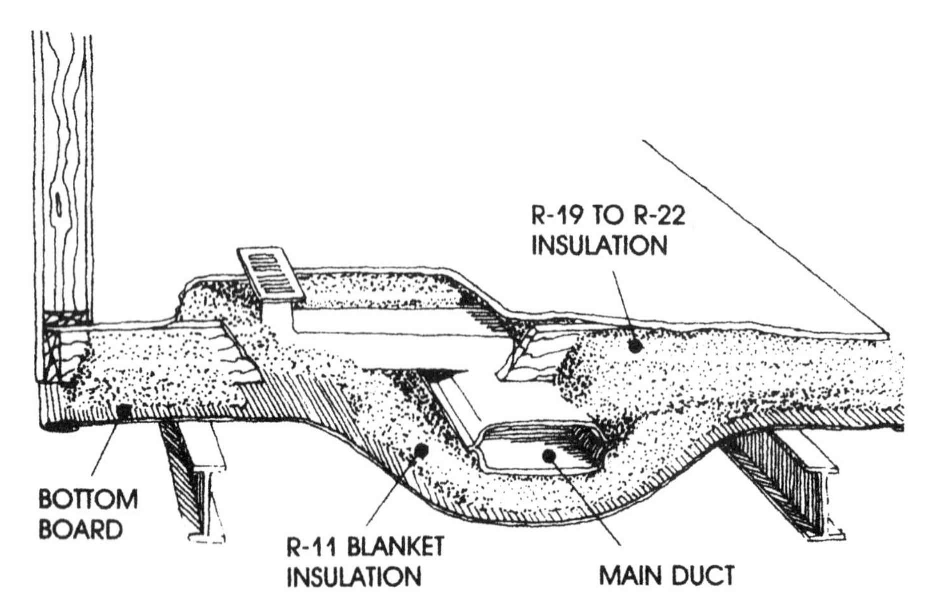

FIGURE 31.6 ■ Batt insulation installed between floor joists

	Approx R/Inch	R11	Approximate inches needed for: R19	R22	R30	R38	R49
Loose-Fill Machine-Blown							
Fiberglass	R2.25	5	8.5	10	13.5	17	22
Mineral Wool	R3.125	3.5	6	7	10.0	12.5	16
Cellulose	R3.7	3	5.5	6	8.5	10.5	13.5
Loose-Fill Hand-Poured							
Cellulose	R3.7	3	5.5	6	8.5	10.5	13.5
Mineral Wool	R3.125	3.5	6	7	10.0	12.5	16
Fiberglass	R2.25	5	8.5	10	13.5	17	22
Vermiculite	R2.1	5.5	9	10.5	14.5	18	23.5
Batts or Blankets							
Fiberglass	R3.14	3.5	6	7	10.0	12.5	16
Mineral Wool	R3.14	3.5	6	7	10.0	12.5	16
Rigid Board							
Polystyrene Beadboard	R3.6	3	5.5	6.5	8.5	10.5	14
Extruded Polystyrene	R4–5.41	3–2	5–3.5	5.5–4	7.5–5.5	9.5–7	12.5
Urethane (Styrofoam)	R5.4–6.2	2	3	3.5	5.0	6.5	8
Fiberglass	R4.0	3	5	5.5	7.5	9.5	12.5

FIGURE 31.7 ■ R-values for various types of insulation products *(Courtesy: Air Conditioning Contractors of America, Manual J)*

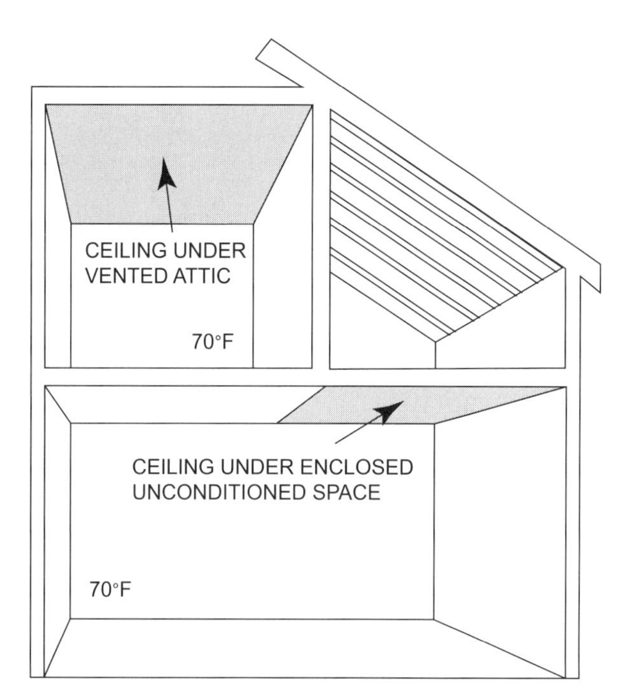

FIGURE 31.8 ■ There is no heat transfer between the first-story ceiling area that is below the second-story floor. There is heat transfer between the first-story ceiling area that is below the unconditioned space *(Courtesy: Air Conditioning Contractors of America, Manual J)*

Wall	R-11
Ceiling	R-30
Foundation	R-6

FIGURE 31.9 ■ Insulation schedule found on some blueprints

31.10 CEILING INSULATION

Common insulation materials are loose-fill cellulose or fiberglass and fiberglass or mineral wool batts. The thickness or R-value will be determined from the blueprints (Figure 31.9). The R-value will be converted to a U-value to be used in the load calculations. The U-value is the amount of heat that transfers through a wall section. The U-value is measured in BTUH per degree difference across 1 square foot of a building assembly (Figure 31.10). In this example, it includes the insulation and ceiling finish. Some blueprints provide the U-value information, so no conversion is required.

The total R-value of the ceiling is R-19.5.

The U-value is the reciprocal of the R-value.

$$\text{U-Value} = \frac{1}{\text{R-value}} \text{ or } \frac{1}{19.5} = 0.051 \text{ BTUH/hour/degree difference/square foot of ceiling}$$

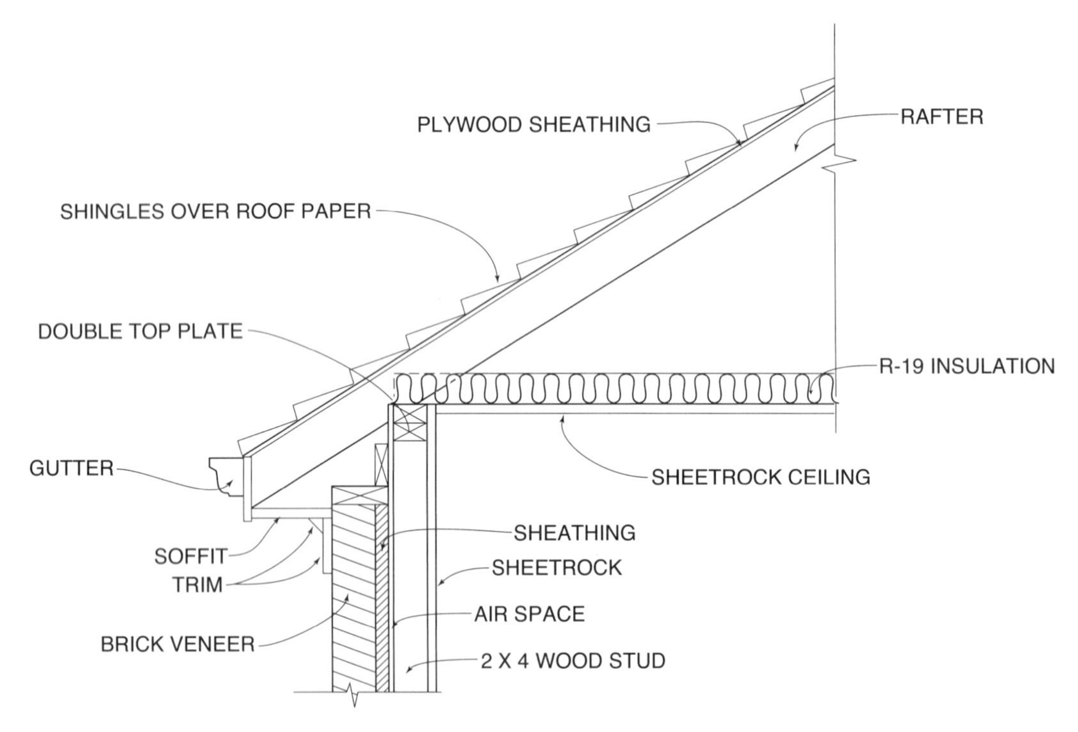

FIGURE 31.10 ■ The R-value of 1/2" Sheetrock is R-0.5 and 6 1/4" fiberglass batt is R-19

31.11 NET WALL AREA

You will need to know how to calculate the net wall area. The net wall area is calculated by subtracting the window and door area from the gross wall area.

To determine the net wall area, first calculate the gross wall area of the structure. The gross wall area can be calculated by multiplying the length and height of each outside wall. Add these wall areas together. Next, add the area of windows and doors together and subtract this total from the gross wall area. This will be the net wall area. See Figure 31.11 for an example of calculating the net wall area.

$A = L \times W$
$12' \times 8'$ $A = 96$ square foot gross area

Calculate the area of the doors and windows, and deduct this from the gross area.

Net Area = Total Area – (the area of the doors + the area of the windows)
96 square feet – (21 square foot door +12 square foot window)
Net Area = 63 square feet

(Net Wall Area = Gross Wall Area – Area of Windows and Doors)

31.12 WALL INSULATION

Determine the wall insulation by reviewing the blueprints (Figure 31.12). The wall cavity may have batt material. Foam insulation or other insulation material may be attached to the outside of the batt material. Add the R-value of the batt and external insulation together to determine the final insulation value.

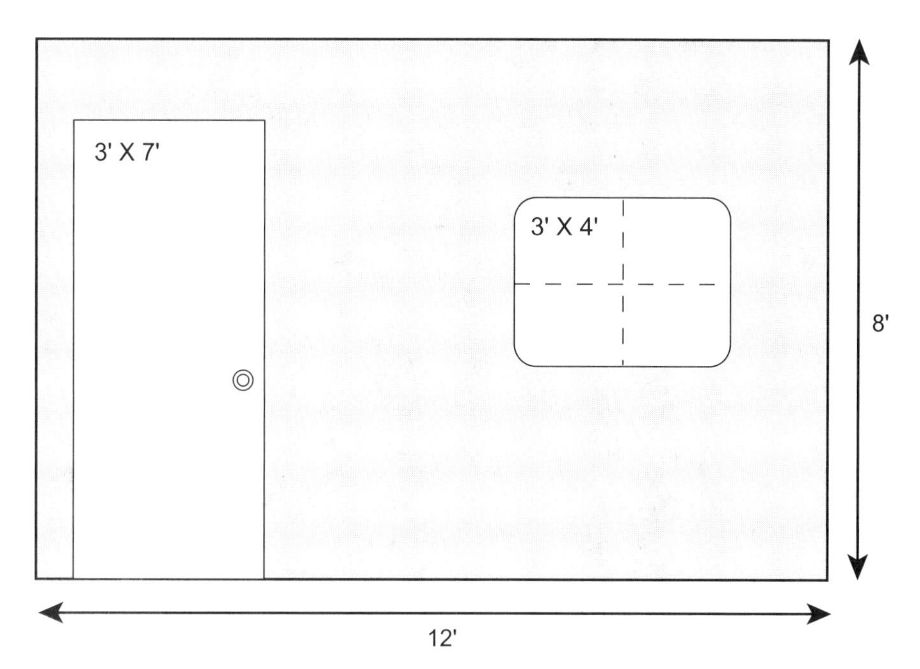

FIGURE 31.11 ■ To find the net wall area, first calculate the gross wall area by multiplying the length by the width

The inside and outside construction will also affect the final insulation value of the wall assembly. There are tables that can aid in calculating the final thermal resistance of the wall assembly (Figure 31.7). In some instances, this can also be found on the blueprints (Figure 31.9). The R-value will be converted to a U-value to be used in the load calculations.

31.13 WINDOW AREA

Calculate the window area by measuring the length and width of the window and sash. This will include all the area inside the frame but not the window molding (Figure 31.13). Measure all external patio or sliding glass doors. Include the glass area of any exterior door that makes up more than 20 percent of the door area.

31.14 WINDOW CONSTRUCTION

Windows are generally a big source of heat gain and loss within a structure. It is important to identify the type of window assembly. Using the blueprints or a site visit, determine the type of window that is or will be installed. Is it single-pane (or single-glazed) glass? Is it double-pane glass? Does it have a storm window or special tinting treatments? These are all important questions that can be answered from the blueprints or by conducting an audit of the building.

For an example of this, review the textbook blueprint A-5, *Transverse Section 12*, located at the bottom of the print. The window construction is 1/4" solar bronze-tempered glass with a bronze-anodized aluminum frame. The doors located in this section are glass. This window and door section can be combined. The scale is 1/4" to the foot. Measuring with a ruler, this section is 2'-7/8" × 4'-0". This includes the bronze framing material. Convert these measurements by multiplying the height and width by 4 because the scale is 1/4" to the foot. The calculated measurement will be 11.9 feet by 16 feet.

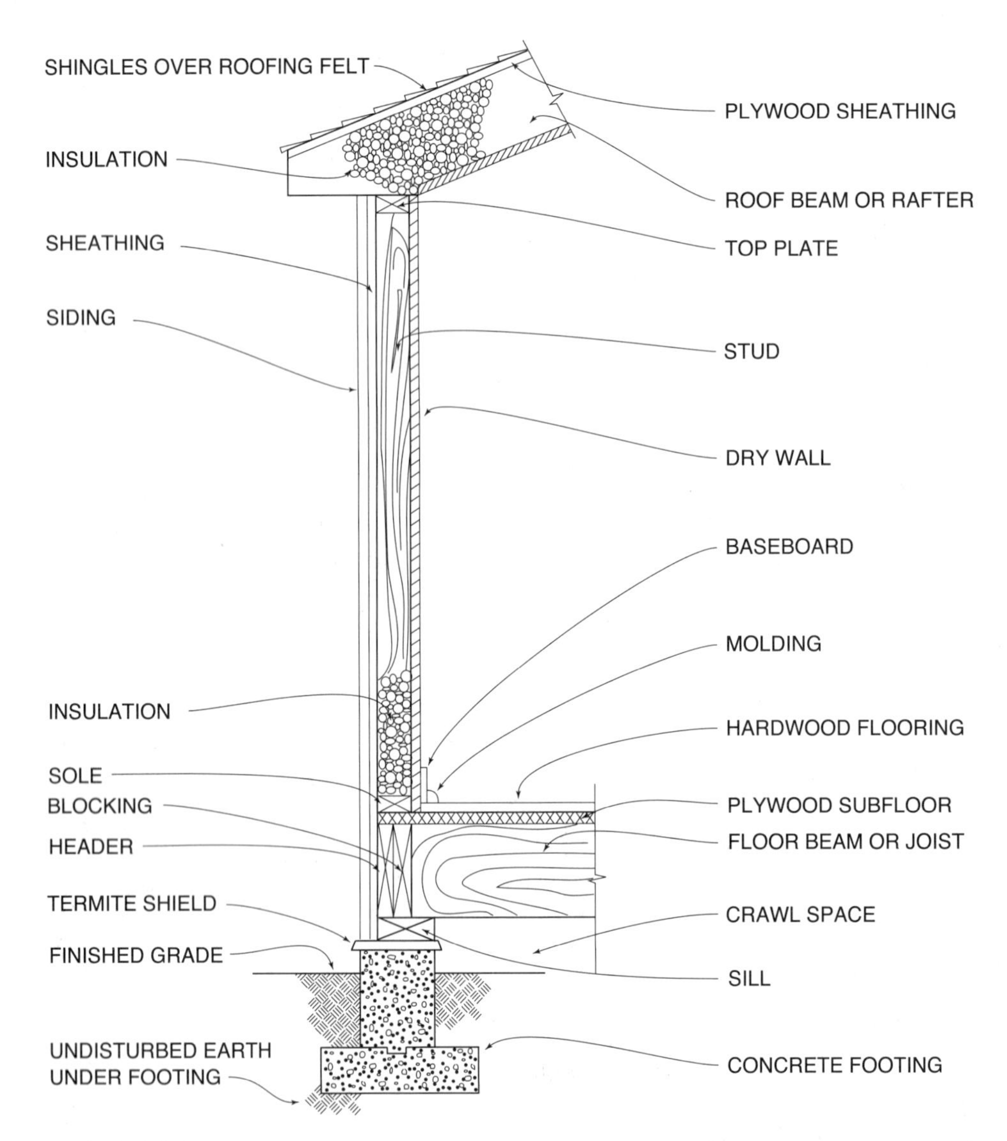

FIGURE 31.12 ■ In this cross-sectional view, add the R-value of all the wall assembly components. The assembly is made of drywall, insulation, sheathing, and siding

There are tables that will aid in determining the heat gain (Figure 31.14) or loss through glass (Figure 31.15). The heat transfer multipliers (HTM) are stated as BTUH per square foot of glass area. You must determine the cooling and winter temperature design conditions in order to use this chart.

31.15 SHADING FACTOR

Shading factor is used to figure how many BTUs are gained or lost through the window and window treatment (curtains, blinds, etc). Windows that face east and west will have the greatest solar heat gain in the summer, while north-facing glass will have the least. South-facing glass will have limited direct solar heat gain with a properly sized overhang. The overhang should be designed to allow heat gain for the winter season.

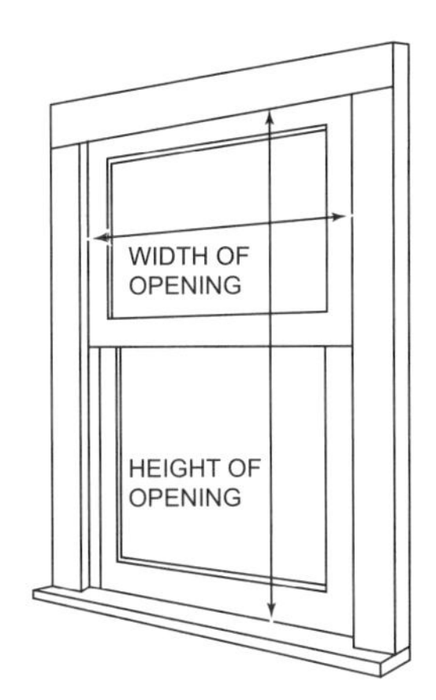

FIGURE 31.13 ■ Measuring the window glass and frame area *(Courtesy: Air Conditioning Contractors of America, Manual J)*

Glass Heat Transfers Multipliers (Cooling)
No External Shade Screen
Clear Glass

	Single Pane						Double Pane Single Pane & Low e Coating						Triple Pane Double Pane & Low e Coating					
Design Temp. Difference	**10**	**15**	**20**	**25**	**30**	**35**	**10**	**15**	**20**	**25**	**30**	**35**	**10**	**15**	**20**	**25**	**30**	**35**
Direction Window Faces							**No Internal Shading**											
N	23	27	31	35	39	43	19	21	23	25	27	29	17	18	19	20	21	22
NE and NW	56	60	64	68	72	76	47	49	51	53	55	57	43	44	45	46	47	48
E and W	81	85	89	93	97	101	68	70	72	74	76	78	62	63	64	65	66	67
SE and SW	70	74	78	82	86	90	59	61	63	65	67	69	53	54	55	56	57	58
S	40	44	48	52	56	60	34	36	38	40	42	44	30	31	32	33	34	35
							Draperies or Venetian Blinds											
N	14	18	22	26	30	34	12	14	16	18	20	22	10	11	12	13	14	15
NE and NW	33	37	41	45	49	53	29	31	33	35	37	39	25	26	27	28	29	30
E and W	48	52	56	60	64	68	42	44	46	48	50	52	37	38	39	40	41	42
SE and SW	41	45	49	53	57	61	37	39	41	43	45	47	32	33	34	35	36	37
S	24	28	32	36	40	44	21	23	25	27	29	31	18	19	20	21	22	23
							Roller Shades—Half-Drawn											
N	17	21	25	29	33	37	16	18	20	22	24	26	14	15	16	17	18	19
NE and NW	41	45	49	53	57	61	38	40	42	44	46	48	34	35	36	37	38	39
E and W	60	64	68	72	76	80	55	57	59	61	63	65	49	50	51	52	53	54
SE and SW	52	56	60	64	68	72	47	49	51	53	55	57	42	43	44	45	46	47
S	30	34	38	42	46	50	27	29	31	33	35	37	24	25	26	27	28	29
							Awning, Porches, or Other External Shading											
ALL DIRECTIONS	23	27	31	35	39	43	19	21	23	25	27	29	17	18	19	20	21	22

FIGURE 31.14 ■ Sample table from *Manual J*. This table is used to determine the heat gain through three different types of window assemblies. The heat transfer multiplier (HTM) is used to calculate the heat gain in the cooling load calculation *(Courtesy: Air Conditioning Contractors of America, Manual J)*

Heat Transfer Multipliers (Heating)
No. 1 Single Pane Window

Winter Temp. Difference	20	25	30	35	40	45	50	55	60	65	70	75	80	85	90	95	U
	HTM (BTUH per sq. ft.)																
Clear Glass																	
Wood Frame	19.8	24.8	29.7	34.7	39.6	44.6	49.5	54.5	59.4	64.4	69.3	74.3	79.2	84.2	89.1	94.1	.990
T.I.M. Frame	20.9	26.1	31.4	36.6	41.8	47.0	52.3	57.5	62.7	67.9	73.2	78.4	83.6	88.8	94.1	99.3	1.045
Metal Frame	23.1	28.9	34.7	40.4	46.2	52.0	57.8	63.5	69.3	75.1	80.9	86.6	92.4	98.2	104.0	109.7	1.155
Low Emittance Glass, e = 0.60																	
Wood Frame	18.4	23.0	27.5	32.1	36.7	41.3	45.9	50.5	55.1	59.7	64.3	68.9	73.4	78.0	82.6	87.2	.918
T.I.M. Frame	19.4	24.2	29.1	33.9	38.8	43.6	48.5	53.3	58.1	63.0	67.8	72.7	77.5	82.4	87.2	92.1	.969
Metal Frame	21.4	26.8	32.1	37.5	42.8	48.2	53.6	58.9	64.3	69.6	75.0	80.3	85.7	91.0	96.4	101.7	1.071
Low Emittance Glass, e = 0.40																	
Wood Frame	16.4	20.5	24.6	28.7	32.8	36.9	41.0	45.0	49.1	53.2	57.3	61.4	65.5	69.6	73.7	77.8	.819
T.I.M. Frame	17.3	21.6	25.9	30.3	34.6	38.9	43.2	47.5	51.9	56.2	60.5	64.8	69.2	73.5	77.8	82.1	.865
Metal Frame	19.1	23.9	28.7	33.4	38.2	43.0	47.8	52.6	57.3	62.1	66.9	71.7	76.4	81.2	86.0	90.8	.956
Low Emittance Glass, e= 0.20																	
Wood Frame	14.2	17.8	21.3	24.9	28.4	32.0	35.6	39.1	42.7	46.2	49.8	53.3	56.9	60.4	64.0	67.5	.711
T.I.M. Frame	15.0	18.8	22.5	26.3	30.0	33.8	37.5	41.3	45.0	48.8	52.5	56.3	60.0	63.8	67.5	71.3	.751
Metal Frame	16.6	20.7	24.9	29.0	33.2	37.3	41.5	45.6	49.8	53.9	58.1	62.2	66.4	70.5	74.7	78.8	.830

FIGURE 31.15 ■ Sample table from *Manual J*. This table is used to determine the heat loss through twelve different types of window assemblies. The heat transfer multiplier (HTM) is used to calculate the heat loss in the heating load calculation *(Courtesy: Air Conditioning Contractors of America, Manual J)*

In addition to considering the overhang or awnings, it is valuable to know the inside shading factor. Will the structure have curtains, blinds, or other inside window treatments? Do not consider the inside window treatment if it is not known or not stated on the blueprints. During the cooling season, outside shading has a greater impact on reducing heat transfer compared to an inside window treatment. The opposite is true of winter conditions.

31.16 DOOR AREA

Measure the length and width of all doors. Add the area together. Deduct the excess window area from the total door area. The excess window area in a door is considered when glass covers more than 20 percent of the door area. In this example, the window area should be included in the glass area tabulations and not in the door area. This information will be on the blueprints (Figure 31.16).

31.17 DOOR CONSTRUCTION

The blueprints should list the door construction characteristics. See blueprint plan A-8, *Door Schedule*. The door may be solid or hollow wood or metal with various types of foam insulation materials. The insulation value of the door will vary according to its construction features. The *Manual J* table in Figure 31.17 indicates that 23.0 BTUH per square foot will be lost through a solid-core wood door that has a temperature difference of 50° F.

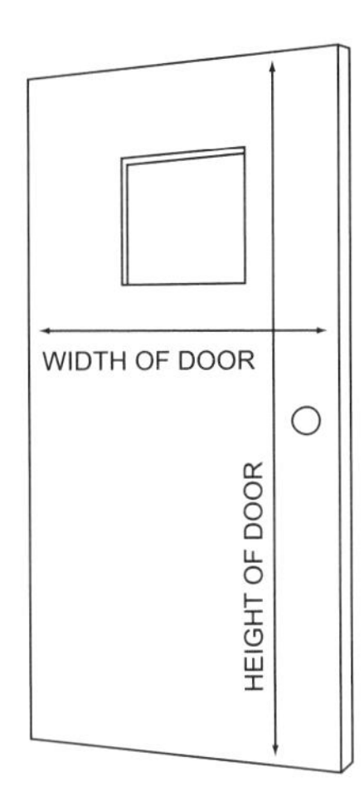

FIGURE 31.16 ■ Measuring the door area *(Courtesy: Air Conditioning Contractors of America, Manual J)*

No. 10 Wood Doors																	
Winter Temp. Difference	**20**	**25**	**30**	**35**	**40**	**45**	**50**	**55**	**60**	**65**	**70**	**75**	**80**	**85**	**90**	**95**	**U**
								HTM (BTUH per sq. ft.)									
A. Hollow Core	11.2	14.0	16.8	19.6	22.4	25.2	28.0	30.8	33.6	36.4	39.2	42.0	44.8	47.6	50.4	53.2	.560
B. Hollow Core & Wood Storm	6.6	8.3	9.9	11.6	13.2	14.9	16.5	18.2	19.8	21.5	23.1	24.8	26.4	28.1	29.7	31.4	.330
C. Hollow Core & Metal Storm	7.2	9.0	10.8	12.6	14.4	16.2	18.0	19.8	21.6	23.4	25.2	27.0	28.8	30.6	32.4	34.2	.360
D. Solid Core	9.2	11.5	13.8	16.1	18.4	20.7	23.0	25.3	27.6	29.9	32.2	34.5	36.8	39.1	41.4	43.7	.460
E. Solid Core & Wood Storm	5.8	7.3	8.7	10.2	11.6	13.1	14.5	16.0	17.4	18.9	20.3	21.8	23.2	24.7	26.1	27.6	.290
F. Solid Core & Metal Storm	6.4	8.0	9.6	11.2	12.8	14.4	16.0	17.6	19.2	20.8	22.4	24.0	25.6	27.2	28.8	30.4	.320
G. Panel	13.4	16.8	20.1	23.5	26.8	30.2	33.5	36.9	40.2	43.6	46.9	50.3	53.6	57.0	60.3	63.7	.670
H. Panel & Wood Storm	7.2	9.0	10.8	12.6	14.4	16.2	18.0	19.8	21.6	23.4	25.2	27.0	28.8	30.6	32.4	34.2	.360
I. Panel & Metal Storm	8.2	10.3	12.3	14.4	16.4	18.5	20.5	22.6	24.6	26.7	28.7	30.8	32.8	34.9	36.9	39.0	.410

FIGURE 31.17 ■ Sample heat loss table for doors *(Courtesy: Air Conditioning Contractors of America, Manual J)*

Let us do an example using this information:

Example: Determine the Door Heat Loss at a 50° F Temperature Difference.

Review blueprint plan A-8, *Door Schedule,* to locate door number 102a.

According to the plan, the door measures 2'-8" × 8'-10".

The area of the door is calculated by converting the inches to decimal form and multiplying 2.67 feet by 8.83 feet to equal 23.58.

The area is 23.58 square feet.

The total heat loss through this door will be 23.0 BTUH × 23.58 square feet = 542.34 BTUH.

31.18 DUCT DESIGN

There are several factors to consider when reviewing the duct layout. If the duct is located in an unconditioned space, such as an attic area, it will gain heat in the summer and lose heat in the winter. This is called the duct load. Ducts that are located below the ceiling insulation or in a furred-down location will not experience this heat transfer. An energy-efficient design will dictate that ductwork is installed where it will not transfer heat. In commercial establishments, ducts are sometimes exposed in the condition space. Normally, ducts installed in this fashion do not need insulation. Ducts may sweat or condense moisture if exposed and suspended 12 feet or higher in an area that is not conditioned above 8 feet. The warm, moist air from the conditioned space will rise due to natural convection. The dew point will be reached on the cold duct surface and water droplets will form, falling on the structure below. Twelve feet is an arbitrary height. Condensation problems may develop at lower duct levels because of cold duct surfaces or excess indoor moisture conditions.

Ducts located outdoors or in unconditioned spaces will need insulation and a vapor barrier. Ducts located outside will need an approved weather barrier. Common duct insulation R-values are R-4, R-6, and R-8. The vapor barrier is usually an aluminum foil–based or an U-V–resistant plastic product.

A duct chart is used to determine the heat gain or heat loss of a duct system. Figure 31.18 shows that an exposed supply air duct, installed in the outdoor ambient air with no insulation, will increase the duct load by .25 or 25 percent. An uninsulated return air duct in an unconditioned space will add another .15 or

Duct Gain and Loss Factors
(Ducts Tightly Sealed with Tape)

Location	Insulation Level	Cooling Gains: Supply	Cooling Gains: Return	Heating Losses: Supply Below 120	Heating Losses: Supply Above 120	Heating Losses: Return Ducts
Exposed to outdoor ambient, roof, open or vented space, exterior wall, uninsulated unconditioned space	None	0.25	0.15	0.25	0.30	0.15
	R2	0.15	0.08	0.15	0.20	0.08
	R4	0.10	0.06	0.10	0.15	0.06
	R6	0.05	0.03	0.05	0.10	0.03
Ducts located in enclosed, insulated, unconditioned unvented spaces	None	0.20	0.12	0.20	0.25	0.12
	R2	0.10	0.06	0.10	0.15	0.06
	R4	0.05	0.03	0.05	0.10	0.03
	R6	0.03	0.00	0.03	0.05	0.00
Ducts located in a return air-ceiling plenum	None	0.10	0.00	0.10	0.15	0.00
	R2	0.05	0.00	0.05	0.10	0.00
	R4	0.03	0.00	0.03	0.08	0.00
	R6	0.00	0.00	0.00	0.03	0.00
Ducts buried in or under a concrete slab	None	0.10	0.06	0.10	0.15	0.06
	R2	0.05	0.03	0.05	0.10	0.03
	R4	0.03	0.00	0.03	0.08	0.00
	R6	0.00	0.00	0.00	0.03	0.00
Ducts in conditioned space	na	0.00	0.00	0.00	0.00	0.00
Ducts "stubbed" from unit to space with negligible run	Any	0.00	0.00	0.00	0.00	0.00

Notes:
(1) Increase multipliers by 0.10 if ducts are not taped.
(2) Use a "weighted average" multiplier if all the entire supply duct system or the entire return duct system is not classified in the same location category. The example below shows how to calculate this "weighted average" multiplier.

Example:
Thirty percent of the supply system is on the roof, 50 percent of the supply system in an unconditioned, unvented space, and 20 percent of the supply system is the conditioned space. Ducts taped and insulated @ R-4. Hot water coil heating supply temperature is 135 degrees.

FIGURE 31.18 ■ Duct gain and loss factors *(Courtesy: Air Conditioning Contractors of America, Manual N)*

15 percent to the total load. In this example, ambient exposed ducts with no insulation can increase the total load up to 40 percent.

Notice that ducts in the conditioned space will not be subject to heat transfer. Insulation will be required on the supply duct to prevent condensation problems.

31.19 NUMBER OF OCCUPANTS

The number of persons in a conditioned space can increase the cooling load on a building. This is considered the internal heat load of a building. A normally occupied residential structure will not add much heat to the area. A heavily occupied commercial space may contain enough people to be a significant portion of the cooling load. Occupants who are involved in exercise can produce several times more BTUs of heat than an audience simply watching a movie.

The human body generates heat in two forms: sensible heat and latent heat. Sensible heat is simply heat that is released from a person through conduction, convection, and radiation. Latent heat is the heat associated with the amount of moisture that must be removed by the air-conditioning system. Moisture is generated when we breathe and perspire. Tables are available that indicate the amount of sensible and latent heat generated by a person under varying activity conditions (Figure 31.19). Data are available that prescribe the

Heat Gain from Occupants

Application	Degree of Activity	Sensible BTUH	Latent BTUH
Assembly Hall, Church or School Auditorium, Theater	Seated, at rest	210	140
Funeral Parlor	Seated, very light work	230	190
Bank	Standing, slow walk, light work	315	325
Barber Shop	Seated, very light work	230	190
Beauty Parlor	Seated, very light work	230	190
Bowling Alley	Bowling or medium-heavy work	635–1000	1165–2000
Conference Room	Moderately active work	255	255
Cocktail Lounge, Bar, Tavern	Standing, slow walk, light work	315	325
Department Stores, Retail Shops	Standing, slow walk, light work	315	325
Drugstore			
Pharmacist's Work Area	Light work	345	435
Sales Area	Standing, slow walk, light work	315	325
Dormitory	Seated, very light work	230	190
Factory	3 mph walk, moderate work	565	1035
Food Services			
Dining Room	Seated, eating	255	325
Cafeteria, Short-Order, Drive-In	Seated, eating	255	325
Kitchen	Light work	345	435
Garage (repair)	3 mph walk, moderate work	565	1035
Gymnasium			
Spectators	Seated, very light work	230	190
Participants	Medium-heavy work	635–1000	1165–2000

FIGURE 31.19 ■ This table represents the sensible and latent heat gain of various types of occupants activities *(Courtesy: Air Conditioning Contractors of America, Manual N)*

Occupancy Estimates

Application	Sq. Ft. Floor Area per Person	Application	Sq. Ft. Floor Area per Person
Assembly hall, church or school or auditorium, theater or funeral parlor	7	Drug stores	
Bank	50	Pharmacist's work area	30
Barber shop	40	Sales area	50
Beauty parlor	20	Dormitory	50
Bowling alley	Seating + 6 per alley	Factory	See note (2)
Conference room	14 to 17	Food services:	
Cocktail lounge, bar, tavern	10	Dining room	14
Department stores, retail shops:		Cafeteria, short-order, drive-in	14
Basement and first floor	33	Kitchen	See note (5)
Other floors	50	Garage	
		Parking	See note (2)
		Repair	See note (2)
		Gymnasium	Seating + 10 per basket

FIGURE 31.20 ■ Sample table from *Manual N*. This table is used to estimate the occupancy load in a commercial business. Local codes may recommend a different occupancy load *(Courtesy: Air Conditioning Contractors of America, Manual N)*

occupancy load of a structure. For example, the occupancy load of a residence is two people per bedroom. The occupancy load of an assembly can be as low as seven square feet per person (Figure 31.20). Therefore, a 7000 square foot auditorium can have an occupant load of 1000 people. Building codes set the occupancy load depending on the occupancy use and type. Check with your local code authority.

31.20 INFILTRATION RATE

Infiltration is unwanted and uncontrolled air that enters into a building. Unconditioned air enters the building as it pushes out indoor air that has been conditioned. It is easier to control the infiltration rate in a commercial building because it is pressurized. Residential structures are not usually pressurized; therefore, infiltration has a greater impact on these buildings.

The infiltration rate is divided into three general categories:

- Poor or loose construction
- Average construction
- Best or tight construction

The infiltration rate is a judgment call. New construction built with energy-conservation measures may be considered tight construction. The air-conditioning contractor sizing the equipment may use the average construction factors because they are not able to inspect the construction practices. A structure can lose its tight construction features if regular maintenance is not performed on the building.

31.21 VENTILATION RATE

The ventilation is also known as fresh air makeup, outdoor air makeup, or outside air intake. The ventilation rate is the amount of outdoor air that is ducted into the negative pressure return air stream. See Figure 25.15 to locate the outside air intake. The purpose of the ventilation air is to reduce the indoor air pollutants that are recirculated in the building. The mechanical code requires a minimum amount of ventilation air in commercial buildings. The amount of outside air depends on the type of occupancy (Figure 31.21). Environments that have greater indoor air pollution will require more outside air. Residential structures are not required to have ventilation air because they rely on air infiltration to improve indoor air quality.

31.22 MISCELLANEOUS LOADS

A miscellaneous load is any heat-generating item not covered in the previously discussed topics. Along with the occupancy load, this is considered an internal heat gain because the heat is developed within the structure. It will include heat generated by motors, computers, lights, or cooking equipment. This information is available from professional sources, such as the ASHRAE and the ACCA (Figure 31.22, Figure 31.23, and Figure 31.24).

Occupancy Type	CFM per Person
Office	15 cfm
Dancehall (smoking)	30 cfm
Sports Auditorium	20 cfm
School	15 cfm

FIGURE 31.21 ■ Minimum required outside air requirements for specific occupancy type

Heat Gain from Appliances (BTUH)

	Electric				Gas				Steam			
	Without Hood			Hood	Without Hood			Hood	Without Hood			Hood
Type of Appliance	Sensible	Latent	Total	All Sensible	Sensible	Latent	Total	All Sensible	Sensible	Latent	Total	All Sensible
Broiler-Griddle 31" x 20" x 18"					11,700	6,300	18,000	3,600				
Coffee brewer/warmer												
per burner	770	230	1,000	340	1,750	750	2,506	500				
per warmer	230	70	300	90								
Coffee urn												
3 gallon	2,550	850	3,400	1,000	3,500	1,500	5,000	1,000	2,180	1,120	3,300	1,000
5 gallon	3,850	1,250	5,100	1,600	5,250	2,250	7,500	1,500	3,300	1,700	5,000	1,600
8 gallon (twin)	5,200	1,600	6,800	2,100	7,000	3,000	10,000	2,000	4,350	2,250	6,600	2,100
Roaster	1,700	1,100	2,800	900								
Deep-fat fryer:												
15# fat	2,800	6,600	9,400	3,000	7,500	7,500	15,000	3,000				
21# fat	4,100	9,600	13,700	4,300								
Dry food warmer per sq. ft. top	320	80	400	130	560	140	700	140				
Griddle, frying per sq. ft. top	3,000	1,600	4,600	1,500	4,900	2,600	7,500	1,500				
Griddle-Grille	6,600	3,600	10,200	3,200								
Hot plate (two heating units)	5,300	3,600	8,900	2,800								
Short-order stove (open grates) per burner					3,200	1,800	5,000	1,000				
Steam table per sq. ft. top					750	500	1,250	250	500	325	825	260
Toaster												
Continuous												
360 slices per hour	1,960	1,740	3,700	1,200	3,600	2,400	6,000	1,200				
720 slices per hour	2,700	2,400	5,100	1,600	6,000	4,000	10,000	2,000				
Pop-up (4 slice)	2,230	1,970	4,200	1,300								

FIGURE 31.22 ■ Sensible and latent heat gain from commercial kitchen appliances *(Courtesy: Air Conditioning Contractors of America, Manual N)*

Heat Gain from Electric Motors (Continuous Operation) (BTUH)

			Location of Motor and Driven Equipment with Respect to Conditioned Space or Air Stream		
			A	B	C
Motor Nameplate or Rated Horsepower	Motor Type	Full Load Motor Efficiency in Percent	Motor In, Driven Equipment in BTU/hr	Motor Out, Driven Equipment in BTU/hr	Motor In, Driven Equipment in BTU/hr
0.05 (1/20)	Shaded Pole	35	360	130	240
0.08 (1/12)		35	580	200	380
0.125 (1/8)		35	900	320	590
0.16 (1/6)		35	1160	400	760
0.25 (1/4)	Split Phase	54	1180	640	540
0.33 (1/3)		56	1500	840	660
0.50 (1/2)		60	2120	1270	850
0.75 (3/4)	3 Phase	72	2650	1900	740
1		75	3390	2550	850
1.5		77	4960	3820	1140
2		79	6440	5090	1350
3		81	9430	7640	1790
5		82	15500	12700	2790
7.5		84	22700	19100	3640
10		85	29900	24500	4490
15		86	44400	38200	6210
20		87	58500	50900	7610
25		88	72300	63600	8680

FIGURE 31.23 ■ Partial list of motors that will increase the heat gain of a conditioned space *(Courtesy: Air Conditioning Contractors of America, Manual N)*

Lighting Heat Gains (BTUH)

Location and Type of Lighting	Lighting Load
Lighting entirely in conditioned space	
Incandescent lights	3.4 x Installed Watts
Fluorescent lights	4.1 x Installed Watts
Lighting recessed into a dead air-ceiling space	
Incandescent lights	3.4 x Installed Watts
Fluorescent lights	4.1 x Installed Watts
Lighting recessed into a return air-ceiling space	
Incandescent lights	
Load to conditioned space	0.60 x 3.4 x Installed Watts
Load to plenum space	0.40 x 3.4 x Installed Watts
Fluorescent lights	
Load to conditioned space	0.60 x 4.1 x Installed Watts
Load to plenum space	0.40 x 4.1 x Installed Watts

FIGURE 31.24 ■ Lighting will add heat to a commercial space. This can be a significant cooling load. Lighting can also reduce the heating load requirements *(Courtesy: Air Conditioning Contractors of America, Manual N)*

31.23 ZONING

Zoning simply means dividing the structure into more than one conditioned space or zone. For example, a two-story house should always be divided into an upper and lower zone, with separate systems serving upstairs and downstairs locations. This design will create comfort zones for both locations. A single piece of equipment that serves two stories does not provide the intended comfort. Hot air naturally rises to the upstairs level, and this creates a warmer upstairs zone in both winter and summer (Figure 31.25). The lower floor is cooler in both seasons. This creates a dramatic temperature imbalance and dissatisfaction from the customer. However, a single system can be used in a two-story structure with the use of specially designed duct-damper systems.

A single-story building may need to be zoned if it requires over 5 tons of air conditioning. Zoning can be installed on smaller cooling requirements. A customer may want separate systems for the sleeping areas and for the general living areas, such as the living room, den, and kitchen.

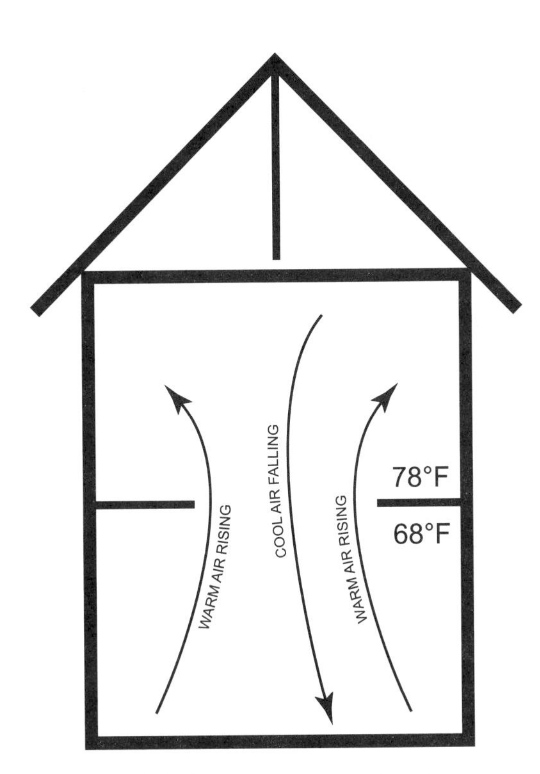

FIGURE 31.25 ■ Notice the stack effect between stories. The lighter, warm air will rise and the heavier, cooler air will fall

SUMMARY

You have been introduced to the information you will need in order to complete a cooling and heating load calculation. Residential load calculations require less information than commercial load calculations. To become proficient in load calculations, you must practice using blueprints as well as compiling information on existing buildings. It is best to learn by using the handwritten load calculation form method first. Once you become experienced with the written load calculation form method, you may begin using computer software for this application. Learning the written form first will help you understand mistakes that you may make when using a software program.

The most important part of developing an accurate load calculation is extracting the correct measurements from blueprints or from an existing building. If this information is not correct, the air-conditioning and heating equipment will not be properly sized for the customer. If the load calculation is used to submit a bid, its accuracy is very important. Over-sizing the equipment may cost the contractor the job because the installation cost will be higher. Under-sizing the equipment will lead to an unsatisfied customer and a possible lawsuit. Understanding blueprints and applying them to load calculations is an important technical skill.

REVIEW QUESTIONS

1. What is the term for correctly sizing air-conditioning equipment?
2. A 10 KW heater will generate how much heat?
3. Where can you obtain load calculation forms and instructions?
4. What are the three methods of heat transfer? Give an example of each.
5. Which form of heat transfer is the most difficult to evaluate?
6. What are the summer and winter indoor design conditions for most climates?
7. What is the total floor area of a two-story building that measures 60' × 80'?
8. Where is most of the heat loss in a house that has concrete slab construction?
9. Name three types of floor insulation.
10. Name three common types of ceiling insulation.
11. Determine the net wall area of a wall section with the following measurements:

    The wall perimeter is 90 feet.

    The wall height is 10 feet.

    There are two doors each measuring 3' × 7'.

    There are three windows each measuring 5' × 5'.
12. What is the area of a window that measures 4' 9" by 3' 4"?
13. Which window orientation has the most heat gain in the cooling season? Select the best answer:

    A. South

    B. East

    C. West

    D. North

    E. East and West

    F. North and South
14. What are the two reasons attic-exposed air-conditioning ducts are insulated?
15. Name three general sources of internal heat gain.
16. Why does a two-story building need two separate air-conditioning systems?

ADDITIONAL STUDENT EXERCISES

Using Provided Blueprints on Load Calculations

1. From plan A-8, *Door Schedule*, what is the total area of doors #102A and #103? Round off to the nearest tenth of a foot.
2. From plan A-8, *Plan of "B" Wall in Public Lobby*, what is the perimeter of this wall section? Round off to the nearest tenth of an inch.
3. From plan A-4, *Details of Fascia*, what is the type and depth of roof insulation?
4. From plan A-4, *Typical Wall Section*, what type of insulation is located directly above the suspended ceiling?
5. From plan A-4, *Typical Wall Section*, what type of insulation is located on the slab foundation?
6. From plan 2, *Foundation Plan*, what is the perimeter of the east-facing wall?
7. From plan M-1, name the type and thickness of the metal duct insulation.
8. From plan M-1, name the type and thickness of the flexible duct insulation.
9. Sketch a simple residential floor plan to scale. Show the interior spaces, including at least three bedrooms, two bathrooms, a kitchen, family room, living room, and a hallway or foyer. Show exterior and interior measurements to scale. Determine the total area of each room.
10. Draw a wall section (length) of the house in question #9. Include all windows and doors as shown on the plan. Calculate the gross wall area and the net wall area. Show all the measurements on the drawing.

APPENDIX A

Geometric Figures and Formulas

RECTANGLE

$A = L \times W$
$P = 2L + 2W$

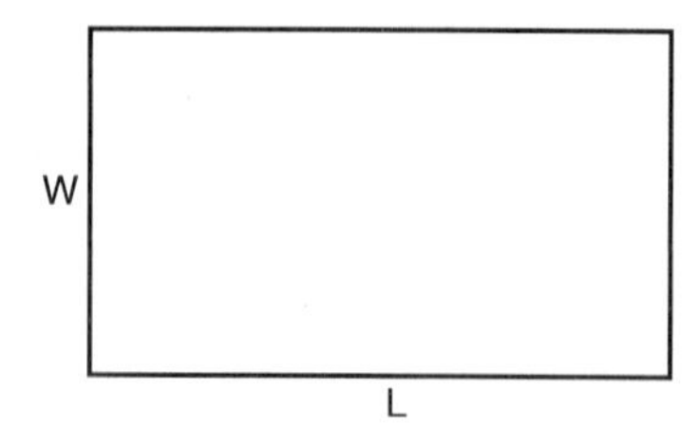

RECTANGULAR TANK

$A = 2 (W \times H) + 4 (L \times H)$
(TOTAL OUTSIDE AREA)
$V = W \times H \times L$

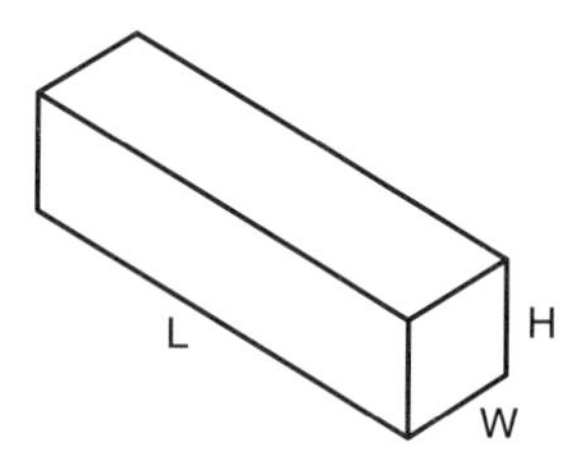

SQUARE

$A = L \times W$
$P = 2L + 2W$

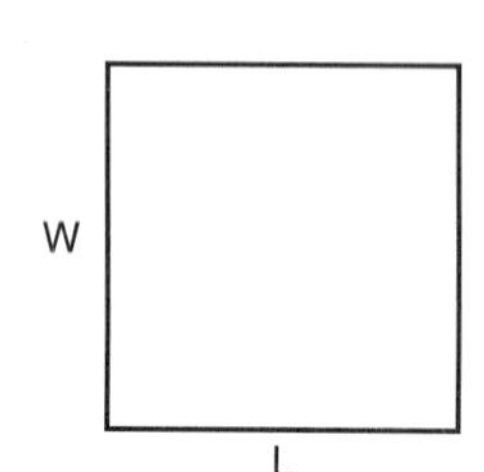

CYLINDER

$S = 6.283 \times R \times H + 6.283 \times R \times R$
(SURFACE AREA OF TANK)
$V = 3.14 \times R \times R \times H$

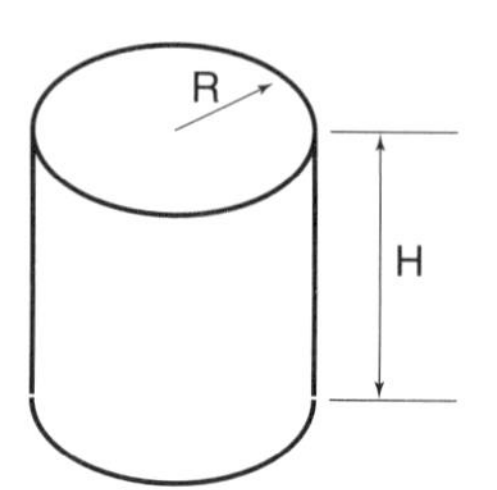

PARALLELOGRAM

$A = H \times L$

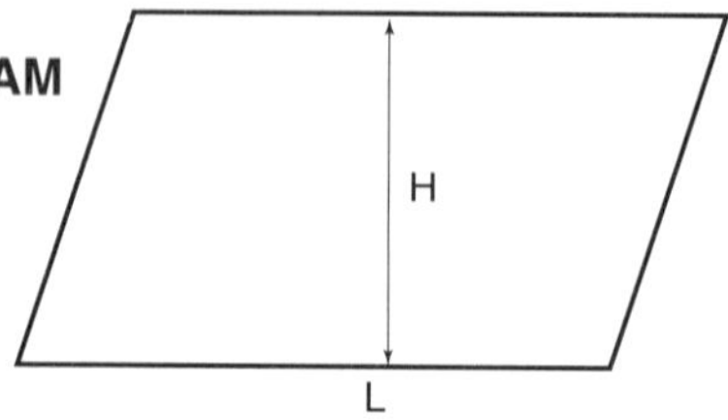

ELLIPTICAL TANK

$V = 3.14 \times Z \times Y \times H$
$S = 6.283 \times \sqrt{\frac{Y^2 + Z^2}{2}} \times H + 6.283 \times Y \times Z$
(S = OUTSIDE AREA)

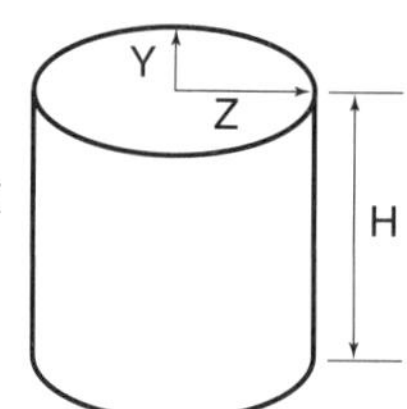

TRAPEZOID

$A = H \times \frac{L_1 + L_2}{2}$

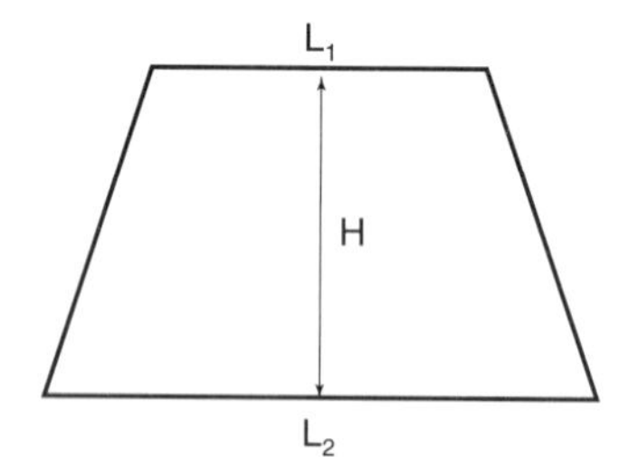

TRIANGULAR STRUCTURE

$V = \frac{W \times H \times L}{2}$

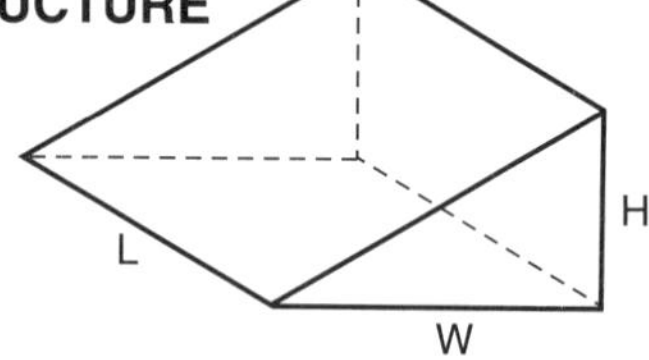

TRIANGLE

$A = \frac{L + H}{2}$

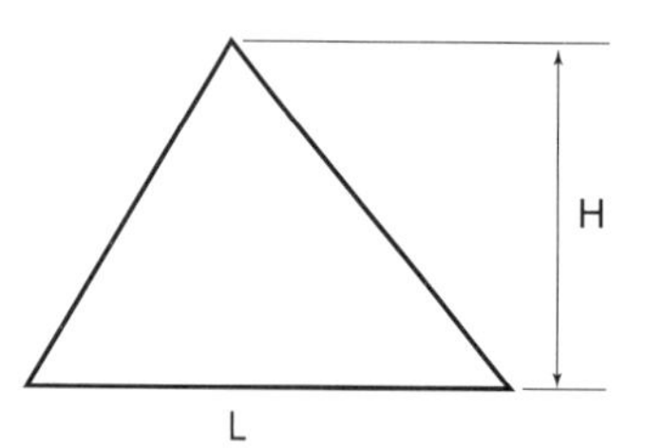

CONE

$S = 3.14 \times R \times S + 3.14 \times R \times R$
$V = 1.047 \times R \times R \times H$

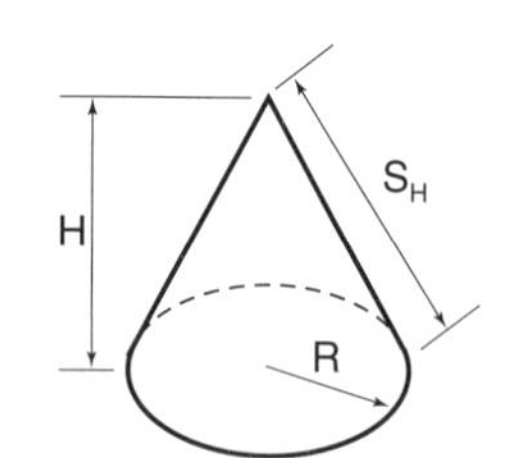

CIRCLE

$A = 3.14 \times R \times R$
$C = 3.14 \times D$
$R = \frac{D}{2}$
$D = R + R$

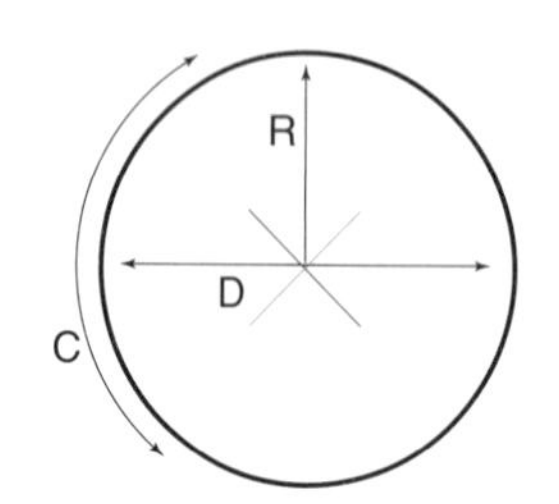

KEY

S = SURFACE
A = AREA
V = VOLUME
Y = SHORT RADIUS
Z = LONG RADIUS
S_H = SLOPE
C = CIRCUMFERENCE
P = PERIMETER
L & W = LENGTHS OF SIDES
H = HEIGHT
R = RADIUS
D = DIAMETER

APPENDIX B

Conversion Tables

TABLE 1 METRIC SYSTEM TO METRIC SYSTEM

10 millimeters	=	1 centimeter
10 centimeters	=	1 decimeter
10 decimeters	=	1 meter
10 meters	=	1 dekameter
100 dekameters	=	1 kilometer

TABLE 2 METRIC SYSTEM TO ENGLISH SYSTEM

1 millimeter	=	0.03937 inch
1 centimeter	=	0.3937 inch
1 meter	=	39.37 inches
1 kilometer	=	0.6214 miles

TABLE 3 ENGLISH SYSTEM TO METRIC SYSTEM

1 inch	=	25.4 millimeters	=	0.0254 meters
1 foot	=	304.8 millimeters	=	0.3048 meters
1 yard	=	914.4 millimeters	=	0.9144 meters
1 mile	=	1.6093 kilometers	=	1609.3 meters

TABLE 4 ENGLISH SYSTEM—CONVERTING FRACTIONS TO DECIMALS

Inch-Fractions		**Inch-Decimals**
1/16	=	0.0625
1/8	=	0.125
3/16	=	0.1875
1/4	=	0.25
5/16	=	0.3125
3/8	=	0.375
7/16	=	0.4375
1/2	=	0.5

TABLE 4 ENGLISH SYSTEM—CONVERTING FRACTIONS TO DECIMALS (CONTINUED)

Inch-Fractions		Inch-Decimals
9/16	=	0.5625
5/8	=	0.625
11/16	=	0.6875
3/4	=	0.75
13/16	=	0.8125
7/8	=	0.875
15/16	=	0.9375

TABLE 5 CONVERSION TABLE—CONVERTING INCHES TO FEET

Inch		Foot
1	=	0.083
2	=	0.17
3	=	0.25
4	=	0.33
5	=	0.42
6	=	0.5
7	=	0.58
8	=	0.67
9	=	0.75
10	=	0.83
11	=	0.92
12	=	1

APPENDIX C

Abbreviations

Abbreviation	Description
A	area
AC	alternating current
AD	access door or area drain
ADD	addendum
AGA	American Gas Association
AIEE	American Institute of Electrical Engineers
AIR COND	air conditioning
AL	aluminum
ALM	alarm
AMP	ampere
AP	access panel
APPD	approved
APPROX	approximate
ARCH	architect or architectural
ASB	asbestos
ASHRAE	American Society of Heating, Refrigeration, and Air-Conditioning Engineers
ASME	American Society of Mechanical Engineers
ASPH	asphalt
ASRE	American Society of Refrigeration Engineers
ASSOC	associate or association
AVE	average
AWG	American wire gauge
B	bath or brine
BB	baseboard
BD	board
BET	between
BF	back feed
BH	boiler house
BK SH	bookshelves
BKR	breaker
BLD LIN	building line
BLD or BLDG	building
BLO	blower
BLR	boiler house
BO	blow-off
BOT	bottom
BP	bypass
BR	bedroom or boiler room
BRK	brick
BS	Bureau of Standards
BSMT	basement
BT	bathtub
BTR	better
BTU	British thermal unit
BVL	beveled
BWV	back water valve
C	thermal conductance
CB CEM	catch basin cement
CEM FL	cement floor
CEM MORT	cement mortar
CEM PLAS	cement plaster
CFLG	counter flashing
CFM	cubic feet per minute
CI	cast iron
CIR	circle
CIR BKR	circuit breaker
CKT	circuit
CL	center line or closet
CLG	ceiling
CLR CND	clear conduit
CO	clean-out
CONC	concrete
COND	conductor
CONN	connection
CONT	contact
CONTR	contractor
CONV	convector
CONV ENCL	convector enclosure
COP	copper
COV	cover
COV PL	cover plate
CP	cesspool
CPM	cycles per minute
CR PL	chrome plated
CSG	casing
CTR	center
CU FT	cubic feet
CU IN	cubic inch
CUR	current
CV	check valve
CW	cold water
CWP	circulating water pump
CYL	cylinder
D R	dining room
D S or DS	downspout

Abbreviation	Description
DC	direct current
DEG or °	degree
DF	drinking fountain
DIA or DIAM	diameter
DIAG	diagram
DIF	diffuser
DIM	dimension
DISC	disconnect
DMPR	damper
DO	ditto
DR	drain
DW	dry well
DWG	drawing
DWN or DN	downspout
E	east
EDR	equivalent direct radiation
ELEV	elevator
EME	emergency
ENCL	enclosure
ENGR	engineer
ENT	entrance
EQUIP	equipment
EST	estimate
EXIST	existing
EXT	exterior
F DR	fire door
F EXT	fire extinguisher
FAB	fabricate
FBRK	firebrick
FD	floor drain
FDR	feeder
FDT	foundation
FHY	fire hydrant
FIG	figure
FIN	finish
FIN FL	finished floor
FL	floor
FLG	flange
FPM	feet per minute
FPRF	fireproof
FPS	feet per second
FR	front
FT or '	feet or foot
FTG	footing
FURN	furnish

Abbreviation	Description
GA	gauge
GALV	galvanized
GL	glass
GPH	gallons per hour
GPM	gallons per minute
GR	grade
GT	grease trap
GTV	gate valve
HB	hose bibb
HP	horsepower
HWH	hot water heater
ID	inside diameter
IN or "	inch
INFO	information
INS	insulate or insulation
INT	interior
INV	invert
J	junction box
JCL	janitor's closet
K	kitchen
KS	kitchen sink
KW	kilowatt
KWHR	kilowatt hour
L	left
LAD	ladder
LAU	laundry
LAV	lavatory
LB or #	pound
LB/CU/FT	pounds per cubic foot
LDG	landing
LEV	level
LH	left hand
LIB	library
LIN FT	linear feet
LKR R	locker room
LR	living room
LV	louver
LWC	lightweight concrete
M	meter or motor
MEZZ	mezzanine
MIN or '	minute
MLDG	molding
MN	main
MRR	men's restroom
MT	men's toilet

Abbreviation	Description
MTG	mounting
N	north
NEC	*National Electric Code®*
NFPA	National Fire Protection Association
OFF	office
OVLD	overload
PAN or PAL PC	panel pull chain
PH	phase or power house
PLUMB	plumbing
PORT	portable
PREFAB	prefabricated
PROP PRV	proposal pressure-reducing valve
PS PTD	pipe shaft painted
PWR	power
R	radius
RAD RD	radiator road or roof drain
RECIR or RECIRC	recirculate
RED REF	reducer refrigerator
REG REM	register remove
REP	repair
RET REV	return revision
RF	roof
RH	right hand
RM	room
RT	right
S	south
SEW	sewer
SH	shower
SL	slate
SLD	solder
SOV	shutoff valve
SP	static pressure
SPKR, SPR	sprinkler
SQ	square
SQ FT	square foot
SQ IN	square inch
SS	service sink
SST	stainless steel
STD	standard
SUP	supply
SV	safety valve
SW	switch
SWBD SYS	switchboard system
T	toilet
TEMP	temperature
THERM	thermometer
TYP	typical
UH	unit heater
UL	Underwriters Laboratories
UNFIN	unfinished
USSG	U.S. Standard Gauge
V	vent or volt
VAC	vacuum
VB	valve box
VEL	velocity
VERT	vertical
VEST	Vestibule
VOL	volume or volt
VS	vent shaft
W	west or watt or wire
WC	water closet
WH	hot water heater
WH	wall hydrant or weatherhead
WP	waterproof
WRR	women's restroom
WT	weight or women's toilet
X SECT	cross-section
YD	yard
YDI	yard drain inlet
YH	yard hydrant
YR	year

APPENDIX D

Symbols

ARCHITECTURAL SYMBOLS

NOTE: The following are architectural symbols that are frequently used on construction drawings. Note that the architectural symbols for the same material are often different in the elevation view, plan view, and section view.

Building Materials

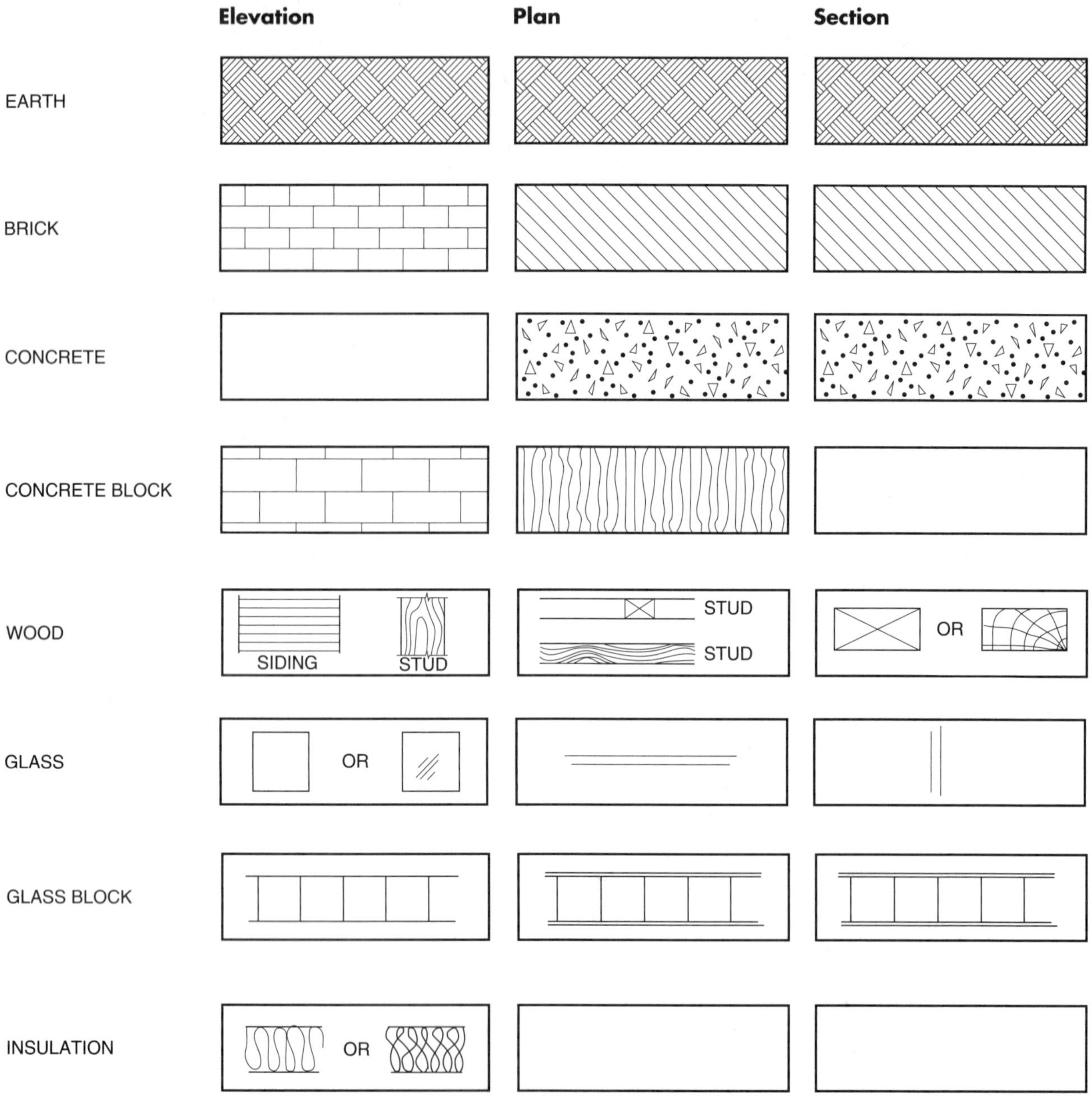

	Elevation	**Plan**	**Section**
EARTH			
BRICK			
CONCRETE			
CONCRETE BLOCK			
WOOD	SIDING STUD	STUD STUD	OR
GLASS	OR		
GLASS BLOCK			
INSULATION	OR		

Exterior Walls

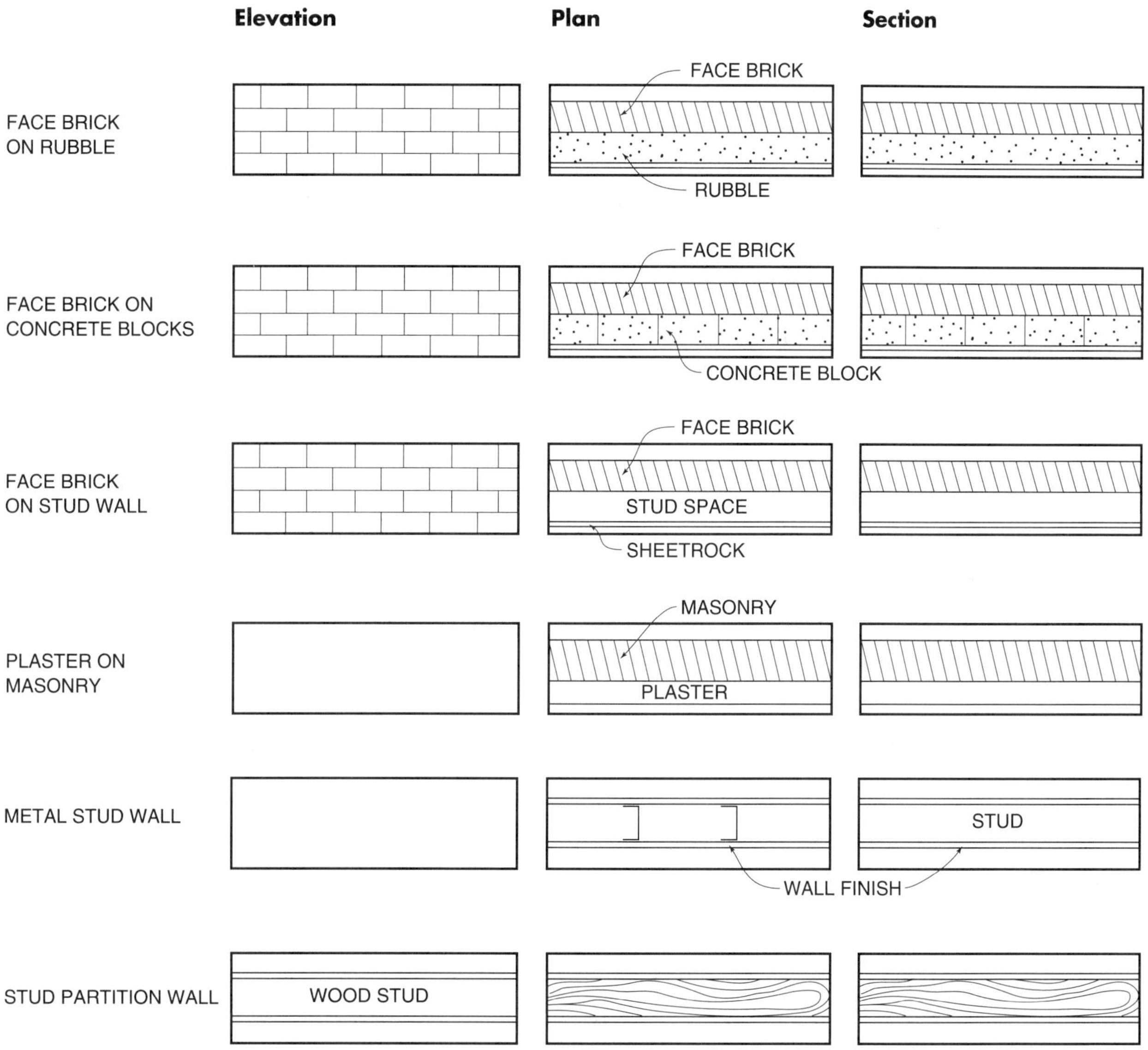

Elevation **Plan** **Section**

FACE BRICK ON RUBBLE

FACE BRICK

RUBBLE

FACE BRICK ON CONCRETE BLOCKS

FACE BRICK

CONCRETE BLOCK

FACE BRICK ON STUD WALL

FACE BRICK

STUD SPACE

SHEETROCK

PLASTER ON MASONRY

MASONRY

PLASTER

METAL STUD WALL

STUD

WALL FINISH

STUD PARTITION WALL

WOOD STUD

SITE PLAN SYMBOLS

Description	Symbol	Description	Symbol
NORTH ARROW	N	WALK	
GRADE POINT		PROPERTY LINE	
FIRE HYDRANT		FUEL GAS LINE	G
MANHOLE	MH	CONTOUR LINE	
TREE		POWER POLE WITH GUY	
NATURAL GRADE (CONTOUR LINE) NUMBER INDICATES ELEVATION IN FEET ABOVE SEA LEVEL	60	ELECTRICAL SERVICE LINE	E
		WATER LINE	W
		FIRE LINE	F
		FINISHED GRADE LINE	60

MECHANICAL SYMBOLS

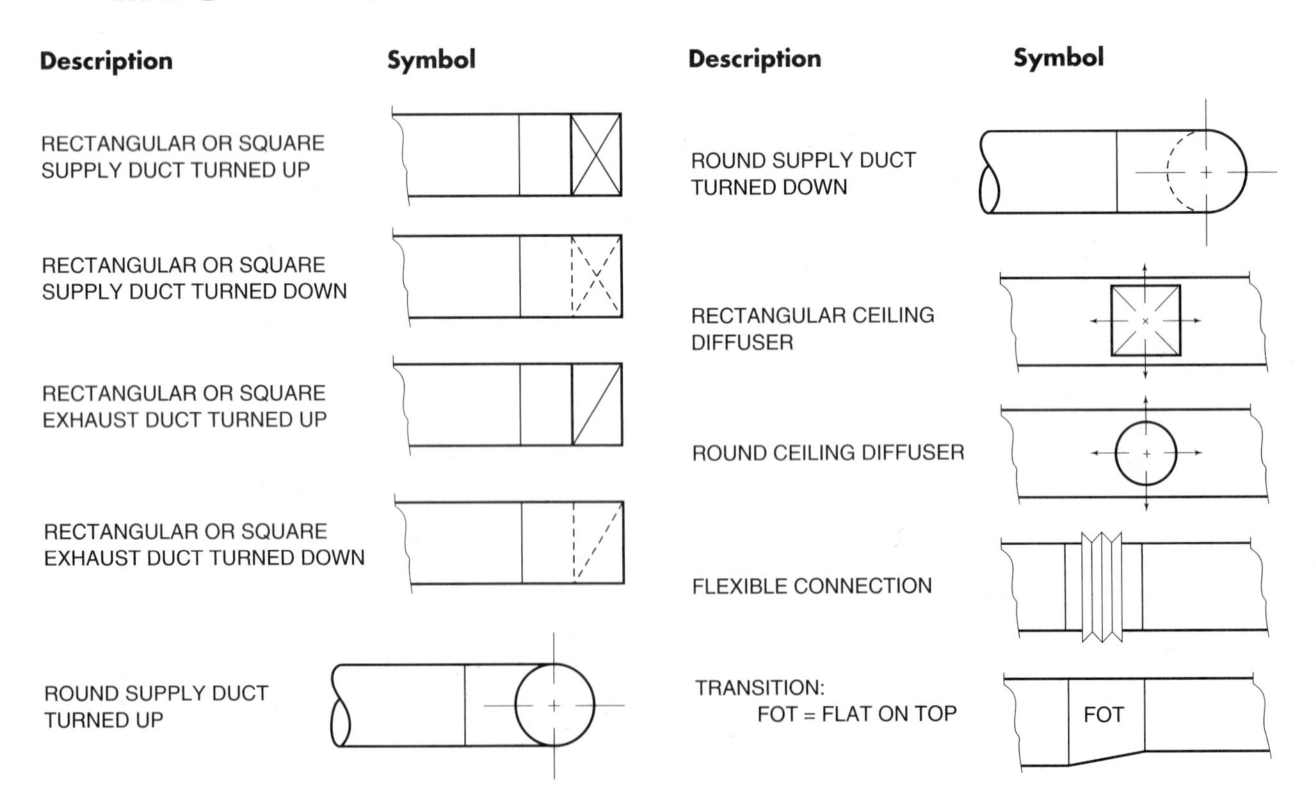

MECHANICAL SYMBOLS (continued)

Description	Symbol
ELECTRIC OPERATED DAMPER	 E.O.D.
FIRE DAMPER	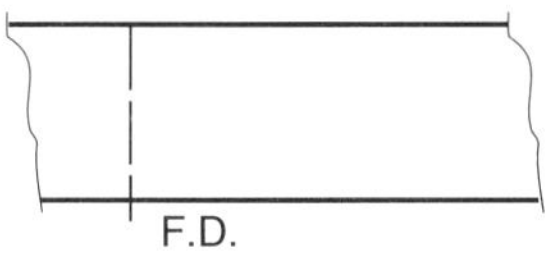 F.D.
SMOKE DAMPER	 S.D.
ELECTRIC HEATER IN DUCT	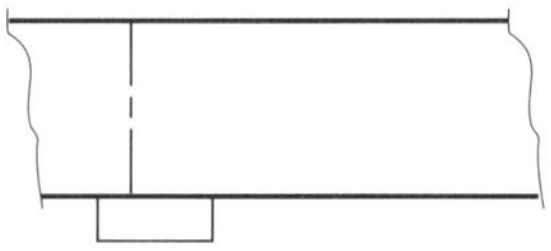
SUPPLY OUTLET WITH SIZE AND AIR QUANTITY SHOWN	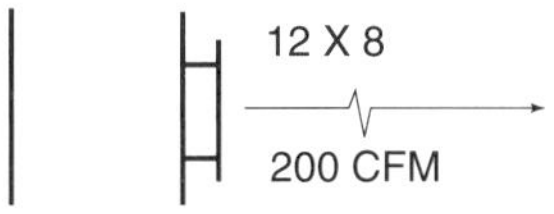 12 X 8 200 CFM
DEFLECTOR IN DUCT BEHIND REGISTER OR GRILLE (ARROW INDICATES DIRECTION OF FLOW)	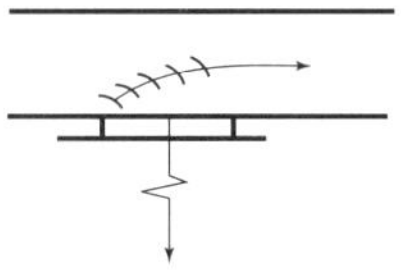
TURNING VANES IN A SQUARE THROAT ELBOW	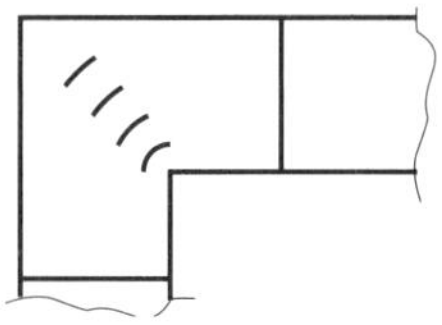
TURNING VANES IN A ROUND THROAT ELBOW	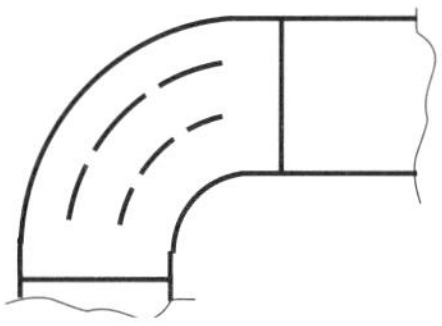
PLAN VIEW OF TRANSITION	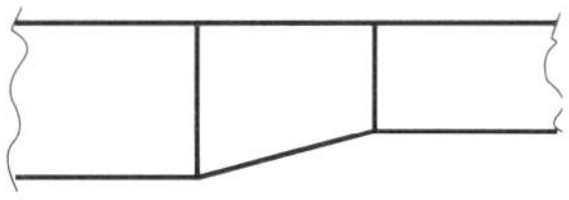
OFFSET UP IN DIRECTION OF ARROW	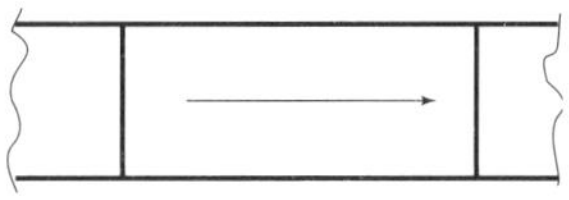
DUCT DIMENSIONS—FIRST FIGURE IS THE SIDE OF DUCT SHOWN (12 X 10)	

Description	Symbol
ACOUSTICAL LINING INSIDE INSULATION	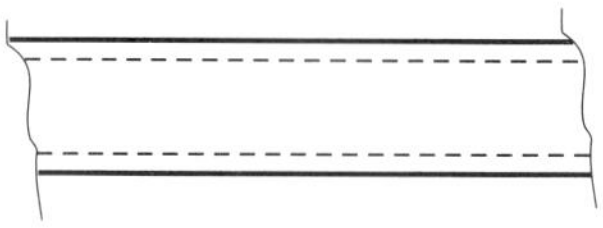
BRANCH TAP IN DUCT	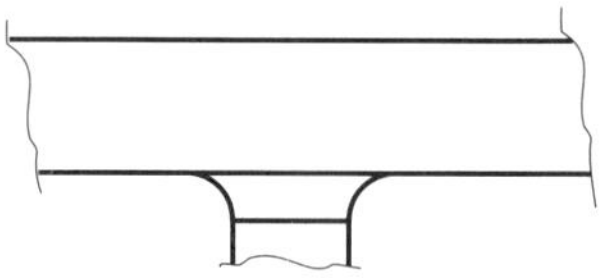
SPLITTER FITTING WITH DAMPER	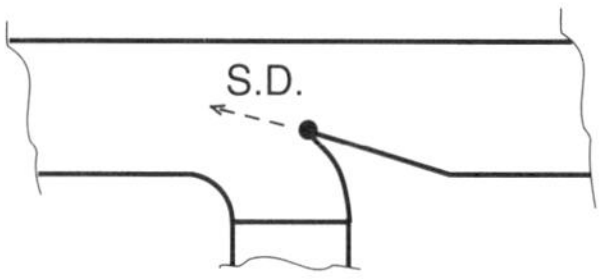 S.D.
VOLUME DAMPER	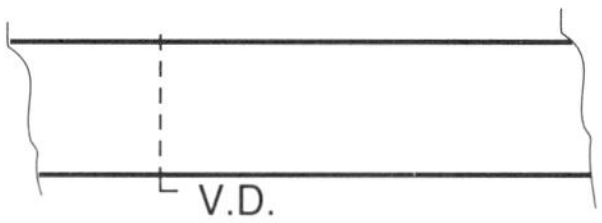 V.D.
BACKDRAFT DAMPER	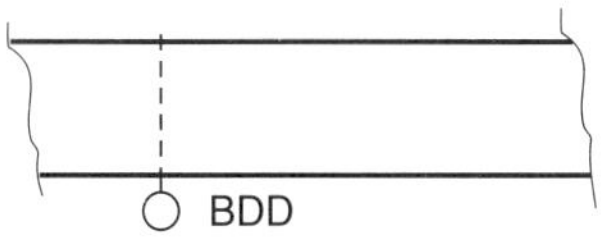 BDD
ACCESS DOOR IN DUCT 10" X 10" SIZE	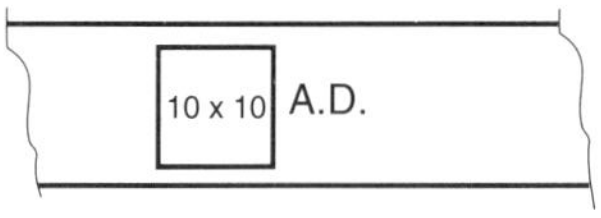 10 x 10 A.D.
PNEUMATIC OPERATED DAMPER	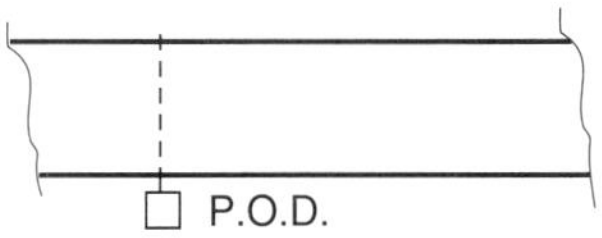 P.O.D.
THREE-WAY VALVE	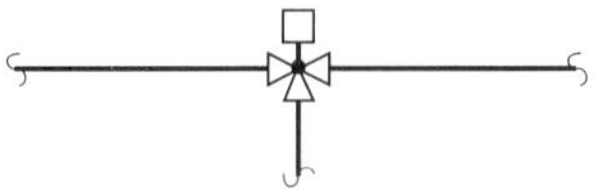
PRESSURE REDUCING VALVE	
PRESSURE RELIEF VALVE OR SAFETY VALVE	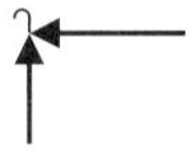
SOLENOID VALVE	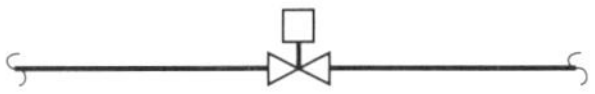
PIPE TURNED UP (ELBOW)	
PIPE TURNED DOWN (ELBOW)	
TEE (OUTLET UP)	

MECHANICAL SYMBOLS (continued)

Description	Symbol
TEE (OUTLET DOWN)	
BACKFLOW PREVENTER	B.F.P.
UNION	
REDUCER	
CHECK VALVE	FLOW
GATE VALVE	OR
GLOBE VALVE	OR
BALL VALVE	
BUTTERFLY VALVE	
DIAPHRAGM VALVE	
ANGLE GATE VALVE	
ANGLE GLOBE VALVE	
PLUG VALVE	
FLEXIBLE CONNECTOR	
FLOW SWITCH	FS
PRESSURE GAUGE AND COCK	
PRESSURE/TEMPERATURE PLUG	
STRAINER, BLOW DOWN	
THERMOMETER	
THERMOMETER WELL	TW

PIPING SYMBOLS

Description	Symbol
LOW PRESSURE STEAM	LPS
LOW PRESSURE CONDENSATE	LPC
PUMPED CONDENSATE	PC
FUEL OIL SUPPLY	FOS
FUEL OIL RETURN	FOR
HOT WATER SUPPLY	HWS
HOT WATER RETURN	HWR
COMPRESSED AIR	A
REFRIGERANT SUCTION	RS
REFRIGERANT LIQUID	RL
REFRIGERANT HOT GAS	RHG
CONDENSATE DRAIN	CD
FUEL GAS	G
CHILLED WATER SUPPLY	CWS
CHILLED WATER RETURN	CWR

PLUMBING SYMBOLS

Description	Symbol
METER	M
SPRINKLER PIPING	S
SPRINKLER HEAD	
FLOOR DRAIN	F.D.
CLEAN-OUT	C.O.
TUB	
TANK-TYPE WATER CLOSET	
WALL-MOUNTED LAVATORY	
URINAL	

Description	Symbol
SHOWER	
WATER HEATER	WH
MANHOLE	MH
WALL HYDRANT	
YARD HYDRANT	Y.H.
FLUSH VALVE WATER CLOSET	
COUNTER-TYPE LAVATORY	
KITCHEN SINK (DOUBLE BOWL)	

PLUMBING PIPE SYMBOLS

Description	Symbol
SOIL, WASTE OR DRAIN LINE	
PLUMBING VENT LINE	
COLD WATER (DOMESTIC)	
HOT WATER (DOMESTIC)	
HOT WATER RETURN (DOMESTIC)	
FIRE LINE	F
FUEL GAS LINE	G

Description	Symbol
ACID WASTE LINE	AW
VACUUM LINE	V
COMPRESSED AIR LINE	A
BACKFLOW PREVENTER	BFP
GATE VALVE	OR
GLOBE VALVE	OR
CHECK VALVE (ARROW INDICATES DIRECTION OF FLOW)	

PLUMBING PIPE SYMBOLS (continued)

Description	Symbol	Description	Symbol
UNION		TEE OUTLET DOWN	
PIPE TURNED DOWN		TEE OUTLET TO SIDE	
PIPE TURNED UP		REDUCER	
TEE OUTLET UP		PIPE SLEEVE	

ELECTRICAL SYMBOLS

Description	Symbol	Description	Symbol
SINGLE CONVENIENCE OUTLET	1	FUSED SWITCH	F
DOUBLE CONVENIENCE OUTLET		GANG OUTLET	R
CONVENIENCE OUTLET OTHER THAN DUPLEX, TRIPLEX, OR DOUBLE DUPLEX, ETC.	3, 4	SWITCH AND CONVENIENCE OUTLET	S
SPECIAL PURPOSE OUTLET (DESCRIBE IN SPECIFICATIONS)		JUNCTION BOX	J
FLOOR OUTLET		LAMP HOLDER	L
FAN OUTLET	F	WALL-MOUNTED LAMP HOLDER	L
GENERATOR	G	CLOCK OUTLET	C
SINGLE-POLE SWITCH		LIGHTING PANEL	
DOUBLE-POLE SWITCH	2	POWER PANEL	
THREE-WAY SWITCH	3	BRANCH CIRCUIT	
FOUR-WAY SWITCH	4	HOME RUN TO PANEL (NUMBER INDICATES THE NUMBER OF THE CIRCUIT BREAKERS IN PANEL)	1
KEY-OPERATED SWITCH	K	BRANCH CIRCUIT (HASH MARKS INDICATE NUMBER OF WIRES)	
SWITCH WITH PILOT LIGHT	P	MOTOR	OR M
WATERPROOF SWITCH	WP	ISOLATING SWITCH	
		BELL	
		OUTSIDE TELEPHONE OUTLET	
		FIRE ALARM BELL	F

ELECTRICAL SYMBOLS (continued)

Description	Symbol
FIRE ALARM STATION	F
TRANSFORMER	T
FUSED ISOLATING SWITCH	F
INSIDE TELEPHONE OUTLET	
TELEPHONE SWITCHBOARD	
BATTERY	
GROUND POINT	

GLOSSARY

Access Door
A door used to enter an area that conceals equipment. A panel used to gain entrance into a piece of equipment or ductwork.

Air Conditioner
The component in a forced-air system that cools and dehumidifies the air.

Air Filter
A porous device that removes particles or contaminants from the air.

Air Handler
An air-moving device used to distribute conditioned air to building spaces.

Architect
A qualified, licensed person who creates and designs drawings for a construction project.

Architect's Scale
A type of ruler that uses smaller units to represent larger ones, such as using 1/2 inch or 1/4 inch to represent 1 foot. The units are used so all building measurements are in proportion to the actual measurements but on a smaller scale for the purpose of creating drawings.

Architectural Plans
Drawings that show the design of the project (also called architectural drawings).

Assembly
A collection of items or equipment needed to complete a specific working unit.

Assembly Drawing
A drawing that shows how to assemble a complex product. It also shows the relationship of the individual parts to the object as a whole.

Battery Limits
The outside perimeter of a building or project.

Beam
A large, horizontal structural member made of concrete, steel, stone, wood, or other structural material to provide support above a large opening.

Blueprints
Architectural or working drawings with dimensions and materials for a structure or building project (also called construction plans). The term is derived from the old reproduction process, which produced a drawing copy sheet with a blue background and white lines. The term "print" is preferred today.

Branch Circuit
In electrical work, the circuit conductors between the final breaker panel and outlet.

Building Code
Regulations adopted by a federal, state, county, or city government for the construction of buildings in a safe and structurally sound manner. Includes standards for the structure, plumbing, mechanical, and electrical components. Codes protect the health, safety, and general welfare of those within or near the buildings and are enforceable by law.

Building Lines
Boundary on a building site that establishes the faces of exterior walls and within which the walls of the building must be confined.

Building Permit
Legally required authorization for construction work. Permits are issued by the code-enforcing agency.

CAD (Computer-Aided Drafting)
Graphic representation of designs using a computer (also called Computer-Aided Design).

CAD Compatibility
The ability of an estimating software package to directly transfer quantity takeoff information from a CAD-generated set of plans.

Catalog Drawings
Drawings of plumbing fixtures that are found in manufacturers' catalogs.

Chase
A recess on the inside of a wall used for piping, mechanical, or electrical lines.

Chiller
The component in a hydronic air-conditioning system that cools water, which cools and dehumidifies the air.

Circuit Breaker
An overcurrent protection device with a mechanism that automatically opens the circuit when an overload condition or short circuit occurs.

Civil Plans
Overhead view of the site. Drawings that show the location of the building on the site from an aerial view, including contours, trees, construction features, and dimensions.

Compressor
A mechanical device that compresses refrigerant and creates a flow of refrigerant in a refrigeration or air-conditioning system.

Computer-Aided Drafting (CAD)
The making of a set of blueprints with the aid of a computer.

Condenser
A heat exchanger coil that removes heat from high-pressure refrigerant vapor. The heat is ejected into the surrounding air or water passing through it.

Conductor
Wire that is used to conduct the flow of electrons in an electrical circuit.

Conduit
A hollow pipe that supports and protects electrical conductors or wires.

Contour Lines
Solid or dashed lines showing the height or elevation of the earth on a civil drawing. Lines on a plot plan are drawn to pass through points having the same elevation. Dashed lines represent natural finish grade.

Contractor
Individual or company responsible for the performance of construction work, including labor and materials, according to plans and specifications.

Convention
A simplified way of representing a building component on prints.

Coordination Drawings
Elevation, location, and other drawings produced for a project by the individual contractors for each trade to prevent a conflict between the trades regarding the installation of their materials and equipment. Development of these drawings evolves through a series of review and coordination meetings held by the various contractors.

Cut List
An information sheet that is derived from shop drawings. It is the shop guide for fabricating duct runs and fittings.

Cutaway Drawing
Section drawing that shows the construction elements of a particular part of the building or of a fixture.

Damper
A movable set of wings that controls and balances airflow in a forced-air system.

Detail Drawings
Enlarged views of part of a drawing used to show an area more clearly or provide more elaborate information than is available on a plan.

Detail Takeoff
An estimating practice in which a takeoff of each individual construction component is done.

Diffuser
An air outlet in the HVAC system that directs air in a wide pattern through a conditioned area.

Dimension Line
A line on a drawing with a measurement indicating length.

Dimensions
Measurements such as length, width, and height shown on a drawing.

Disconnect
An electrical switch that disconnects electrical circuits from motors and machines.

Duct
1) In HVAC, a large round or rectangular pipe used for carrying air. 2) In electrical work, a rectangular-shaped trough that serves as a wireway.

Ductwork
The distribution system for forced-air heating or cooling systems.

Easement
A designated right-of-way, such as the access guaranteed to utility companies for repair of utilities that are located on, or cross over, private land.

Electrical Drawing
Drawing that shows the locations of outlets, switches, and electrical fixtures.

Electrical Metallic Tubing (EMT)
A lightweight, tubular steel pipe or raceway used to carry electrical wiring.

Electrical Plans
Engineered drawings that show all electrical supply and distribution. Shows placement of electrical fixtures, appliances, and circuits. Exact placement of conductors is not specified. Commonly noted with a capital "E" preceding the sheet number.

Elevation (EL)
Height above sea level or other defined surface (usually expressed in feet).

Elevation Drawing
View of a building or object, showing height and width. An orthographic view without any allowance for perspective.

Elevation View
A view that depicts a vertical side of a building, usually designated by the direction that side is facing (right, left, east, west, etc.).

Engineer
A person who applies scientific principles in design and construction.

Engineer's Scale
A straight-edge measuring device divided uniformly into multiples of ten divisions per inch, which allows drawings to be made with decimal values.

Estimating Cost
The calculation of construction costs of a project.

Estimating Method
The approach used by an estimator to perform project analysis, takeoff, and pricing in a consistent and organized manner.

Estimating Practice
The system used to integrate all parts of the estimating process in a cohesive, consistent, reliable manner to ensure an accurate final bid.

Evaporator
A heat exchanger that absorbs heat and moisture from the surrounding air by evaporating refrigerant.

Expansion Tank
A tank that allows the water in a hydronic heating system to expand without raising the water pressure to dangerous levels.

Expansion Valve
A device that reduces the pressure on liquid refrigerant by allowing the refrigerant to expand. It is located in the liquid line near the evaporator.

Extension Line
A line used on a drawing to locate a dimension away from the actual points of the dimension. This method is used when a drawing would be too crowded or cluttered if the dimension were shown within the two points.

Fibrous Insulation
Insulation composed of small diameter fiberglass (or slag) fibers.

Fire Break
The space of fire-resistant materials between structures or groups of structures to impede fire from spreading.

Fixture Drawing
A diagram that shows the components of a fixture in detail.

Floor Plan
A drawing that provides an aerial view of the layout of each room. A building drawing indicating a plan view of a horizontal section at some distance above the floor, usually midway between the ceiling and the floor.

Flue
A heat-resistant passage or chimney that conveys smoke or other byproducts of combustion.

Forced-Air System
A system that uses blowers to circulate warm or cool air to condition building spaces.

Form Lines
Contour lines drawn from visual observation and without accurate elevation information.

Foundation Footing
1) The part of a foundation resting on bearing soil and supporting the foundation wall. 2) The base for a column.

Foundation Plan
A drawing in a set of prints that gives a plan view of a building's foundation. A drawing that shows the layout and elevation of the building foundation.

Furnace
A self-contained heating unit fueled by fuel oil, natural gas, or electricity that includes a blower, burner(s), heat exchanger or electric heating elements and controls and heats air in a forced warm air heating system.

Fuse
An overcurrent protection device with a fusible link that melts and opens the circuit when an overload condition or short circuit occurs.

General Contractor
A contractor or contracting company that agrees to fulfill an entire building agreement with various items or types of work to be completed, such as carpentry, electrical, mechanical, and plumbing work. Subcontractors, working under the supervision of the contractor, may be used to do some of the actual building work.

Grille
A device that covers the opening of supply-air or return-air ductwork.

Ground Fault Circuit Interrupt (GFCI) Receptacle
A device that interrupts the flow of current to the load when a ground fault exceeding a predetermined value of current occurs.

Hazardous Material
A material capable of posing a risk to health, safety, or property.

Hidden Line
A dashed line showing an object obstructed from view by another object.

High Pressure Boiler
A boiler that has a maximum allowable working pressure above 15 pounds per square inch (psi) and over 6 brake horsepower (bhp).

HVAC
Heating, ventilation, and air conditioning.

HVAC Drawing
A construction drawing that shows the placement of furnace and air-conditioning equipment and the locations of ducts, registers, pipes, and/or radiators. Labeled as "M" on the bottom part of a drawing.

Hydronic System
A system that uses water, steam, or other fluid to condition building spaces.

Internet Capability
The ability of estimating software to link and transfer information from the World Wide Web into estimating software.

Isometric Drawing
A three-dimensional drawing in which all horizontal lines are drawn at a 30-degree angle from horizontal and all vertical lines are drawn at a 90-degree angle from horizontal. Not drawn to scale but used to provide clarity.

Job Cost Analysis
The study of the final costs of building a project as compared to the original estimate budget.

Leader
In drafting, the line on which an arrowhead is placed and used to identify a component.

Legend
A description of the symbols and abbreviations used in a set of drawings.

Line Number
A group of abbreviations that specify size, service, material class/specification, insulation thickness, and tracing requirements of a given piping segment.

Longitudinal Section
Sectional view created by passing a cutting plane through the long dimension of a house.

Louver
A ductwork cover containing horizontal slats that allow the passage of air but prevent rain from entering.

Low-Pressure Boiler
A boiler with a maximum allowable working pressure of up to 15 psi.

Low-Temperature Limit Control
A temperature-activated electrical switch that energizes a damper motor and shuts the damper if the ventilation air temperature drops below a predetermined setpoint.

Magnetic Motor Starter
A control device that has overload protection and uses a small control current to energize or de-energize the load (motor) connected to it. A contractor with overload protection.

Makeup Air
Air that is brought into a building to replace air that is lost to exhaust.

Material Safety Data Sheet (MSDS)
Document that describes the components of various substances on a construction site, including their dangers, proper personal protective equipment worn during exposure, disposal procedures, and necessary actions in case of an emergency.

Mechanical Compression Refrigeration
A refrigeration or air-conditioning process that produces a refrigeration effect with mechanical equipment.

Mechanical Plans
Engineered drawings that show the mechanical systems, such as plumbing and process piping.

Metric Scale
A straight-edge measuring device divided into centimeters, with each centimeter divided into 10 millimeters.

National Electric Code
Nationally accepted electrical code designed to protect persons and property from hazards arising from the use of electricity. Sponsored and published by the National Fire Protection Association.

Negotiated Construction Work
Construction work for which a qualified contractor submits a bid to an owner based on the plans and specifications without a competitive bid process.

North Arrow
A directional arrow or mark on a plan locating the north direction for orientation.

Notes
Brief comments, instructions, or information on a construction drawing or other document.

Oblique
Neither perpendicular nor parallel. Having an axis not perpendicular to the base. A drawing in which one face of an object is shown in a flat plane, with the receding lines projecting back from the face.

Oblique Drawing
A pictorial drawing that shows the shape of an object. It shows the front of the object with the body of the object at a slight angle.

Occupational Safety and Health Administration (OSHA)
A federal government agency established under the Occupational Safety and Health Act of 1970, which requires all employers to provide a safe environment for their employees.

Orthographic
Method of projecting planes at right angles.

Orthographic Drawing
Construction drawings that show straight-on views of the different sides of an object. Used for elevation drawings.

Orthographic Projection
The projection of a single view of an object. A drawing in which each face of an object is projected onto flat planes at 90 degrees to one another.

Overhead
The cost of doing business that is not related to a specific job.

Perspective Drawing
The representation of an object as it appears when viewed from a given point.

Pictorial Drawing
A drawing that shows a three-dimensional view of an object.

Piping and Instrumentation Drawing (P&ID)
A schematic flow diagram of a complete system (or systems) that shows function, instrument, valving, and equipment sequences.

Plan Figure
A flat figure.

Plan View
1) A view looking down. 2) In orthographic projection, a top view. 3) In architecture, a floor plan, roof plan, or plan view of a cabinet. 4) The overhead view of an object or structure.

Plot Plan
Scaled view that shows the shape and size of the building lot and the location, shape, and overall size of the house on the building lot.

Plumbing
A general term used for water supply and liquid waste disposal.

Plumbing Drawings
Construction drawing that shows the location of fixtures and pipe runs and gives the size and type to be installed. Marked with a "P" on the bottom of the plans.

Plumbing Plans
Engineered drawings that show the layout for the plumbing system. Marked with a "P" on the bottom of the plans.

Pressure-Reducing Valve
A valve that limits the maximum pressure at its outlet, regardless of the inlet pressure. A valve that sets a maximum operating pressure level and protects the piping or vessel from overpressure.

Pressure Switch
An electrical switch operated by the amount of pressure acting on a diaphragm or bellows element.

Process Piping
Piping installed in industrial facilities for compressed air, vacuum gas, or fuel.

Property Lines
Recorded legal boundaries of a piece of property.

Raceway
An enclosed channel for wiring conductors.

Receptacle
A device used to connect equipment to an electrical system with a cord plug.

Riser Diagram
A vertical and/or horizontal schematic depicting the layout, components, and connections of a piping system.

Rock Wool Insulation
Lightweight heat and sound insulating material made by blowing steam through molten rock or slag.

Roof
The covering for the top exterior surface of a building or structure.

Roof Plan
A drawing of the view of the roof from above the building.

R-Value
Measure of the effectiveness of a material to provide thermal insulation. Higher R-values indicate greater insulating capabilities.

Safety Switch (disconnect)
A device used periodically to remove electrical circuits from the supply source.

Scale
The ratio between the size of an object when it is drawn and its actual size.

Schedules
Tables or documents containing information that describes and specifies the types and sizes of items required for the construction of a building.

Schematic
A one-line drawing showing the flow path for electrical circuitry.

Schematic Drawing
A single-line drawing of a plumbing system or electrical wiring routing or circuit.

Section Drawing
A cross-sectional view of a specific location showing the object. A drawing that depicts a feature of a building as if a cut was made through the middle of it.

Sectional View
Scaled view created by passing a cutting plane through a portion of a building.

Sepia (drawing)
A print or construction drawing with dark reddish-brown lines on a light background. Used to make blueline prints.

Service
Conductors and equipment used to deliver electrical energy from the secondary distribution to the wiring system of the location being served.

Service Conductors
Electrical wiring extending from an underground or overhead transformer to the service equipment at the location served.

Setback
The distance a local code requires a building to be set back from the street.

Shop Drawing
A drawing that indicates how to fabricate and install individual components of a construction project. May be drafted from the blueprints of a project or provided by the manufacturer.

Side Yards
Space along the sides of a structure that provides access to rear yards, reducing the possibility of fire jumping from one building to the next and promoting ventilation around the structure.

Single-Line Drawing
A plumbing drawing that uses a single line to represent the centerline of a pipe. Single-line drawings can be used to represent a pipe of any diameter.

Site Plan
A construction drawing that indicates the location of a building on a land site.

Sketch
A drawing representing the primary features of an object, usually a rough draft or freehand drawing.

Skylight
A roof opening covered with glass or plastic designed to let in light.

Slab
A horizontal (or nearly horizontal) layer of concrete.

Slab-on-Grade
A concrete slab that is placed directly on the ground.

Sliding Door
A horizontal-moving door suspended on rollers that travel in a track that is fixed at the top of the opening or on rollers mounted in the bottom of the door.

Specifications (specs)
Written requirements included with the drawings or blueprints of a construction project. They provide more details or descriptions of the technical standards that must be met during construction.

Spool Sheet
A plan, section, or isometric representation of a segment of piping that is to be fabricated before erection.

Structural Plans
A set of engineered drawings used to support the architectural design.

Subcontractor
Person or party that performs part of the work on a construction project under an agreement with the general contractor.

Superintendent
Contractor's field representative responsible for supervision of an entire project.

Supply Piping
Piping that delivers water from the source to the point of use.

Suspended Ceiling
A ceiling hung on wires from structural members.

Switch
A device that is used to start, stop, or redirect the flow of current in an electrical circuit.

Symbol
A drawing that represents a material or component on a plan. A mark or drawing used to indicate a specific object, material, class, or entity. A legend shows the symbols used on a drawing and their meanings.

Takeoff
Estimating the amount of material required for a job. The process of surveying, measuring, itemizing, and counting all materials and equipment needed for a construction project, as indicated by the drawings.

Terminal Unit
A device that transfers heat or coolness from the water or steam in a piping system to the air-building spaces.

Thermostat
A temperature-activated electrical switch that controls heating and/or cooling equipment.

Title Block
A block on the front of a construction drawing or blueprint, located in the lower right corner, that provides identification and revision information about the drawing.

Topographic
Type of drawing used to describe the physical features and elevation of an area.

Traditional Estimating Method
An estimating method in which quantities are calculated and entered onto ledger sheets for pricing.

Transformer
An electric device that uses electromagnetism to change the voltage from one level to another.

Transverse Section
A section drawing where the "cut" is made along the short dimension of the building.

Underwriter's Laboratories, Inc.
A not-for-profit organization that examines and evaluates devices, systems, and materials to determine their degree of safety.

Valve
A device that controls the pressure, direction, or rate of fluid flow.

Vent Pipe
A pipe that removes odors and gases from the waste piping and exhausts them away from inhabited areas. Approved piping to remove flue gases from a gas or oil furnace.

Wall Bearing Construction
Walls or partitions that support the roof or ceiling members of a structure.

Watt
Measurement of electrical power determined by multiplying the volts and amps. Designated as "P" in the power formula.

Zero Line
A line connecting points on a topographic map where existing and planned elevations are equal.

INDEX

Note: *Italicized pages refer to illustrations and tables.*

A

Abbreviations, 45–48, 270–272
  acronyms, 45
  examples of, 48
  prefixes, metric, 18
  schedules, 61–62, *105*
  using with symbols, 47, 274–281
  variations in, 46
  See also Symbols
ACCA (Air Conditioning Contractors of America), 240
Access door in duct, symbol for, 277
Accidents, avoiding, 13–15
Accoustical lining, symbol for, 277
Acid waste line, symbol for, 279
Acronyms, 45
Acute angle, 26
Addition, 4–5
Air Conditioning Contractors of America (ACCA), 240
Air conditioning equipment, 190–*193*
  energy conservation, 244
  gas furnace connections to, *74*
  grounding, 14
  mechanical plans, 226
  sections of, 190–193
  sizing, 239
Air Distribution Schedule, 226
Air-handling equipment, *182*, 190–*193*
Air requirements, outside, *257*
Alphabet of lines, *105*
American National Standards Institute (ANSI), 45
American Society of Heating, Refrigeration, and Air Conditioning Engineers (ASHRAE), 240
Angle gate valve, symbol for, 278
Angle globe valve, symbol for, 278
Angles
  acute, 26
  bisecting, 189
  obtuse, 26
  right, 25
  sides of, 24
  straight, 25–26
  of triangles, 26
  types of, 25–26
  vertex of, 24
Angular measurements, 23–28
  angles in, 24–26
  in circles, 24–25
  overview of, 23–24
  using protractors in, 27
ANSI (American National Standards Institute), 45
Appliances, heat gain from, *257*
Architects, 31–36, 97–98
  in preparing working drawings, 56
  responsibilities and liabilities of, 56
  seals, 56, 92, 98–*99*
  title blocks, 97
Architect's scale, 31–36
  commonly used scales on, 34–36
  in oblique drawings, 127
  one-half inch scale in, 34
  one-inch scale in, 33
  See also Engineer's scale
Architectural details, 174–175, 204–205
Architectural plans, 203–205
  details, 204–205
  floor plans, 61, 204–205
  foundation plans, 204
  interior elevations, 205
  location plans, 203
  longitudinal sections, 204
  roof framing plans, 204
  schedules, 205
  sections, 205
  site plans, 203
  transverse sections, 204
  wall sections, 70–*71*, 204
  See also Plans
Architectural symbols, 274–275
Architectural woodwork, 90
Area
  calculating, 6–8
  gross, *10*, 248–*249*
  net, 248–*249*
Arrows or special markings, 69
As-built drawings, 87
ASHRAE (American Society of Heating, Refrigeration, and Air Conditioning Engineers, 45
Assembly hall, heat gain from occupants, *255*
Atmospheric pressure, 13
Auditorium, heat gain from occupants, *255*

B

Backdraft damper, symbol for, 277
Backflow preventer, symbol for, 278–279
Ball valve, symbol for, 278
Banks, heat gain from occupants, *255*
Barber shops, heat gain from occupants, *255*
Bars, heat gain from occupants, *255*
Basement plans
  in general construction section, 60
  in mechanical plan section, 63
  in plumbing section, 62
  in structural section, 61
Battery, symbol for, 281
Batts, *246*
Beauty parlors, heat gain from occupants, *255*
Bells, symbol for, 280
Belt-driven equipment, 14, *106*
Benchmark, 203
Bid bonds, 86

Bidding, 86
Blinds, 250–252
Blueprint process, 77–*79*
Blueprint room, safety rules for, 13
Blueprints. *See* Construction drawings
Body heat, *242*, 255
Boilers, *140*, *183*, 193–194, *196*
Bolt/nut and rivet schedule, 62
Bookcases, 174
Border lines, *105*, 109
Bowling alleys, heat gain from occupants, *255*
Branch circuit, symbol for, 280
Branch tap in duct, symbol for, 277
Break lines, *105*, *107*–109
Bricks
  architectural symbol for, 47, 274–275
  symbol for, 46–47, 163, 274–275
British thermal unit (BTU), 45, 239–240
Broilers, heat gain from, *257*
Broken lines, description or use of, *105*–106, 115–116, 232
BTU (British thermal unit), 45, 239–240
Building codes, 256
Building construction review agencies, 86
Building materials, symbols for, 274
Buildings
  front elevation of, 60, *72*, 111–*112*, 120, 169–170
  infiltration rate in, 256
  inspection of, 87
  number of occupants in, 255–256
  orientation of, 243–244
  perimeter of, 6
  triangular shapes in, 23
  ventilation rate in, 256
  working drawings for, 120
  zoning, 259
Butterfly valve, symbol for, 278

C

Cabinets, built-in, 90, 170–171, 174
CAD (Computer-Aided Design)
  advantages of, 54, 57
  building construction agencies, 86
  construction lines, 107
  contractors, 86
  description or use of, 54
  Diazo process, 79–81
  drawings, 54
  electrostatic printing, *79*, 81–82
  equipment suppliers, 86
  with microfilms, 83
  in oblique drawings, 125–128
  in orthographic projection, 111–120
  overview of, 54
  owners, 87, 97–99
  in printing construction, *79*
  record drawings, 87
  review and approval of, 85–86
  saving, 83
  symbols in, 46
  users of, 54
  wording notations for, 153–154
  See also Drawings; Working Drawings
  See also Lines
Cafeterias, heat gain from occupants, *255*
Carpentry, 90, 93
Carpeting, 61, 91
Cast-in-place concrete, 90
Catch basin details, 174
CD-ROM, *55*
Ceiling area, 245
Ceiling diffusers, symbols for, 276
Ceiling insulation, 247
Ceiling suspension systems, 91
Center lines, *105*, *106*, 108, 188
Centimeter, 18
Ceramic veneers, 90
CFM (cubic feet per minute), 45
Chair rails, 170
Chalkboards, 91
Check valve, symbol for, 278–279
Cheeks (ductwork), 190–*191*
Chemicals, 14
Chilled water return, symbol for, 278
Chilled water supply, symbol for, 278
Chillers, 171
Churches, heat gain from occupants, *255*
Circles
  angular measurements in, 24–25
  area of, 7–8
  circumference of, 8
  drawing, 189
  equations for, 8, 264
  locating the center of, 189
  quadrants of, 24
  segments, 189, *191*
Circuits, 232
Clay backing tiles, 90
Clean-out, symbol for, 279
Clean rooms, 91
Clear glass, *251*–*252*
Clock outlet, symbol for, 280
Coal heating, 240
Cocktail lounges, heat gain from occupants, *255*
Coffee brewer/warmer, heat gain from, *257*
Coffee urn, heat gain from, *257*
Coil mounting details, 175
Cold water (domestic), symbol for, 236, 279
Columns, 60–62, 90, 206–207
Column schedules, 62
Combustible materials, 14
Compass, 53, 188
Completion log, 158–159
Compressed air, 92, 235
Compressed air line, symbol for, 278–279
Compressors, 171
Computer-Aided Design. *See* CAD (Computer-Aided Design)
Computer drafting machine, 54, *55*, 161
Computer monitors, 54–*55*
Concrete
  in specifications, 90
  symbol for, 162, 274
Concrete block, symbol for, 162, 274
Condensate drain, symbol for, 278
Condensate, symbol for, 278
Condensation, 254–255
Condensing unit, *76*
Conduction, 240–*242*
Conduit schedules, 64
Cones, equations for, 264
Conference rooms, heat gain from occupants, *255*
Connectors, 9, 86
Construction agencies, building, 86
Construction contractors, 59–60, 86–87
Construction drawings, 51–57, 77–83, 85–87, 169–176
  as-built drawings, 87
  blueprint process for, 77–*79*
  CAD (Computer-Aided Design), 82–83
  dates on, 98
  equipment suppliers, 86
  infiltration rates, 256
  in load calculations, 239–259
  overview of, 51
  plan review and approval of, 85–86
  plans for contractors, 86–87
  pricing contract bids, 86
  printing, 79–83
  record drawings, 87
  See also Working drawings
Construction plans, 60–61, 86–87
Contour line, symbol for, 276
Contraction joints, 90
Contractors, 60, 86
Convection, 240
Convenience outlets, symbols for, 280
Conversion tables, 18, 266–267
Conveying systems, in specifications, 91
Cooling towers, 171
Core drilling, 90
Cornices, 67–*68*, 174
Costing, for ductwork, 219, 221
Counter-type lavatory, symbol for, 279
Cranes, 91
Crosshatching, 161–163, 181–*183*
  combining with poché, 162–163
  at different angles, 161–163
  of different materials, 163
  in freehand sketching, 181–*183*
  See also Shading
Curb details, 174
Curtains, 250–252
Cutting plane lines, 69, 71, 108
Cutting planes, 67
Cylinders, equations for, 8, 264

D

Damper details, 175
Dampers, symbols for, 277
Dates, on working drawings, 98

- Decimals
  - converting from fractions, 20
  - converting to fractions, 20
- Decimeter, 18
- Deep-fat fryer, heat gain from, *257*
- Degrees, of an angle, 25–26, 28
- Demolition site plans, 60
- Department stores, heat gain from occupants, *255*
- Design conditions, 243
- Designers, 54, 56–57, 97–98, 157–158, 173–174
- Details, 72, 173–176
  - architectural, 72, 174–175
  - drawing, 72, 176
  - electrical, 176
  - in general construction section, 61, 173
  - locating on sheets, 158, 176
  - mechanical, 175
  - not to scale, 72
  - plumbing, 63
  - site, 174
  - structural, 62, 174
  - walls, 204–205
- Dew point, 254
- Diaphragm, symbol for, 278
- Diazo printing, 79–81
- Diazo printing machine, *80*
- Diffuser schedules, *48*, 63
- Dimension lines, *105*–107
  - description or use of, 106–107
  - in freehand sketching, *105*–107, 181
  - in orthographic projection, 116
  - *See also* Lines
- Dimensions, 17, 39, *105*–107, 116, 181, 194
- Dining rooms, heat gain from occupants, *255*
- Directional signage, 91
- Disks, 54–*55*, 83
- Distribution entrance panel, 14
- Dividers, 53, 188
- Division, in mathematics, 4–5
- Divisions, in specifications, 89–92
- Domestic cold water piping, 235–237
- Doors, 252–253
  - access in duct, symbol for, 277
  - area, *253*
  - construction, 60–61, 91, 169, 249, 252–253
  - details, 169, 174, 205
  - heat loss, 252–*253*
  - load calculation, 243
  - schedules, 61, 205
  - in specifications, 169–170
- Dot matrix printers, 83
- Double convenience outlet, symbol for, 280
- Double-pole switch, symbol for, 280
- Drafters
  - abbreviations and symbols, 45–47, 270–272
  - average job time, 5
  - details and, 176
  - drawing lines, 104
  - freehand sketching, 179–181
  - instruments for, 53–54, 187–195
  - notations, 72
  - reading degrees, 27
  - scales and, 31–33,
  - working drawings and, 56
- Drafting
  - Computer-Aided. *See* CAD (Computer-Aided Design)
  - machine, 187
  - manual instruments, 52, 187–195
  - scales in, 31–32
  - time-saving procedures in, 189–190
  - *See also* Drawings
- Drafting equipment
  - computer, 54, *55*
  - manual, 52, 187–195
- Drafting ink, 56
- Drafting pencils, 188
- Drafting pens, 103–104, 188
- Drafting room, safety rules for, 13
- Drafting scales, 31–32
- Drafting tables, 52
- Drain line, symbol for, 279
- Drawing film, 56
- Drawing paper, 56
- Drawings
  - as-built, 87
  - details, 61–64, 72–*76*, 90, 116, 140–*142*, 158
  - freehand sketching in, 103–104
  - isometric, 140–*142*
  - lines on, 103–109
  - locating on pages, 181
  - oblique, 125–*129*
  - original, 51
  - record, 87
  - scales in, 31–32
  - sepia, 56
  - sheet number in, 158
  - three-dimensional, 72, 125, 161, 164–165, 181
  - *See also* Working drawings
- Drawing sheets, 56, 78, 109, 157–158
  - locating details on, 158
  - numbering, 158
  - organizing, 158
  - in organizing jobs, 158
  - schedules of plans, 157
  - titles, 158
- Drug stores, heat gain from occupants, *255*
- Drums, in electrostatic printing machines, 81
- Dry food warmer, heat gain from, *257*
- Duct deflector symbol for, 277
- Duct Factor Chart, *212*–213
- Duct insulation schedules, 63, 254
- Ducts, 254–255
  - access door, symbol for, 277
  - connections, 190–192, *215–216*
  - design, 254
  - gain and loss factors, 254–255
  - insulation factors, 254
  - liner, 219
  - symbols for, 276
  - take-off form, 215, 218, 221–222
- Ductwork, 140–*142*, 190–194, 211–222
  - costs of labor, 219
  - costs of materials, 219, 221
  - details, *141–142*
  - elbow, 190–192, *215–216*, 277
  - fittings, *194, 216*
  - insulation, 219–221
  - insulation factors, 219–*220*, 254
  - linear feet of, 214–215
  - in mechanical plans, 63, 91–92, 171, 175, 226
  - plans, 226
  - sheet metal, 213
  - splitter, 192–*193*, *214*, *217*, 277
  - take-off factors, *212*–213, *220*–221
  - take-off form, 215, 218, 221–*222*
  - total weight of metal, *212–213*, 217–219
  - required for, 217
  - transition, 192, *194*, *215*
  - types of, 211
- Dumbwaiters, 91

E

- Earth, symbol for, 46, 274
- Elbows (ductwork), 190–192, 277
- Electric heaters, 227
- Electric heating, *107*, 239–240
- Electric operated damper, symbol for, 277
- Electrical details, 176
- Electrical division, in specifications, 92, 232
- Electrical elevations, 171
- Electrical energy, 14, 63–64, 92, 97
- Electrical heaters in duct, symbol for, 277
- Electrical meter, 171–*172*
- Electrical outlets, 63, 170, 171
- Electrical panel details, 176
- Electrical plans, 63–64, 231–232
  - electrical wiring in, 232
  - lighting fixture schedules in, 64, 232
  - reading, 231–232
  - specifications in, 232
  - symbol schedules in, 232
  - symbols for, 280–281
  - *See also* Plans
- Electrical section, 63–64
- Electrical service line, symbol for, 276
- Electrical specifications, 232, 280–281
- Electrical weather head, *172*
- Electrical wiring, 232
- Electrostatic printing, *79*, 81–82
- Electrostatic printing machine, 81–82
- Elevation plans
  - in architectural section, 204–205
  - in general construction section, 60, 169–172
  - in structural section, 62
- Elevation points, 203
- Elevations, 69–72, 169–*172*
  - in architectural plans, 69–72, 204–205

electrical, 171
exterior, 169–170
front, 169, *72*
in general construction section, 169–172
interior, 70, 170–171, 205
left, 111, 120
mechanical, 171
plumbing, 63, 171
rear, 169–170
right, 60, 69, 111, 120
sides, 169
in structural section, 62
Elevation views, 111, 120
Elevators, 271
Elliptical tank, equation for, 262, 264
End view
base lines for, *135*–139
constructing, 125–127, 133–137
in isometric drawings, 133, 136
in oblique drawings, 125
viewing planes and, 111, *114*
Engineers, 39–42, 97–98
in preparing working drawings, 56–57
responsibilities and liabilities of, 56
seals, 56, 92, 98–*99*
title blocks, 97
Engineer's scale, 39–42
commonly used scales on, 41–42
units in, 40
using, 8, 56, 62, 85–87, 97–99
See also Architect's scale
English system, 17–20
conversion to metric system, 17–18, 266
converting fractions to decimals in, 20, 266–267
converting inches to feet, 267
one-foot graduation in, 18
one-inch graduation in, 18
one-yard graduation in, 19
Enlarged floor plans, 205
Equipment
safety rules in handling, 14
schedules, 61, 63
in specifications, 89, 91
suppliers, 86
Equipment connection details, 175
Equipment mounting bases, *141*
Equipment room plans, 193–194
Erasing shields, 188
Existing contour lines, description and use of, *105*
Expansion joints, 90
Expansion tanks, *140*, *183*, 193–194, 196
Extension lines
description or use of, 70, 106–107
in freehand sketching, 70
in orthographic projection, 116
See also Lines
Exterior clean-out details, 175, 236
Exterior elevations, 169–172
Exterior insulation, 219
Exterior walls, symbol for, 275
Extruded polystyrene, *246*

F

Face brick, symbol for, 275
Fahrenheit, 239
Fan details, *69*, *76*, *107*, 140–*141*, 175
Fan outlet, symbol for, 280
Federal safety programs, 15
Fences, 90
Fiberglass, 211, 219, *246*–*247*
Fiberglass batts, 245, 248
Film, drawing, 35, 54, 56
Film lead, 104
Finishes, in specifications, 61, 91, 170
Fire alarm bell, symbol for, 280
Fire alarm station, symbol for, 281
Fire bricks, 90
Fire damper, symbol for, 277
Fire extinguishers, 171
Fire hydrant, 276
Fire line, symbol for, 276
Fireplaces, 91
Fire protection systems, 92, 235–236
Fixture schedules, 63–64, 232, 236
Flagpoles, 91
Flat scale rule, 31, 40
Flexible connection, symbol for, 276
Flexible connector, symbol for, 278
Floor area, 241, 244
Floor drain details, 175, 194, 271
Floor drain, symbol for, 279
Floor framing plans, 61
Floor insulation, 245
Floor outlet, symbol for, 280
Floor plans, 203, 205
architectural, 203, 205
calculating floor area from, 244
developing elevations from, 61, 169–170
in electrical plan section, enlarged, 205
in general construction section, 61
in mechanical plan section, 63, 226
plan views as, 111, 120
in plumbing section, 62
See also Plans
Floppy disks, 54, 83
Flow switch, symbol for, 278
Flush valve water closet, symbol for, 279
Foam insulation, 248, 252
Folding rule, 17–18, *20*
Food service equipment, 91
Food services, heat gain from occupants, *255*
Food warmer, heat gain from, *257*
Foot (English system), 18
Foundation plans, 204
in architectural section, 204
in electrical plan section, 63
in general construction section, 60
in load calculations, 245
in mechanical plan section, 63
in plumbing plan section, 62
in structural section, 61
Foundation type, 245
Fountains, 174, 271
Four-way switch, symbol for, 280
FPM (feet per minute), 45
Fractions, 5–6, 151–*152*
lettering of, 151–*152*
Freehand sketching, 103–104, 179–184
crosshatching in, 181–182
dimension lines in, 106–107, 181
in drawings, 181–182
extension lines in, 70, 106–107, 181
lettering for, 180
lines used for, 180
locating notes on, 181
locating on a page, 181
poché, 181, 183
scales in, 31–32, 180
shading in, 181, 183
special treatment of surfaces, 182
types of, 179–180
See also Drawings
Front elevation, 169–170
Front View, 133–136
base lines for, 135, 137–138
constructing, 107
in isometric drawings, 133–143
in oblique drawings, 125–130
selecting, 116–120
Fuel gases, 62, 92
Fuel gas line, symbol for, 276, 278–279
Fuel oil return, symbol for, 278
Fuel oil supply, symbol for, 278
Full scale drawings, 31
Fumes, 15
Funeral parlors, heat gain from occupants, *255*
Furnishings, in specifications, 91
Fused isolating switch, symbol for, 281
Fused switch, symbol for, 280

G

Galvanized steel sheetmetal, 213
Gang outlet, symbol for, 280
Garages, heat gain from occupants, *255*
Gas heating, 239
Gaskets, *163*–*164*
Gate valve, symbol for, 278–279
Gates, 90
Gauge metal, 213
General Conditions, 90
General construction section, 60–61
General requirements, in specifications, 89–90
General use template, 46
Generator, symbol for, 280
Geometric figures and formulas, 264
GFCI (ground fault circuit interrupters), 14.
Glass bricks, 274
Glass, symbol for, 274
Glazing, 91
Globe valve, symbol for, 278–279
Gloves, 13
GPM (gallons per minute), 45
Grade point (symbol), 276
Grading plans, 60
Graduations, 18–21

Graphite lead, 56, 104
Griddles, heat gain from, *257*
Grille schedules, 47, 63
Gross area, *10*, 248–*249*
Ground fault circuit interrupters (GFCI), 14
Ground point, symbol for, 281
Guardrails, 90
Guidelines, for lettering, *150–152*
Gutter details, *68*, 174, *248*
Gymnasium, heat gain from occupants, *255*
Gypsum wallboard, 91

H

Hand tools, 14
Hangers, 90, 175, 219
Heating coils
- drawing with projection, *128–129*
- oblique drawing of, 128–*129*
- piping connection to, *142*
- three-way control valve, *75*, 277

Heating equipment
- drawing components for, 190–193
- safety rules in handling, 14
- sizing, 239, 260

Heat pumps, 226–227, 240
Heat transfer multiplier (HTM), 250, *251–253*
Heat transfers, 240–241
- between floors, *247*
- from inside to outside of house, *242*
- from outside to inside of house, *241*
- gain from, *251*
- loss from, *252*
- overview of, 240

Heels (ductwork), 190–191
Hidden lines, 106
- in orthographic projection, 114–116

Hoists, 91
Home run to panel, symbol for, 280
Horizontal lines, 25
Hospitals, 91, 235
Hot plates, heat gain from, *257*
Hot rolled steel sheetmetal, *213*
Hot water (domestic), symbol for, 279
Hot water heater, *73*
Hot water piping, 194
Hot water return (domestic), symbol for, 278, 279
Hot water supply (domestic), symbol for, 278
HTM (heat transfer multiplier), 250, *251–253*
Humidity, 243
HVAC technicians
- job site safety rules for, 13–14
- self-test for, 3–*11*

Hydronic heating, 240

I

Inches, 18
Incinerators, 91
Indoor temperature, 243
Infiltration rate, 256
Ink, drafting, 56
Inking pens, 54, 56, 103–104, 188
Inside telephone outlet, symbol for, 281
Instrumentation rooms, 91–92
Insulation, 219–*222*
- accoustical lining, symbol for, 277
- batt, *246*
- ceiling, 243, 247–*248*
- cellular, 219
- chart, *220*
- costs of labor for, 221
- costs of materials, 221
- details, 175
- for ductwork, 63, 219–221, 219–*220*, 254
- exterior, 219
- floor, 245
- interior, 219
- R-values for, *246*, *248*
- specifications, 91–92
- symbol for, 274
- take-off form, *220*–221
- take-off procedure, 219
- wall, *70*, 243, 248–249

Insulation details, *68*, 175
Insulators, 240
Insurance companies, 86
Interior clean-out details, 175
Interior elevations, 70, 170–171, 205
Interior insulation, 219
Invited bidding, 86
Iron salts, 78, 83
Irrigation site plans, 60, 90
Isolating switch, symbol for, 280
Isometric drawings, 133–143
- creating, 133–137
- overview of, 133
- of piping diagrams, 139–140
- shading, 164
- using, 140–*142*
- variations in, 137–139
- *See also* Drawings

J

Job sites, safety in, 13–14
Joists, *68*, *70*, 90, 204, 208, 245–*246*, *250*
Junction box, symbol for, 280

K

Key-operated switch, symbol for, 280
Keyboards, 54
Kilowatts, 240
Kitchen appliances, heat gain from, *257*
Kitchen sink, double bowl, symbol for, 279

L

Labor costs, for ductwork, 219, 221
Lamp holder, symbol for, 280
Landscaping site plans, 60
Laser printers, 83
Latent heat, *257*
Laundry equipment, 61, 91
Lavatory, symbol for, 279
Leader lines
- description or use of, 153–154

Lead pencils, 33, 54, 104, 188
Left elevations, 111
Left side view, 111, 113–*114*, 116, 120, *138*
Legend and Symbol Schedules, 47, 61, 274–281
Lettering, 149–152
- for freehand sketching, 180
- guides for, *150–152*
- lowercase, 150–151
- numerals, 151–*152*
- overview of, 149
- size and spacing of, 151
- uppercase, 149–150

Liabilities, 86, 92, 99–100
Liability, in specifications, 92
Liability notes, 99
Lifts, 91, 206
Lighting fixtures,
- details, 231–232
- heat gain from, *258*
- schedules, 64, 232

Lighting panel, symbol for, 280
Linear measurement, 17–20
Linear measurement instruments, 17, 20
Lines, 103–109
- baselines, 137–*138*
- border, 108
- break, 108–109
- center, 108
- construction, 107, 114–115
- contour, 276
- cutting plane, 108
- dimension, 106–107, 116
- in drafting, 103–104
- drafting pens, 103–104, 188
- drawing with pencils, 33, 104, 188
- electrical service, 276
- existing contour, *105*
- extension, 70, 106–107, 116
- finished grade, 276
- fire, 276, 279
- in freehand sketching, 103–104
- fuel gas, 276, 279
- hidden, 106, 115
- intersection of, 189
- leader, *105*, 153
- long break, *105*
- miter, 115

new contour, *105*
object, 104, 114
in oblique drawings, 127–128
parallel, 136
projection, 107–108, 114–116
property, 276
schedule of, 103–*105*, 109
segments, 190
short break, *105*
water, 276
weights of, 103–104
on working drawings, 188
Load calculations, 239–260
building orientation in, 243–244
ceiling area in, 245–*247*
ceiling insulation in, 247–248
design conditions in, 243
door area in, 252
door construction in, 252–253
duct design in, 254–255
factors in, 254
floor area in, 244
floor insulation in, 245
foundation type in, 245
heat transfer basics, 240–241, 257–*258*
infiltration rate in, 256
information needed for, 241–243
miscellaneous loads in, 257–*258*
net wall area in, 248
number of occupants in, 255–256
shading factor in, 250–252
ventilation rate in, 256
wall insulation in, 248–249
window area in, 249
window construction in, 249–250
zoning in, 259
Loading dock equipment, 91
Location plans, 203
Lockers, 91
Lockout/Tagout procedure, 14
Long break lines, description or use of, *105*
Longitudinal sections, 204
Loose-fill cellulose, *246*–247
Louver details, 175
Louvers, 91
Low emittance glass, *252*
Lowercase lettering, 149–152
Low pressure condensate, symbol for, 278
Low pressure steam, symbol for, 278

M

Maintenance equipment, 91
Manhole details, 174–175
Manholes, symbol for, 276, 279
Manual drafting equipment, 52–53
Manual drafting instruments, 52–53, 187–188, 195
Manual J load calculation form, 240
Manual N load calculation form, 240
Masonry, in specifications, 90
Material costs, for ductwork, 219
Material handling systems, 91
Material Safety Data Sheets (MSDS), 14
Mathematics, 9
self-test for basic, 3–*11*
Measurements
angular, 24–25
English system, 17–20, 266–267
instruments, 20
linear, 17–20
metric system, 17–18, 266
standards, 17
Measuring instruments, graduations for, 18–19
Mechanical details, 175
Mechanical division, in specifications, 91–92
Mechanical elevations, 171
Mechanical equipment room, 195
Mechanical pencils, 188
Mechanical plans, 63, 225–227
abbreviations and symbols on, 270–272
ductwork in, 226
other information shown on, 227
plans, 63
reading, 225–227
specific systems in, 226
specifications in, 227
supply and return air outlets in, 226, 277
written notes and instructions in, 226
See also Plans
Mechanical plan section, 63
Mechanical schedules, 63
Mechanical specifications, 227
Mechanical symbols, 276–278
Mechanical unit details, 175
Medical equipment, 91
Medical gases, 62, 92, 235
Metal doors and frames, 91
Metal fabrications, 90
Metal gauges, 213, 221
Metals
ductwork, 215–219
in specifications, 90–91
total weight required for, *212*–213
Metal stud wall, symbol for, 275
Meter, symbol for, 279
Metric Conversion Act, 18
Metric system, 17–18
conversion to English system, 17–18, 266
Microfilms, 83
Mile, 18
Millimeter, 18
Mineral batts, 245, 247
Mineral wool, 245–247
Minutes, of an angle, 28
Mirror, 171
Miscellaneous loads, 257
Miter lines
in orthographic projection, 115
plan view, 116–*118*
See also Lines
Moisture protection, in specifications, 91
Moldings, *70*, 170, 174, 249–*250*
Motors
heat gain from, *258*
symbol for, 280
Mounting pads, 175–176
Mouse (computer), 54
Moving stairs and walks, 91
MSDS (Material Safety Data Sheets), 14
Multiplication, 4–5
Mylar® film, 56, 104

N

National Bureau of Standards (NBS), 17
National Institute of Standards and Technology (NIST), 17
Natural grade (symbol), 276
Near scale, defined, 31
Net area, 248–*249*
Net wall area, 248–*249*
New contour lines, description or use of, *105*
Night set-back thermostats, 227
North arrow (symbol), 276
Notations, for blueprints, 153–154
Not to Scale, 72
Nuclear reactors, 91
Numerals, 151–*152*

O

Object lines, 104–*105*
description or use of, 104
in orthographic projection, 114–116
Oblique drawings, 125–130
creating, 125–127
lines used in, 127–128
overview of, 125
shading, 164–*165*
using architect's scale for, 31–32, 127
variations in, 128–*129*
See also Drawings
Obtuse angle, 26
Occupancy load, in commercial business, *256*
Occupants, heat gain from, 255–*256*
Occupational Safety and Health Administration (OSHA), 15
Off-set fitting, *194*
Oil heating, 239–240
One-foot graduation, 18
One-foot scale, 20
One-half inch scale, in architect's scale rule, 34
One-inch graduation, 18
One-inch scale, in architect's scale rule, 33
One-yard graduation, 19
Open bid list, 86
Operating rooms, 91
Original drawings. *See* Working drawings
Organizing, drawing sheets, 157–159

Orthographic projection, 111–120
  in constructing basic views, 116, *119*
  in freehand sketching, 179
  lines used in, 114–116
  in oblique drawings, 125
  overview of, 111
  selecting proper views in, 116–120
  shading of, *128–129*, *164–165*
  viewing planes, 111–114
  and working drawings, 120
OSHA (Occupational Safety and Health Administration), 15
Outdoor temperature, 243
Outlets, special purpose, symbol for, 280
Outside, air requirements, *257*
Outside telephone outlet, symbol for, 280

P

Painting, 91
Panel board schedules, 64, 231
Paper, drawing, 35, 56
Parallel bars, 52
Parallel lines, 136
Parallel surfaces, 136
Parallelogram, equation for, 264
Parking lot lighting, 176
Partial wall section, *71*
Pencils, 104, 188
Pen plotters, *55*, 83
Pens, drafting, 103–104, 188
Perimeter, calculating, 6
Pest control, 91
Pilot light switch, symbol for, 280
Pipe elbows, symbols for, 277
Pipe flanges, *163–164*
Pipe sleeve, symbol for, 280
Pipe (turned down), symbol for, 280
Pipe (turned up), symbol for, 280
Piping schedules, 63
Piping symbols, 278
Piping system, 62–63, 92, 139–140
  connections, *142*, 194
  details, 175
  diagrams, 175
  domestic cold water, 235
  foundation plans and, 62
  hot water, *196*
  irrigation, 60
  isometric diagram, 139–140
  and mechanical elevation, 171
  in plumbing plan section, 62–63, 92, 236
  refrigeration, 227
  roof drainage, 236
  sanitary drainage, 236
  schedule, 63
Planes, 111–*114*
Planning booklet, 158
Plans
  architectural, 203–205
  basement, 60–63
  blueprints, 51, 77–79
  construction, 60–61
  demolition, 60
  ductwork, 211–221
  electrical, 63–64
  elevation, 169–172
  floor, 62–63, 169–170, 204–205, 226
  floor framing, 61
  foundation, 60–63, 204
  grading, 60
  irrigation site, 60
  landscaping site, 60
  mechanical, 63
  plumbing, 62–63, 235–237
  roof, 61, 204
  roof framing, 61, 204
  schedules, 64, 157
  site, 60
  site location, 60
  soil boring site, 60
  structural, 61–62
  utilities site, 60
  viewing planes, *112–114*
  written notes and instructions, 226
Planters, 174, 203, 205
Plan view, 111–120
  constructing, 116
  in isometric drawings, 133–137
  selecting, 116, 120
  viewing plane, 111–*114*
Plaster on masonry, symbol for, 275
Plastic drawing film, 56
Plastic templates, *53*, 188
Plastics, in specifications, 90–91
Plotters, 54–55, *79*, 82–83
Plug valve, symbol for, 278
Plumbing, 62–63, 235–237
  details, 63
  elevations, 63
  in specifications, 62
  symbols, 62, 279–280
Plumbing Fixture Schedule, 236
Plumbing fixture template, 46
Plumbing pipe symbols, 235–236, 279–280
Plumbing plans, 235–237
  reading, 236
  specific systems in, 62–63, 235–236
Plumbing section, 62–63
Plumbing vent line, symbol for, 236, 279
Plumbing vent system, 236
Pneumatic operated damper, symbol for, 277
Pneumatic tube systems, 91
Poché, 162–*163*
  combining with crosshatching, 163
  in freehand sketching, 162–*163*
  See also Shading
Points, 125–126
Polystyrene beadboard, *246*
Pools, 91, 174
Potassium dichromate, 78
Power panel, symbol for, 280
Power pole with guy, symbol for, 276
Power riser, 232
Power service diagrams, 176
Prefabricated structural wood, 90
Pressure, 13
Pressure gauge, symbol for, 278
Pressure reducing valve, symbol for, 277
Pressure relief valve, symbol for, 277
Pressure/temperature plug, symbol for, 278
Pricing, 86
Printers, 54, 79–81, 83
Process piping systems, 62, 92
Projection lines
  description or use of, 107–108
  in orthographic projection, 114–116
  See also Lines
Property lines, description or use of, *105*
Property lines, symbol for, 276
Protective gear, 13
Protractor, 27, 188
Pruning, 90
Pulleys, *106*, *108*
Pump details, 175
Pumped condensate, symbol for, 278

Q

Quadrants, of circles, 24

R

R-values, *246*, *248*
Raceway schedules, 64
Radiation, 240, *242*
Radiation treatment rooms, 91
Radius, 7–8, 46, 190–192
Rag content drawing paper, 56
Rankine, 46
Rear elevations, 111
Recast concrete, 90
Record drawings, 87
Rectangles
  area of, 6
  equations, 264
  perimeter of, 6
Rectangular tank, equation for, 264
Reduced scales, 18
Reducer, symbol for, 278, 280
Refrigerant hot gas, symbol for, 278
Refrigerant liquid, symbol for, 278
Refrigerant suction, symbol for, 278
Refrigerants, 13–14
Refrigeration piping, 227
Register and Grille Schedule, 63
Register schedules, 47, 63
Relative humidity, 243
Responsibility, in specifications, 92
Retail shops, heat gain from occupants, *255*
Revisions, on working drawings, 95, 98, 100
Right angle, 25
Right elevations, 60, 69, 111, 120
Right side view. *See* End view
Riser diagrams, 175, 237

Roadway paving details, 174
Roaster, heat gain from, *257*
Rock wool batts, 245
Roof drainage piping, 235
Roof drain details, 236
Roof framing plans, 61, 204
Roofing materials, 91
Roof-mounted equipment, 175
Roof plans, 61, 204
Room finish schedules, 61
Rotating equipment, 14
Rulers, 17–18

S

Safes, 171
Safety, 13–15
  in blueprint room, 14
  in drafting room, 14
  federal law, 15
  on job sites, 13–14
Safety glasses, 13
Safety valve, symbol for, 277
Sanitary drainage piping, 90, 236
Scales, 31–36, 39–42
  architect's, 31–36
  defined, 31
  in drawings, 72, 98–100
  engineer's, 39–42
  flat rule, 31–32
  reduced, 18
Schedule plans, 62
Schedules
  abbreviations, 61–62, 270–272
  in architectural plans, 31–36
  bolt/nut and rivet, 62
  color product, 205
  column, 62
  conduit, 64
  diffuser, 63
  door, 61, 205
  duct insulation, 63
  equipment, 61
  grille, 63
  hospital, 61
  kitchen, 61
  laundry, 61
  legends, 47
  lighting fixture, 64
  lines, 103–104, 109
  panel board, 64
  piping, 63
  of plans, 60–64
  raceway, 64
  register, 63
  room finish, 61, 205
  standards, 17–20
  steel or structural members, 62
  symbols, 47, 61–62
  units, 63
  welding symbols, 62
  window, 61
Schools, heat gain from occupants, *255*
Screens, 91, *251*
Sealants, 91
Seals, architect's and engineer's, 56, 92, *98–99*
Seconds, of an angle, 28
Sections, 67–69, 205
  of air-conditioning units, 69
  in architectural plans, 205
  electrical, 63–64
  general construction, 61
  locations of, *71*
  mechanical, 63
  overview of, 67–69
  plumbing, 62–63
  site work, 59–60, 90
  structural, 62
  views of, *68–71*
  of walls, 70–*71*
Section views, of cornice for brick veneer wall, *68*
Security companies, 86
Seismic exploration, 90
Self-test questions, for HVAC technicians, 3–*11*
Sensible heat, *257*
Sensitized print paper, 78–79
Sepias, 56
Shading, 161–*165*
  in freehand sketching, 164
  of isometric drawings, 164
  marking limits for, *165*
  of oblique drawings, 164
  stick-on, 164
  symbols in, 164
  See also Crosshatching; Poché
Shading factor, 250–252
Shafts, *106*, *108*
Sheet Metal Duct Estimating Factors Chart, *212*–213
Sheet metal ductwork, 213
Sheet titles, 158
Shelving, 91, 170–171, 174
Shop drawings, 89
Short break lines, description or use of, *105*
Shower, symbol for, 279
Sides, 111–120
Single convenience outlet, symbol for, 280
Single-pole switch, symbol for, 280
Single splitter fitting, 192–*193*
Site details, 174
Site location plans, 60, 203
Site plans, 203
  and location plans, 60, 203
  symbols, 276
Site work, in specifications, 90
Site work section, 60
Six-foot folding rule, 20
Six-inch scale, 34
Sizing, of air conditioning and heating equipment, 239
Slab construction, 62–63, 245
Slant numerals, 151–*152*
Sloped letters, 150–*152*
Smoke damper, symbol for, 277
Soil, and heat transfer, 245
Soil boring site plans, 60
Soil line, symbol for, 279
Soil testing, 60
Solar radiation, 240
Soldering, 14, 272
Solenoid valve, symbol for, 277
Sound and vibration rooms, 91
Speakers, symbol for, 208
Special construction, in specifications, 91
Special purpose outlets, symbol for, 280
Specifications, 89–92
  concrete work, 90
  conveying systems, 91
  doors and windows, 91
  electrical, 92
  equipment, 91
  finishes, 91
  furnishings, 91
  general requirements, 89–90
  masonry, 90, 275
  mechanical, 91
  metals, 90
  overview of, 89
  plumbing, 92
  responsibility and liability, 92
  site work, 90
  special construction, 91
  specialties, 91
  thermal and moisture protection, 91
  woods and plastics, 90
Splitter fitting, 191–*193*
Splitter fitting with damper, symbol for, 277
Spray-on cellulose, 245
Sprinkler head, symbol for, 279
Sprinkler piping, symbol for, 279
Sprinklers, 60, 92, 235
Squares
  area of, 6
  equations, 264
  perimeter of, 6
Stairs, 90–91, 174
Standards agencies, 17, 45
Steam heating, 63, 240, *257*
Steam, low pressure, symbol for, 278
Steel tape rule, 20
Stick-on shading, 164
Stones, 90, 174
Storm drainage pipe installation details, 174
Straight angle, 25–26
Straight line, 24
Strainer, blow down, symbol for, 278
Structural details, 62
Structural section, 61–62
Stud partition wall, symbol for, 275
Studs, 183, 275
Styrofoam, *246*
Submittals, 89
Subtraction, 4–5
Sun control devices, 91
Supplemental General Conditions, 90
Supply and return air outlets, 226
Supply outlet, symbol for, 277
Swimming pools, 91
Switch and convenience outlet, symbol for, 280
Switch with pilot light, symbol for, 280

Symbols, 45–48, 270–272
  architectural, 274–275
  arrows, and special markings, 69, 276
  building materials, 274
  in CAD (Computer-Aided Design), 46
  cutting planes, 69, 108
  electrical, 231–232, 276–277, 280–281
  examples of, 48
  mechanical, 226, 276–278
  metric system, 18
  nonstandard, 48
  piping, 236, 278
  plumbing, 62, 236, 279
  plumbing pipe, 236, 279–280
  schedules, 47, 61–62
  in sections, 69
  in shading, 46
  site plans, 276
  using with abbreviations, 47, 270–272
  variations in, 46
  walls, exterior, 275
  See also Abbreviations
Symbol schedules, 47, 61–62, 274–281

T

T-squares, 52, 187
Tackboards, 91
Take-off form (ductwork), 215, 218, 221–*222*
Take-off form (insulation), 219–221
Taking off, 86
Tank-type water closet, symbol for, 279
Taverns, heat gain from occupants, *255*
Technicians (HVAC)
  job site safety rules for, 13–14
  self-test for, 3–*11*
Tee (outlet down), symbol for, 278, 280
Tee (outlet to side), symbol for, 280
Tee (outlet up), symbol for, 277, 280
Telephone outlets, 280–281
Telephone switchboard, symbol for, 281
Templates, 46
Terrazzos, 91
Theater, heat gain from occupants, *255*
Thermal protection, in specifications, 91
Thermal resistance, 245, 249
Thermometer, symbol for, 278
Thermometer well, symbol for, 278
Thermostats, 227
Three-dimensional drawings, 125, 127–128, 130, 161, 164–165, 181
Three-way splitter damper, *193*
Three-way switch, symbol for, 280
Three-way valve, symbol for, 277
Throats (ductwork), 190–191, 277
Timed override switches, 227
Title blocks, 95–100
  dates, 98
  job names in, 97
  liability notes, 99
  locations of, 95–*96*
  names of architects and/or engineers in, 97
  names or initials in, 98
  overview of, 95–97
  revisions, 98
  scales, 98
  seals of architects and engineers on, 56, 92, 98–*99*
  sheet number in, 97–98
  sheet titles in, 97
  stick-on, 97
Toasters, heat gain from, *257*
Toilet fixtures, 171
Toners, 81, 83
Tools, 14
Transformer, symbol for, 281
Transition, symbol for, 276
Transverse sections, 204
Trapezoid, equation for, 264
Tree relocation, 90
Trees, symbol for, 276
Triangles
  angles of, *52–53*, 161, 188
  area of, 7
  equations, 264
Triangles (drafting equipment), *53*, 188
Triangular scale, 31, 39
Triangular shapes, 23
Triangular structure, equation for, 264
Tub, symbol for, 279
Turning vanes in a round throat elbow, symbol for, 277
Two-dimensional drawings, 125

U

U-value, 247
Ultraviolet (UV) light, 79
Union, symbol for, 278, 280
Unit hangers details, 175
Unit schedules, 63
Uppercase lettering, 149–151
Urethane, 246
Urinal, symbol for, 279
Utilities site plans, 60

V

Vacuum line, symbol for, 279
Valves
  symbols for, 277–279
Vapor barriers, 254
Vaults, 91
Vellum® film, 104
Vending equipment, 91
Ventilaton, safety issues, 13
Ventilation rate, 256
Vents, 236
Vermiculite, *246*
Vertex, 24
Vertical letters, 149–150, *152*
Vertical numerals, 150
Viewing planes, 111–120
Views, elevation, 111
Voltage, 14
Volume, 8
Volume damper, symbol for, 277

W

Walks, symbol for, 276
Wall hydrant, symbol for, 279
Wall-mounted lamp holder, symbol for, 280
Wall-mounted lavatory, symbol for, 279
Walls, 204–205, 248–249
  architectural symbol for, 275
  details, 204–205
  insulation, 248–249
  interior elevations, 70–*71*, 205
  net area, 248–*249*
  with poché, 163
  sections, 70–*71*, 204
Waste handling equipment, 91
Water closets, symbols for, 279
Water heater piping details, 175
Water heater, symbol for, 279
Water heating, 235
Water line, symbol for, 276, 279
Waterproofing materials, 91
Waterproof switch, symbol for, 280
Watts, 240, *258*
Weatherhead, 171–*172*, 176
Welding, 62
Welding symbols schedule, 62
Whole numbers, 4–5
Windows, 249–252
  area, 249, *251*
  construction, 249–250
  schedules, 61
  in specifications, 91
Wiring diagrams, 176
Wood
  in specifications, 90–91
  symbol for, 274
Wood flooring finishes, 91
Working drawings, 51–57, 59–64
  CAD (Computer-Aided Design), 54
  creating manually, 52–54, 188–189
  creating with computers, 54
  dates on, 98
  Diazo process, 79–81
  drawing film for, 54, 56
  drawing paper for, 56
  electrical section, 59, 63–64, 176
  electrostatic process, 81–82
  general construction section, 59–62
  lines on, 188–189
  major categories of, 59

mechanical plan section, 63, 226–227
and orthographic projection, 120
overview of, 51
plan size, 54
plumbing section, 62–63
printing, 79–83
responsibilities of architects and engineers in, 56–57
revisions, 98
sepias in, 56
site work section, 60
structural section, 61–62
title blocks, 98
See also Construction drawings; Drawings
Workplace safety. *See* Safety
Wrapped insulation, 219
Wrappers (ductwork), 190–191
Written notes and instructions on plans, 226

Y

Yard (English system), 19
Yard hydrant details, 175
Yard hydrant, symbol for, 279
Yardsticks, 18–19

Z

Zoning, 259